An Amateur Astronomer's Life

Seven Decades Of Enthusiasm

For The Heavens

Geoff Kirby BSc

Acknowledgements

My heartfelt thanks go to my best friend and partner Sandy who has supported me wholeheartedly throughout this project.

Her patience is unending.

I also acknowledge with great gratitude my family who have tolerated my obsessional astronomical activities.

Copyright and References

Because of the labyrinthine laws on copyright and the non-uniformity of these laws across countries and because some copyright holders are aggressively protective of their precious pictures, I have included here pictures taken by me, or which are either copyright-free, free to be used under the Creative Commons License or where the copyright holder has given me permission to use their pictures.

Where pictures cannot be used in this book because of copyright restrictions I give references - typically thus [23] - directing you to websites where the pictures can be viewed.

I urge you to follow the URLs given in the references to fully enjoy this book.

For your convenience the references to websites can be easily reached by going to a website which is located at www.geoffkirby.co.uk/References

There you can click on the number of the reference in this book and be taken to the reference website page.

Simples!

The Author's website is at

www.geoffkirby.co.uk/Books

iv

I enjoyed using these very special binoculars on cloudy nights!

Give me the ways of wandering stars to know.

The depths of heaven above and the Earth below.
Teach me the various labours of the Moon
and whence proceed the eclipses of the Sun.
Why flowing tides prevail upon the main
and in what dark recess they shrink again.
What shakes the solid Earth, what cause delays
the Summer nights and shortens Winter days.

Virgil 70 - 19 BC
Translated by John Dryden 1631 - 1700

Alternatively...

"...heav'n is for thee too high

To know what passes there; be lowly wise:

Think only what concerns thee and thy being:

Dream not of other worlds..."
Angel Raphael in Milton's Paradise Lost Book VIII

You choose...

Contents

xiv

1. Introduction

This book is not an instruction manual for buying or building telescopes.

It is not even a textbook on amateur astronomy.

It is not a 'coffee table' book packed with large glossy pictures of galaxies, planets and their moons, etc., taken with enormously expensive telescopes.

As I write these words a search of Amazon.co.uk for astronomy books produced over 65,000 titles. There is no shortage of books to guide readers interested in pursuing their wonderful hobby of astronomy and space.

So what is this book?

It is unique.

It is my personal story of over seven decades of enthusiasm for enjoying the heavens.

It is about me as a ten-year old boy, whose father had died only a few months previously, being in a painfully uncomfortable bed in a brutal Dickensian-style boarding school dormitory. I watched the Moon turning blood red during a total eclipse. I was desperately home sick and wanted to be at home sharing this sight with my widowed mother.

It is about the pleasures of being out on freezing nights peering down an eyepiece to glimpse some vague shimmering fuzzy patch in the heavens and being in awe that the light had been travelling for millions of years just to hit my retina.

It is about developing 1000 ASA monochrome roll film in a toilet as part of a search for comets and 'novae' - stars which suddenly burst into brilliance for a brief few weeks of glory before fading back to insignificance.

It is about spending two years grinding and polishing a mirror so that I could see the wonders of the universe through my very own telescope.

Then there was the agony of sitting on a needle sharp lava field in total darkness on the summit of Mount Teide in the Canary Islands trying to photograph Halley's Comet in 1986.

Madness!

Not for me the comfort of modern astronomy where observers sit indoors remotely controlling a large and very expensive telescope via a joystick and a laptop screen.

This book is a nostalgic ramble through an age when telescopes were mostly home-made, guided by hand and not by a computer and the light went into my eye and not a camera lens.

Chapter 2 describes how my first interest in the heavens was ignited during the 'blackout' of World War 2 when I sat on top of my family's air raid shelter watching German 'Doodlebug' weapons flying over London with their long orange exhaust flames tracing their deadly path to kill innocence civilians.

One of these missiles exploded scarcely one hundred metres from my family home, destroying it and leaving my parents and me buried in our Anderson shelter under piles of rubble and shards of glass.

Between these raids the Moon, planets and stars shone down, calmly aloof from the carnage below, urging me to learn more about them.

I lived at 34 Savernake Road in North London. At number 14 lived Henry Wildey; one of the greatest British telescope makers of the twentieth century.

Henry fuelled my interest in astronomy by giving me my first little telescope, taking me to look through the magnificent Cooke refractor in the nearby Hampstead Observatory and taking me to meetings of the Royal Astronomical Society.

By the age of ten I was fully signed up to a passion for the heavens.

In chapter 3 I move on to the many frustrating, amusing and exciting entries in my log book.

These entries start in 1957 when I started to build, from scratch, my first telescope and end in 2016 when my highly myopic and cataract scarred eyes finally stopped me from viewing the heavens.

An obsession with astronomy requires unlimited patience from a partner. Relationships are often severely tested.

It means tolerating a huge ugly observatory being built where the flower beds have just been destroyed.

It means a significant element of romantic neglect as illustrated by my logbook entry on February 14th 1976 which reads

'I took my wife to a romantic Valentine's Day dance. However, I drove home in the interval to observe a lunar occultation and got back just in time for the first dance of the second half!'

That's devotion - but whether more to astronomy or my wife I leave you to decide!

On July 2nd 1975 I recorded

"My second son was born at 2 am this morning and I was at the birth. Tonight I am a little tired."

Despite having had no sleep the previous night as I accompanied my wife through her labour I still managed to make many variable star estimates!

In 1959 I joined the British Astronomical Association (BAA) - which is the subject of Chapter 4.

Belonging to the BAA was an 'interesting' experience as in the Chinese curse 'May you live in interesting times'.

I was to meet many fascinating people; Patrick Moore, Heather Couper and Nigel Henbest as well as Bill and Ethel Granger. Bill went everywhere with his cat 'Treacle Pudding' on his shoulder and Ethel still holds the world record for the smallest waist measurement of 13 inches; the same circumference as a one litre bottle of wine.

They and many more characters will appear in Chapter 4.

In Chapter 5 I try to persuade newcomers to astronomy not to rush out with their credit card and buy a huge and complicated telescope which will probably be used less than a dozen times before being consigned to the cupboard under the stairs.

Newcomers should start by learning the skies with the unaided eye.

There is much to see and useful observations to be made.

Meteor observing is an unaided eye activity and I was co-discoverer of a nova in Cygnus in 1975 just by looking up at the sky from my back garden.

Once you are familiar with the sky seen with the unaided eye it is time to move on to using binoculars. Eventually you will want a telescope but do not rush into this acquisition.

By my teenage years I decided that I needed a telescope to make serious observations of the universe. So I made one from scratch as shown above.

OK!

Have a good laugh but I made observations with this little beauty that were published in astronomical journals.

Chapter 6 describes how I made the mirrors for several telescopes. Also how I polished and shaped my telescope mirrors to an accuracy of one-tenth the wavelength of light (about 50 nanometres or two millionths of an inch). The equipment I built to achieve this extreme accuracy cost me less than £1.

When the Hubble Space Telescope was first launched the error on its mirror shape was 2,000 nanometres [1] - forty times the error on my best mirror!

Chapter 7 describes how I built telescopes to hold my homemade mirrors. The telescope mountings were crude and made extensive use of parts from discarded washing machines, TV aerials, bricks used as counterweights and a sawn off scaffolding pole to use as a sighting telescope.

Here you see a typical homemade telescope mounting. The upper bearing is a television aerial pole fitted to the bearing from a washing machine.

The mirror was homemade.

The total cost of the telescope and mounting was about £30.

In much of this endeavour I was inspired by Reg Spry who, in 1978, published a wonderful little book entitled *'Make Your Telescope From Everyday Materials'* [2].

Two years previously he had published an article in the BAA Journal entitled *'A 150 mm From Scrap'* [3]. In that article Reg wrote

"Looking at the completed instrument I list the ingredients as follows: Two larder shelves and various other pieces of timber in garage and garden; old car half-shaft and lifting jack; pieces of plumbing; plastic coffee jar caps; parcel binding tape; lenses from broken binoculars; various screws, nuts and bolts and odd pieces of metal from the workshop 'come-in-handy' box; a few Meccano parts for the eyepiece mount; main mirror and flat mirror bought for £20. The total cost of the telescope was £21.50."

Some readers of this book may be feeling gloomy about having pushed their credit cards to their limits to buy a telescope that performs no better than a home-made one costing about one-tenth of the price.

You will discover as you ramble through this book that Reg Spry and I had much in common when it came to avoiding spending money on astronomical equipment.

Chapter 8 moves on to my homemade observatories.

This was my first and most ugly observatory.

Eight sheets of chipboard were covered in roofing felt and assembled into an octagonal building.

A sloping roof was added and gave stability to the flimsy walls.

The whole contraption rotated on a low circular wall of breeze blocks using eight shopping trolley wheels.

After years of excellent service this was replaced by a more conventional home-made dome; again built entirely from materials available cheaply at my local hardware store.

The following chapters from 9 to 16 describe the major astronomical projects that I have been engaged in over the past seventy years.

The Moon was my first love as it should be for all newcomers to astronomy and Chapter 9 describes my lunar observations.

Here is a world of mountains, ravines, lava flows and craters all visible in very modest telescopes.

More advanced observers curse the Moon's light for preventing them from observing objects outside our Solar System; the so-called 'Deep Sky' objects.

However, the Moon is a wonderful object to observe. Patrick Moore spent his observing lifetime concentrating on the Moon; its allure never faltered for him and it should not be neglected by you.

Chapter 10 introduces you to those much neglected planets Mercury and Venus as well as to the more popular Mars, Jupiter and Saturn

I published more scientific papers about my observations of Venus than any other topic.

In fact I was offered the post of Director of the Mercury and Venus Section of the BAA in 1978.

I declined this offer because the Mercury and Venus Section had then contained a very small number of unpleasant characters who had had the areas of their brains that control charisma and modesty surgically removed at birth.

The Director J Hedley Robinson was a thoroughly nice man who was harassed by this small clique of members and I had no wish to be involved with these people albeit they were a very small group within the BAA Mercury and Venus Section membership.

At a meeting of the BAA in the early 1970s one of these people walked up to me, eyes sparkling and face twisted in rage. He formed his hand into the shape of a gun and 'shot' me three times through the head. He then moved his index finger over his throat and stabbed the finger at me - a clear threat to my wellbeing.

I didn't need this behaviour from someone who was simply disagreeing with me over a technical point relating to observations of the planet Venus!

Meteor observing is the subject of Chapter 11.

This is a great beginner's topic as it is performed with the unaided eye and can be enjoyed as a group activity by an astronomy club.

It can be uncomfortable however.

I once stretched out on a sun lounger on the top of a barren hill in Dorset on the night of January 4th (the peak of the Quadrantid meteor shower). There was a sharp frost and the faint light from polluted skies glistened off my boots.

Around me there were cars whose occupants were engaged in much warmer activities than me judging from the steamed up windows and creaking car suspensions.

I wonder what they thought of me and my two friends on sun loungers staring up at the skies?

However, they were the losers.

Whatever they were getting up to in their cars they would surely have had far more enjoyment joining me watching meteors!

I spent many years recording occultations of stars by the Moon and these observations are described in Chapter 12.

The 1970s was a peak decade for timing the disappearance and reappearance of stars at the Moon's limb.

This was because simple timings of the instant when a star is exactly on the limb of the Moon were used to refine the prediction of the Moon's orbit around the Earth and also to refine the position of stars in space.

In the twenty-first century these observations are of less value as the Moon's orbit is now predictable to a few centimetres thanks in part to the optical retroreflectors left on the Moon by the Apollo astronauts.

Also, the HIPPARCOS space probe greatly improved our knowledge of the positions of the stars.

Despite this reduction of interest by the professional astronomers in observing lunar occultations there is a great interest in observing the occultation of stars by asteroids and this is a current important topic for study.

I have twice observed stars being hidden briefly by asteroids and these observations have greatly improved our knowledge of the two asteroids involved.

Chapter 13 describes the many years I spent hunting for novae; stars which suddenly flare up in brightness.

One of the greatest nova hunters was George Alcock who memorised the positions of 30,000 stars so that he could immediately recognise an interloper - a nova - in a star pattern.

George used only binoculars for this monumental task; an affirmation of my advice to spend time using binoculars rather than rushing out to buy a huge eye-wateringly expensive telescope.

I carried out my nova hunting by photographing the whole of our Milky Way galaxy every clear night using a standard camera on a 'Barn Door' mount - see page 208 - that cost me less than one pound to make.

Did I find a nova? You will have to read Chapter 13 to find out!

Chapter 14 covers the period when I observed variable stars.

There are many stars that vary in brightness and the observations and recordings of these changes in brightness are much sought after by professional astronomers. This is because they do not have the resources to study thousands of variable stars whereas thousands of amateur astronomers can do that task in their spare time.

I observed about one hundred different variable stars but my interests focussed on eclipsing variables.

These are two or more stars that are in orbit around each other. During their celestial dance one star passes in front of another causing a regular change in brightness.

One of these eclipsing stars is easy to follow with the unaided eye - Algol, the 'Demon Star' of mediaeval Arab observers. I provide a chart on page 353 to encourage you to observe this star.

Nearly all of the variable stars that I followed were observed using binoculars and not a telescope.

However, a rare opportunity arose in 1996 to observe a variable star that was suspected of being an eclipsing variable. I was able to make observations with my large Dobsonian telescope which confirmed this as a previously unknown eclipsing star.

I was a joint author of a scientific paper published in 1997 [4].

Chapter 15 covers my extensive work observing artificial satellites and tracking their position so that their orbital characteristics could be deduced.

All this work was performed using binoculars whilst relaxing in my back garden on a sun lounger - newcomers to astronomy please note - large expensive telescopes are not required!

Amongst my exciting experiences was the demise of SKYLAB in July 1979.

My cliffhanging involvement in helping to get SKYLAB safely to crash in the Indian Ocean rather than on a populated area of the Earth is described in chapter 15.

My observations and those of many fellow enthusiasts had helped NASA to set SKYLAB into a controlled tumble thereby increasing its drag so that it crashed safely in an unpopulated area of the Earth [5].

Chapter 16 describes my observations of sunspots for which I used a refracting telescope projecting the images of the Sun onto white card - the only completely safe and cheap way to see the Sun's surface.

The equipment as seen below was simple and made largely from bits of wood found in my garage.

Although the equipment appears crude I made observations and positional measurements of sunspots that were surprisingly accurate.

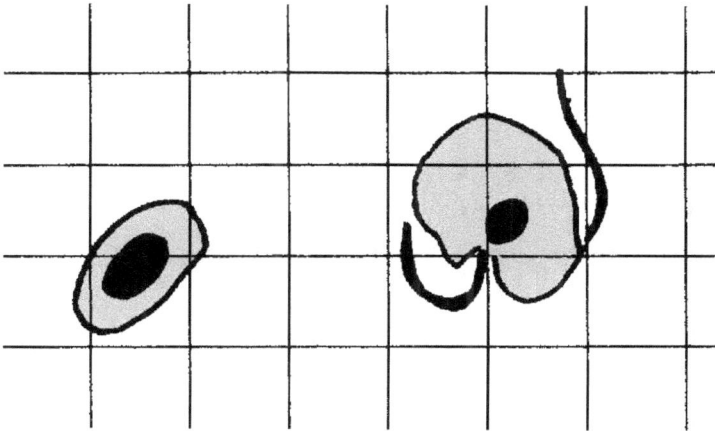

For example, the above drawing shows two dark lines attached to a sunspot which I believe were two prominences seen in visible light. This was the only time I saw this effect so it must have been rare.

The following two chapters describe several non-observational astronomical topics starting with Chapter 17 which describes trips I have made to the USA.

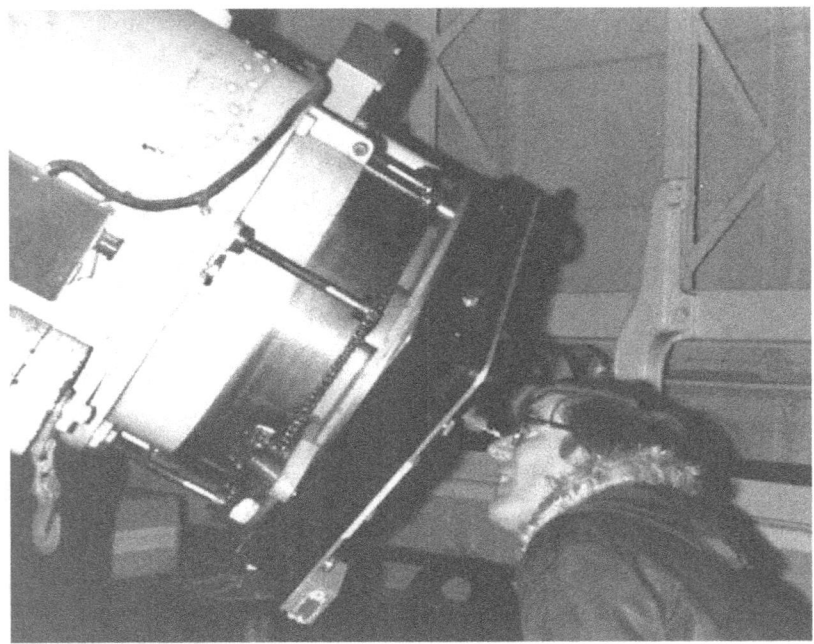

Here I am observing with the 26 inch (660 mm) aperture refractor of the US Naval Observatory in Washington DC [6].

It was through this telescope that Asaph Hall discovered Phobos and Deimos; the two moons of Mars in 1877.

These trips to the USA took in many of the major astronomical observatories and research centres.

Chapter 18 describes visits to the Canary Islands (Tenerife and La Palma) in the 1980s visiting places of astronomical interest.

One of these trips was with my family to see Halley's Comet in 1986.

The comet proved to be a disappointment - can you spot it in my photograph below? - but the visit to the islands and the observatories was very interesting.

Throughout my life as an amateur astronomer I have been plagued by stray light; usually from neighbours' bedroom windows but also a great deal of light from street lights.

Light pollution has been a subject close to my heart as shown by my campaigning efforts described in Chapter 19.

For this I was awarded a Certificate of Appreciation by Bob Mizon of the Campaign for Dark Skies. (This is now known as The Commission for Dark Skies).

Chapter 20 describes my arguments with astrologers.

Of special note was a gathering of professional astrologers who converged on my small rented house on Portland, Dorset in 1989 to debate the workings of their dark art and to try to convince me that astrology is not a steaming pile of poo.

Who won? Read Chapter 20 to find out!

Chapter 21 describes the years in the 1970s and 1980s when I was interested in Unidentified Flying Objects (UFOs).

I travelled around Dorset interviewing people who claimed to have seen a UFO and probing the background to their claims.

In the course of this I met many interesting and eccentric people.

I also travelled extensively around the United Kingdom giving a slide show about UFOs and reproduced above is the title slide.

Maybe this indicated to my audience that I was not taking the subject too seriously?

In October 1981 I published an article in a magazine summarising my personal views on the UFO phenomenon. This is reproduced in Chapter 21.

This drew a line under my involvement in the UK UFO investigating activities.

This book ends with an extensive list of references to sources of information.

Throughout the book I have suggested 'projects' that could be undertaken either individually or by groups in a an astronomy club.

It seems to me that far too many people go along to astronomy clubs as passive members and never aspire to have a telescope or make observations of their own.

My projects are set out on page 539. Of the forty-two listed twelve are suitable for unaided eye observations, seven for binoculars, twelve for telescopes and the rest are non-observational projects.

There are plenty of projects to pursue even without optical aid.

Before ending this introductory chapter I must mention what is probably the best book ever written on the delights of amateur astronomy. This is *'Starlight Nights: the Adventures of a Star-Gazer'* by Leslie C. Peltier [7].

Leslie has rightly been acclaimed as the 'World's Greatest Amateur Astronomer' [8].

If you only read one astronomy book in your life describing the awe and wonder of looking up at the night sky then this must be that book!

2. A Child Looks Up At The Stars

Wartime Survival

I was born a couple of weeks after World War 2 started.

I believe - but cannot prove - that these two events were coincidental and that Hitler was not personally involved in trying to kill me.

Two weeks before my due date of arrival my heavily pregnant mother was taken by bus, along with a large number of similarly encumbered women, to a hospital in the safety of Bedford, a small market town in the English South Midlands.

As soon as my mother and I were fit to travel, a bus loaded down with new mothers and screaming babies travelled back to London.

Not long after this first of many traumatic events for my parents and me the hospital where I had been born was demolished by a Luftwaffe bomb.

Maybe Hitler was out to get me after all but got his timing wrong?

My parents and I lived in North London throughout World War 2.

We suffered enormous deprivations alongside our neighbours; fear, hunger and the close proximity to death and personal loss.

No wonder I look terrified in the picture below!

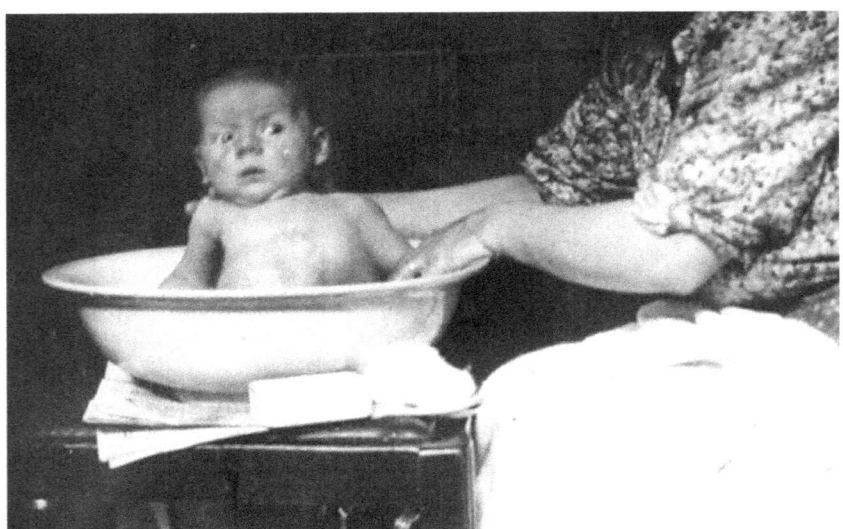

On a personal level our house was all but destroyed by a V1 'Doodlebug' missile and we had to camp in the shell of the house.

Life was so hard that my pet rabbit had to be killed by my father and cooked by my mother because we were so short of food.

When the air raid sirens sounded I was allowed to sit on the roof of our Anderson shelter seen above - although not when deep snow was on the roof! I watched the dogfights of the fighter aircraft, was awestruck by huge waves of Luftwaffe bombers heading west and, later in the war, I watched 'Doodlebugs' with their throbbing engines and tail flame heading in from France.

A V1 'Doodlebug' on its launcher at Duxford Museum

Although only a small child, my parents knew it was better for me to be outside watching than cowering in the claustrophobic environment of an air raid shelter. They also knew that I had enough sense to dive into the shelter if the 'Doodlebugs' or bombers were coming towards us and the wooden door would then be shut tight.

If I heard or saw anything to indicate danger approaching I would give a penetrating whistle and the neighbours then hurried to their shelters. This meant that I had heard or seen German aircraft coming or that a V1 'Doodle Bug' was on its way. The V1 missiles could be heard in the daytime by their unmistakable throaty drone.

I have published an account of these early terrible years of my life in the book *'My Lost Childhood - How Hitler and the Freemasons Stole My Childhood'* [9].

My interest in the night skies was sparked off during this traumatic time.

During the wartime blackout regime it was possible to stand in my north London garden at night and see the Milky Way as clearly as it can now only be seen from the depths of the countryside.

Sky transparency was sometimes poor due to the amount of coal burnt in houses and factories. This got so bad that 'The Great Smog' of 1952 led to the Clean Air Acts and 'pea soup' smogs became a memory.

This was my introduction to the wonders of the universe; a sense of awe and beauty that has never left me.

I recall walking with my father down our street one beautiful evening. The crescent Moon was hanging in the western sky against a darkening blue sky. It impressed me with its beauty even as a six-year old.

"Daddy, why does the Moon change shape from night to night?"

I asked.

He thought for a second and said that a cloud covered it up by a different amount every night we looked at it.

I didn't believe that as there was no cloud in the sky. Perhaps there was a much higher layer of clouds invisible except where it covered the Moon. Perhaps the Moon had clouds on it.

I doubted the whole thing!

I suspect that he didn't believe it either but he was not letting on.

There were few articles in magazines and newspapers that passed on elementary information about the heavens such as the example above from the 1940s.

Note the tag line 'Mainly for Husbands'!

This was written by the world famous comic film star Will Hay.

His fame in many 1930s films came from his depiction of bumbling school teacher, an incompetent railway stationmaster or similar characters [10].

He surprised his worldwide fans by being an intellectual man and a talented amateur astronomer. He became world famous in the astronomical community - and in the British press - when he discovered a white spot on Saturn in 1933 [11]

Despite these occasional snippets of astronomical education, the public knowledge of astronomy was very poor in those days.

For example, my mother believed into the 1960s that it got dark at night because the Sun was hidden behind a thick cloud.

There were few television programmes or popular magazines that would explain that sort of thing and encyclopaedias were out of the financial reach of most people. Patrick Moore's 'Sky At Night' TV programmes started in 1957 and astronomy only then really entered the public consciousness although the 'Sky At Night' was broadcast to a minority audience.

My father was a postman and my mother served behind the counter of the local Co-operative grocery store. Both worked six days a week and had little time for finding out why the Moon appeared to change shape or why it got dark at night.

Anyway, I was not convinced by my father's 'cloud theory' of the Moon's phases.

I went to the local library and looked up the answer.

Simple really.

I was just seeing the bit of the globe lit up by the Sun.

I went home and told my dad - I even think that I got an old tennis ball and showed him phases using the kitchen light. Although poor, my parents encouraged me in everything I wanted to do and discover.

One day when I was seven years old I was struggling to read an encyclopaedia - my favourite reading material as a small child. I couldn't work out how to translate Roman numerals into ordinary numbers. My father sat down and explained to me how they worked despite making himself very late for work. That didn't matter to my father; my education was paramount.

We always went to Broadstairs in Kent for our one week's summer holiday - here I am in a woollen swimming costume knitted by my mother. This was fine until I emerged from the sea with about a gallon of water trapped inside. This caused the groin of the costume to stretch down to my knees.

A Telescope Reduces Me To Tears

On a visit to the nearby Margate pier at the age of seven I spotted a telescope - one (old) penny in the slot for a minute's view. I was so excited. All I wanted to do was to come back after dark and see the Moon through it.

My father took me along the next clear night and the Moon was up there, tempting me to that telescope. I pointed it up, aligned it on the Moon and inserted my one penny coin.

Nothing!

I was bitterly disappointed. My father was at a loss to know what to do - the telescope was broken!

I cried inconsolably.

A kindly old gentleman asked what was wrong and my father explained the problem. The old chap gave me a handful of pennies so that I could play the amusement machines but all I wanted to do was look at the Moon through a telescope.

I cried all the way back to the guest house.

This fascination with the Moon was both scary and comforting in the years to come.

Scary because the features on the Moon appeared to me to be like the face of a man yelling in terror when the illumination was a day or two after Full Moon.

There is more on faces, lobsters and nursery rhymes on the Moon in Chapter 9.

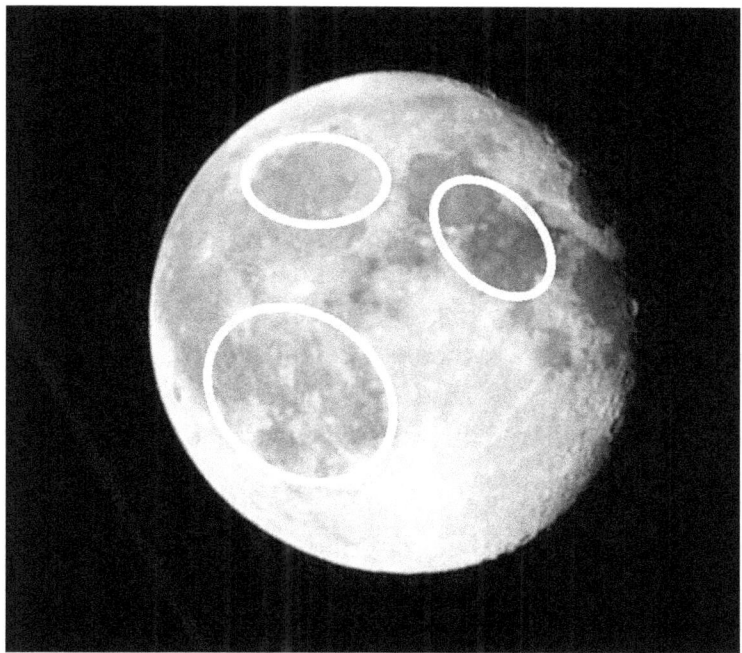

The 'Man in the Moon Screaming' as I saw him.

I would be in my bed at an abusive Freemasons' boarding school after a day of beatings and bullying and watch the clouds drift over the Moon.

At these times I would remember the holidays spent with my parents with the Moon shining down on the sea at Broadstairs.

My urge to explore the heavens was far greater than the normal children's pursuits at that time.

Henry Wildey Inspired Me

Back home I lived at 34 Savernake Road in North London. At number 14 lived a gentleman who was widely regarded in our neighbourhood as an eccentric.

I believe his daytime job was as a builder, plumber and decorator but he had a large gun-like object in his garden. I knew that it was a telescope for looking at the stars but many neighbours believed it was some secret weapon; probably a ray gun for shooting down Luftwaffe bombers.

There were stories that he was employed by the Ministry of War to spot enemy aircraft with this huge device.

His name was Henry Wildey and he was a Council Member of the British Astronomical Association (BAA) and a Fellow of the Royal Astronomical Society (RAS).

I owe my early growing interest in astronomy to this happy coincidence of living ten doors away from an internationally famous amateur astronomer.

Henry made his own telescopes and sold many. There are probably many still in use around the world. Even today a mirror by Wildey is much prized for its workmanship and performance.

The main telescope that Patrick Moore used for most of his life - nicknamed 'Oscar' - was made by Henry Wildey.

I remember the first time I went into Henry Wildey's house. The front room facing the road was supposed to be the family lounge but in Henry Wildey's house that room was a temple to telescope making. His wife Violet was *very* tolerant!

It was full of machines, bags of grinding powder, glass discs and all the paraphernalia of the telescope maker.

Put bluntly, his house was given over to telescope making!

Henry died in 2003 the day after his 90th birthday. There are touching and informative obituaries for Henry Wildey which are well worth reading [12], [13].

My parents took me to see Henry Wildey and told him of my interest and the problems of the seaside telescope. Henry gave me an ex-War Department gun sight elbow telescope. I suppose it was the equivalent of one half of a pair of 7 x 50 binoculars - not much of an optical aid by present standards but I was delighted!

With my little telescope I could project the Sun onto my bedroom ceiling and study its sunspots - although my mother soon forbade this claiming that I would burn the house down!

I could see the main features of the Moon and the larger craters.

I could see the Galilean satellites of Jupiter.

Saturn's rings were a bit beyond my instrument but I imagined that I could see something bulging out on each side of the tiny disc.

Losing My Religion

There was a terrible blot on my landscape.

My family lived just to the north of a very high church - in both the architectural and religious senses. It was so high that from our windows the Sun, Moon, planets and other interesting objects predominantly in the Southern skies were often hidden by that dreadful roof.

How I cursed that church and all who were associated with it!

Look at the enormous monstrosity - like an ugly cathedral dumped down in a quiet residential area in North London.

At the age of nine I suddenly realised that God did not exist and I became an enthusiastic atheist from that time onwards; my antagonism strongly fuelled by that horrible building.

In 1949, at the age of nine, my atheist convictions were

confirmed by four years of hell.

My father died when I was nine years old and I was sent to the Freemasons' boarding school (the Royal Masonic Institution for Boys at Bushey in Hertfordshire).

This was the start of four of the most miserable years in my life which I have published in my book 'My Lost Childhood' [14].

I desperately needed a home life and yet I was being taken into an abusive and terrifying environment just after the death of my father. This was the worst thing that could have happened to me.

Most of the teachers were incompetent abusive bullies.

A friend had an allergic reaction to tomatoes. Despite this he was forced to eat tomatoes which he vomited up.

He was then sent to the Headmaster carrying the plate of vomit and he was forced to eat the regurgitated mess.

One nine year old pupil had projecting ears. He was teased mercilessly - by a teacher! - who called him 'Royal Air Force'.

My love of astronomy was all I had to take my mind away from this hell.

Few people could have come to astronomy as I did and appreciated it from such an early age, with such passion and with so much need.

I survived the bullying from boys and teachers alike through my intense interest in the heavens.

My First Astronomy Book

For my eighth birthday my parents had given me a wonderful little astronomy book.

By today's standards it is uninspiring I suppose but, to me, it was wonderful. I read it over and over and almost knew it by heart. It had wonderful blurry photographs of the Moon and planets.

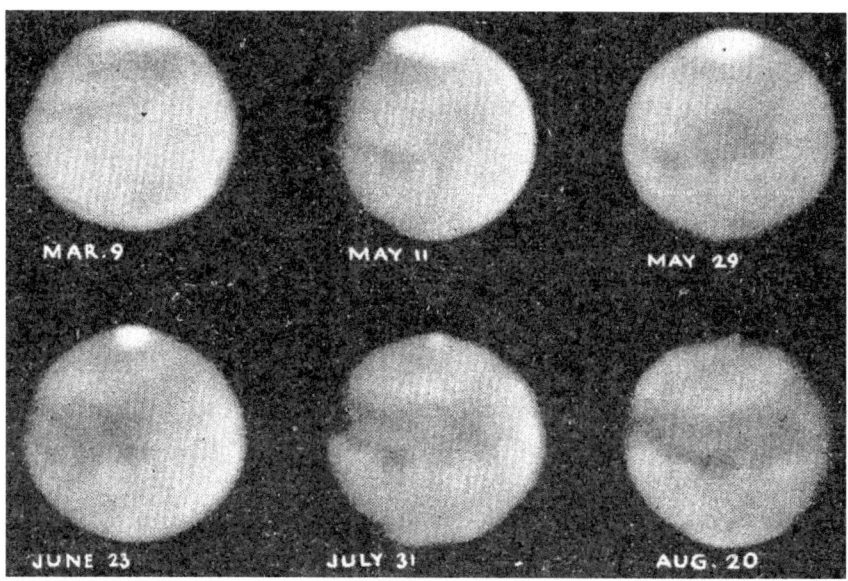

Above we see a display from this book of a sequence of photographs of Mars. These - incredibly - were the best available in the late 1940s.

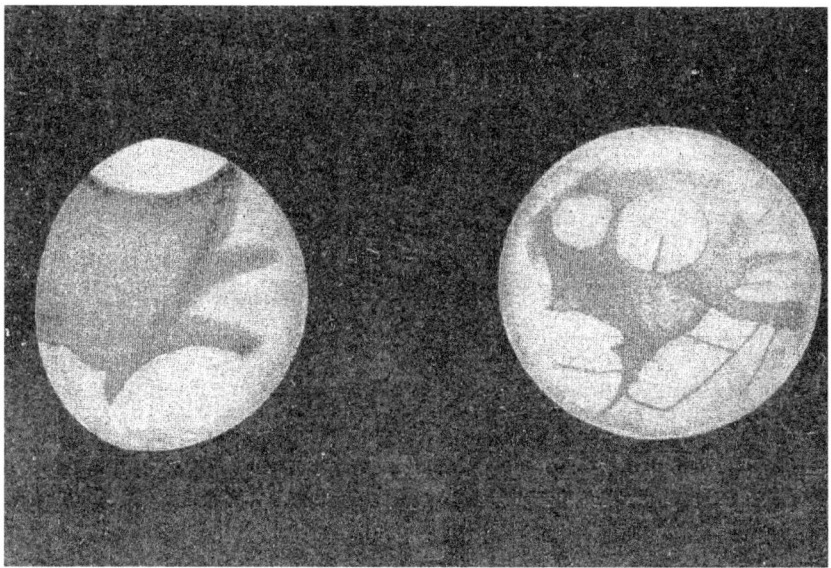

Above we see two visual drawings of Mars from this book captured during moments of good atmospheric seeing with a large telescope.

Note the 'Martian Canals' which were much in evidence well into the 1940s.

Although some astronomers doubted these features were artefacts of Martian engineers as claimed by Percival Lowell, the reality of the 'canals' was not disputed when I was a child.

In the two-volume and magnificent encyclopaedic *'Splendour of the Heavens'* published in 1923 fifteen pages are devoted to describing the 'canals'.

Changes in colour in the 'canals' and nearby large dark areas were widely assumed to be vegetation. Only the 'canals' artificiality was debated; not their reality.

Most wondrous of all there was an artist's impression in my little book of what Martians might look like! I came to believe in my heart that there really were Martians and they looked like this...

The reasoning behind the above illustration was based on the assumption that Martians were developed millions of years beyond humans. Martian brains were huge and their bodies had largely withered away due to all their physical needs being provided by machines and computers although in the late 1940s 'computers' as we now know them were beyond comprehension.

Indeed, in a classic science fiction story written in the late 1940s but set well into the 21st century, space travellers on their way to Mars compute their trajectory using a slide rule! [15]

Ideas about the surface of Venus have also changed widely over the years.

In the early 1950s nothing was known about the surface of

Venus. Some scientists believed it could be a searing dry desert, others that it was covered with oceans of hydrocarbons whilst yet others believed there could be life there in a world made up of steaming jungles.

A view of the surface of Venus depicted as a scorched desert painted by Chesley Bonestell (of whom more later) can be seen at [16].

It was entirely possible that weird creatures - even intelligent beings - lived on Venus. So little was known about Venus that almost anything was possible.

So, in 1950 when the Eagle comic started publishing the adventures of Dan Dare and his battles with the inhabitants of Venus - the Treens who were ruled by the deadly and ruthless Mekon - nobody could prove it was not really like that on Venus - see [17].

Incidentally, I had pristine copies of the first one hundred issues of 'The Eagle' comics but my mother threw them away - they would have been worth a fortune now!

The Inspiring Paintings Of Chesley Bonestell

A great inspiration to my budding interests in astronomy were the paintings of Chesley Bonestell.

Although there are many 'space artists' around today Chesley was the first to inspire the public with his visions of space. Indeed, he is credited with getting the US space programme under way through his beautiful and awe inspiring paintings.

He was born in 1888 and died in 1986 having seen many of the images he painted become reality as space probes visited the Moon and planets.

Although he started publishing paintings in the late 1940s it was the paintings published in 1952 that took the world by storm.

They showed the public what was out in space.

They appeared in advertisements, on children's lunch boxes, on picture cards - they were all over the place.

I saw Chesley's pictures at that time and managed to borrow a copy of Willy Ley's 1949 book 'Conquest of Space' which Chesley illustrated [18].

The one picture that overwhelmed me was the one entitled 'Saturn from Mimas'. You can see it here [19] amongst many of

Bonestell's most famous pictures. Click on each of them in turn and be amazed by his artistic talent and limitless imagination.

The huge globe of Saturn filling so much of the sky was amazing enough but, after studying the picture for a while, one experiences a shock on seeing the tiny figures of astronauts on the surface of Mimas in the bottom left corner. It is this stunning touch that brings home the scale and wonder of the scene.

It is a masterpiece of perspective, scale and imagination.

Chesley performed intricate calculations using the best astronomical data available at the time to show planetary scenes as accurately as possible.

If his pictures are now known to be inaccurate it is only because our knowledge has moved on.

I was so in awe with this picture of Saturn seen from Mimas that I cut it out of a magazine, carried it around and showed my mother, my friends and my teachers. I don't think that any of them appreciated my sense of wonder. Only I knew that there really was a satellite of Saturn called Mimas and that this must be what the view really looks like.

Look again at the figures of human explorers in the bottom left-hand corner. When this picture was originally published, the magazine editor insisted on adding the caveat

"The human figures are included solely to give a scale to the picture"

A young Arthur C. Clarke was furious at this presumption that the scene was beyond expectation of reality and penned a strongly worded letter supporting Chesley's belief that humans would sometime travel to Mimas and view this scene.

Chesley lived to the age of 98 and died leaving an unfinished painting on his easel.

Chesley Bonestell's amazing paintings are widely available to view and buy as posters on the Internet.

I was lucky to be given a copy of the book *'The Exploration of Mars'* printed in 1956.

This book was written by Willy Ley and Wernher von Braun; the latter who was responsible for the Nazi V2 'Vengeance' weapons that rained down on London when I was a child. The book was illustrated by Chesley Bonestell.

The cover illustration shows a manned spacecraft approaching Mars - complete with 'canals' spreading like a spider's web over the planet - see [20].

Mars was still believed to be covered in a web of 'canals' well into the 1950s.

The Film 'Destination Moon' Is Released

In 1950 the Film *'Destination Moon'* was released.

My boarding school teachers raved about it - as did the press.

For details of this film please see [21] and [22].

It was advertised as

'The most accurate representation of a manned journey to the Moon ever filmed, accurate in every scientific detail'.

I nagged my poor mother into taking me and I sat goggled eyed throughout.

It was amazing!

The special effects were 'out of this world'!

Recently it was shown on television and I eagerly waited for it but it was utter rubbish! The rockets were obviously models and the scenery crude inaccurate plaster models.

At a late stage in the preparation for the flight to the Moon one of the astronauts is taken ill. A mechanic is substituted. I think he must have been a car mechanic from a back street garage because he knew very little about spacecraft and travelling in space!

He panics nearly all the time.

Once free flight is reached he floats up from his bench.

"What's happening? Why am I floating?"

he asks, goggle eyed and sweating.

The rocket ship commander explains briefly about zero gravity.

"What's that out there?"

gibbers the mechanic.

The rocket ship commander explains that the illuminated ball hanging in the studio just outside the rocket ship window is the Moon.

"How are we going to get back?"

blubbers the mechanic.

The rocket ship commander explains briefly about landing on the Moon and coming back.

This mechanic is more of a liability than an asset! In fact, at the end of the film (Spoiler Alert!) he is left behind on the Moon.

By the end of the film I am falling off the settee with laughter and a damp stain is spreading from my groin.

How could anyone have raved about this rubbish so few years ago? Because we knew no better and Chesley Bonestell helped us to learn what space was really like. He and Henry Wildey helped to hook me on a lifetime's interest in astronomy.

I was beginning to get restless with my small gunsight telescope and eventually I wrote a letter from boarding school to my mother asking her to buy me a *'parabolic mirror'* so that I could make a bigger telescope.

Now, at the age of ten I knew how a reflecting telescope worked but I was naïve in thinking that astronomical mirrors were there on the shop shelves along with other every day commodities.

My poor mother must have been distraught. She had no idea what I was asking for and could neither had found one or been able to afford one even if she did. What she bought was a thick glass car headlight reflector. It was about 125 mm diameter with a focal ratio of about f/1 - totally useless as a telescope optic.

Although initially disappointed that I was not going to own a 'proper' telescope, the discovery that my headlight reflector could concentrate the Sun's rays and set fire to wood, grass and paper made me a star attraction with my fellow pupils.

My mirror was a vicious weapon in the wrong hands - and mine *were* the wrong hands!

The Hampstead Observatory

During my holidays from school Henry Wildey encouraged my interest in the heavens. He was the Curator of the Hampstead Scientific Society Observatory. This was a dome on the highest point in London adjacent to the Whitestone Ponds.

The first night he took me up there was sheer magic. I was about ten years old. We walked up the long hill and then I saw the dome - just like I had imagined except some vandal had daubed

'It's a pudding!'

on the side.

I was shocked at such sacrilege at this temple to the stars.

Mindless graffiti is not a recent phenomenon.

Henry opened up the door and in I stepped.

There, in the dim light, was a superb 150 mm aperture Cooke refractor. It was all gleaming brass and knobs. Henry wound up the clockwork drive and a large paddle wheel started to silently spin.

That magnificent telescope is still in use and still belongs to the Hampstead Scientific Society. There are details and photographs of the observatory at [23].

On the walls were black and white photographs and diagrams.

No colour pictures from the Hubble Space Telescope in those days! There were very few pictures available and they were rather boring by today's standards. The two pictures I remember were all-time classics. One was Halley's Comet in 1910 with Venus close to the tail. The other was of M31 - the Andromeda Galaxy - with a meteor trail streaking across it.

Henry opened the shutters and the stars were visible. It was the most exciting event in my life so far! I looked at everything he turned the telescope to with amazement and wonder.

He showed me double stars with contrasting colours.

He showed me nebulae.

He showed me the Moon and I was dumbstruck by the myriad of mountains, craters, ravines and 'seas'.

He pointed out the Leibnitz Mountains at the South Pole where there were deep valleys into which the Sun had not shone for billions of years. Was there ice in those black valleys? Were there creatures living there under a thin blanket of air and water?

The wonder of much of the universe has been spoilt by the space probes. In around 1950, so little was known about the bodies in space - even the Moon which Patrick Moore suggested in 1949 might harbour vegetation.

In the 1950s the surface of Venus was unknown and could have been the home to creatures. Mars was still believed to be covered in 'canals' which could have been artificial.

O'Neill's Bridge was a hot topic in the mid-1950s and believed - briefly - to be a twelve mile span bridge possibly constructed by an alien race.

More on this 'bridge' in Chapter 9.

I still remember my first sight of Saturn through that Cooke refractor. It seems now as if it was a huge pink globe almost filling the field of view. The ring system was wide open and I saw the Cassini Division splitting the ring into two.

At that point I decide that I was going to get myself a 'proper' telescope but it would be a few years before I got my first reflector.

In the meantime, I contented myself with viewing through the Hampstead Cooke refractor on every clear night under Henry Wildey's guidance.

'Splendour of the Heavens'

Henry once took me to a meeting of the Royal Astronomical Society. I was about eleven and I was greatly in awe of everything and everybody.

Before the meeting I was introduced to J. G. Porter - see [24] - who was the 'Patrick Moore' of the time.

Porter broadcast regularly on popular astronomy and wrote many books which brought astronomy into the homes of people in Great Britain in the 1950s.

Henry had some business to attend to and he left me in the library with a copy of *'Splendour of the Heavens'*.

Although full of crude black and white photographs and some rather gaudy coloured drawings, it was enchanting and I vowed I would own my own volumes one day. It took forty years but I eventually found them in a second-hand book shop!

One of the many fascinating and charming diagrams from *'The Splendour Of The Heavens'* illustrates the distance to the Moon in terms of how long it would take a shell, an aviator and a cyclist to travel from Earth to Moon.

The cyclist would take 3½ years pedalling at a rate of 186¾ miles each day. That's a long journey each day on a bone shaking bike of a century ago!

An aviator is shown flying in a crude monoplane travelling at 62¼ miles per hour. He or she would take 160 days. Those were the days of leisurely air travel!

When Henry took me into the meeting we stood in silence for a minute in respect for a Fellow who had died. Then we had a talk on *'Interstellar Dust'*. I managed to follow most of the talk and I recall being impressed with the scarcity of interstellar particles.

It was about this time that I met my first 'Independent Thinker' as Patrick Moore has generously called them - see [25]. This amateur cosmologist was outside the Hampstead observatory when I went one evening with my mother. He told us that he didn't believe in the conventional theory of the universe. He believed that the sky was a faceted crystalline shell and that stars were the distant reflections of our Sun.

Perhaps he was right.......?

The 1950s - Change And Inspiration

In 1953 - Coronation Year - I left the Freemasons' abusive boarding school which had ruined my life for the previous four years.

My mother had plans to emigrate to South Africa to join my half-sister and her family. I started to learn Afrikaans and can still remember a few phrases.

However, those plans fell through so my mother and I moved in 1954 to Wembley in Middlesex.

My treasured gunsight telescope was stolen by the removal men.

In 1955 I sat my O-Level examinations and had little time for astronomy.

In 1957 I sat my A-Level examinations and went to Queen Mary College, London University to study Physics.

This left me with quite a lot of spare time and I embarked, in that year, on my first 'real' telescope - a homemade 150 mm (6 inch) aperture Newtonian reflector.

Perhaps I was inspired on this enterprise by the launch of Sputnik 1 in October 1957.

This caused a huge interest - and fear - because nobody believed that the Soviets could achieve such a feat.

Space fever gripped the world and my mother and I would stand for ages in our backyard hoping to see this first space craft.

We saw it several times and it seemed so wonderful at that time. There was this starry point of light gliding across the sky like a silent aircraft and we watched it in wonder.

Also in 1957 there was the appearance of Comet Arend-Roland which was the first unaided eye comet to appear in my lifetime.

I went to a local vantage point one clear evening along with many other interested members of the public. They were pointing up into the sky, buzzing with admiration and excitement. I failed to see that comet and this brought home to me that my eyesight was not good.

In October 1959 the Soviet space probe LUNA-3 passed behind the Moon and transmitted back the first images of the Moon's surface which is permanently hidden from observers on the Earth [26].

There was enormous excitement as the world waited for the pictures to be released, more so than any images later transmitted back from the planets or their satellites.

The fact that humans had never seen nearly half of our Moon was an obsession for millennia for sky watchers as well as poets, philosophers and artists.

On the morning that the first pictures were printed all over the newspaper front pages I grabbed a copy and gazed in wonder at it all the way on the London Underground to my university.

Not only was this an amazing engineering achievement but the blurry picture showed a Moon so unlike the side that we were familiar with covered in large 'lava seas'.

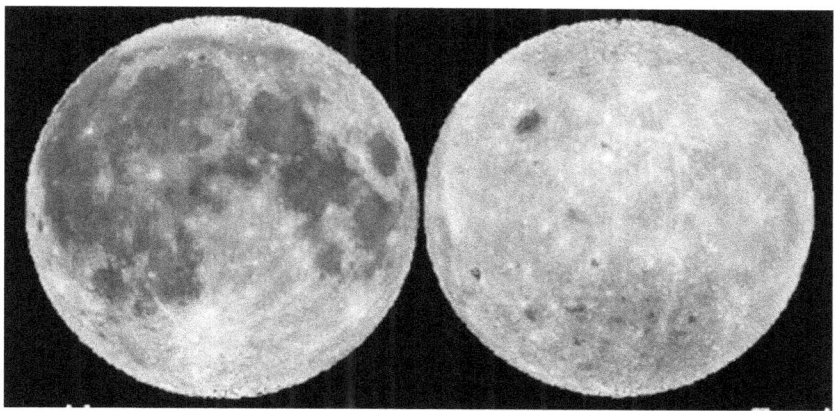

Curiously, Patrick Moore had predicted that the unseen side of the Moon would be very different from 'our' side and he was right.

In 1959 I joined the British Astronomical Association (BAA).

This was a proud moment in my life. I felt that I had arrived in the astronomical world. However, my relationship with the BAA was not always smooth.

The story of the BAA as far as my membership was concerned is the subject of Chapter 4 so I will refrain from further comment until then.

I had joined the BAA in 1959 and then resigned in 1992.

Apparently my letter giving my reasons for resigning was read out at a BAA Council meeting by Patrick Moore - an honour I suppose!

This letter is reproduced on page 139.

The next chapter is a ramble through my observing logbooks. This is a gentle way of introducing you to the varied projects and interests described in the rest of this book.

3. From My Log Books

An Amateur Scientist Experiments

In 1953 the Junior Astronomical Society was formed and later renamed The Society For Popular Astronomy.

I joined soon afterwards and was immensely proud to receive the journal 'Altair'. I felt so honoured to be a member that I put the letters JAS after my name on my school homework.

Between the ages of about nine and fifteen I had a passion for experimental science - much to the alarm of my mother - and astronomy was just one aspect of this interest.

I was often cutting open batteries, heating up coins, and mixing chemicals from a simple chemistry set which my mother eventually threw away believing that I was about to blow up the house.

Many of my experiments were highly dangerous.

Those that weren't were revolting.

For those of you with strong stomachs here are a few of my early teenage experiments.

I made a two-pronged fork out of Meccano and persuaded a friend to hold the 'handle' whilst sticking it into the mains socket. (I may be mad but I am not stupid!)

He was a little nervous but I explained that the electricity would go up one tine of the fork and down the other without going up the handle to his arm. He took the precaution of wrapping paper around the handle.

That probably saved his life.

He poked the fork into the mains socket. There was a bang and a dazzling flash and the main house fuse blew.

My friend was on the floor on the opposite side of the room shaking with alarmingly large staring eyes.

I rushed off and disconnected the mains wire from a bedside lamp, blackened the bare wires with candle soot. I told my mother that I had caught my foot in the cable, the wires had come out and blown the fuse.

She believed it.

By that time my friend was staggering home holding his thumping head and twitching a lot.

I read that manganese mixed with carbon powder and heated could explode. I knew there was manganese in an old fashioned dry battery as well as a carbon rod. I cut open several batteries and ground up the insides. I tried to make them explode by heating the mixture on my mother's gas stove.

Nothing happened.

Very disappointing!

I read that the alchemists believed that the noblest creature (man) must contain the most noble metal (gold). In their searches they found instead white phosphorous.

That sounded interesting to me as I had experience of white phosphorous. During the war a German phosphorous incendiary bomb had fallen near my house and failed to ignite.

I hid it under my bed and often played with it.

That is, until my father discovered it. He went berserk and the house was cordoned off whilst the army bomb disposal team took it away.

I followed the alchemists' prescription. I urinated into one of my mother's best saucepans and set the liquid to boil to a residue. The residue was then to be mixed with powdered carbon and heated at which point white phosphorus should be seen igniting [27].

I had boiled away most of my best and strongest urine when my mother came home from work. Her first reaction was to throw open every window to let out the stench. Her second reaction was to clip me around the ear.

Mothers can be so boring!

However, she probably saved my life by preventing me from inhaling burning white phosphorous.

I have always been fascinated by thunderstorms. I love them and pull back the curtains so that I can have a good look at the uncontrollable violence being unleashed all around me.

In the mid-1950s I used to try plotting the positions of individual flashes. I would set up a small table in the garden covered with a large sheet of paper with a mark in the middle. I would then draw the direction of every flash on the paper and time the delay in the thunder.

Dividing the number of seconds by five gave the approximate distance in miles which I would plot on the sheet. I then added a note of the time and I ended up with a plot showing the progress of the storm as it passed by.

One day I was in the garden plotting away and, as usual, my mother was standing at the back door shouting

"Come in you silly bugger! You'll be killed!"

I ignored her until my hair started to tingle and a stroke of brilliant lightning appeared above me lighting up the entire garden in a vivid electric blue colour and a simultaneous CRASH shook the ground.

I can tell you that, had my time from garden to house been measured, it would have been in the Guinness Book of Records!

A final experiment worth recalling here was a test of psychokinesis - the ability to move objects by only thought power [28]. I had read of people claiming to move quite heavy objects by thought power alone which seemed pretty silly. The effect ought to be tested by looking for very small forces.

I set up a torsion balance consisting of a long bamboo rod with two ping-pong balls at the ends suspended from the ceiling of my bedroom by a fine copper wire. I measured the period of swing and then concentrated on each ball in turn in sympathy with the period of swing in the hope of building up a movement by concentrated thought power alone.

I reckoned that I could have detected a psychokinetic force of well under a dyne (10^{-5} Newton in modern units) which is an extremely small force; about equal to the force exerted on the palm of my hand by a grain of sand.

No oscillations were built up so I decided that I at least had no significant psychokinetic abilities.

A waste of time?

No!

If you do not test for such things you will never know if such an effect exists.

At least this experiment didn't threaten the fabric of the house, my life or the lives of my friends!

My First Telescope

My first true proactive astronomical activity was to build the telescope shown on page 4. This occupied me in the period 1957 to 1959 during which time I was a student studying for a Physics degree at Queen Mary College (as it then was) which was part of London University.

I ground, polished and smoothed the mirror to a near perfect parabola using the simple guidance in the little book *'Making and Using a Telescope'* by H Percy Wilkins and Patrick Moore published in 1956 [29].

There has been an awful lot written about making telescopes and I have no intention of going through all that in this book. What I will do in Chapters 6, 7 and 8 is to describe what I did and you can learn from my successes and failures.

By 1959 I had my first simple homemade telescope in operation and my log book started to fill up.

It may be surprising that I was able to observe anything much with the crude little home-made 150 mm (six inch) aperture reflector shown on page 4.

However, I had a great time with it - partly because I had made it myself from bits and pieces and it had cost me well under £10 - in those days.

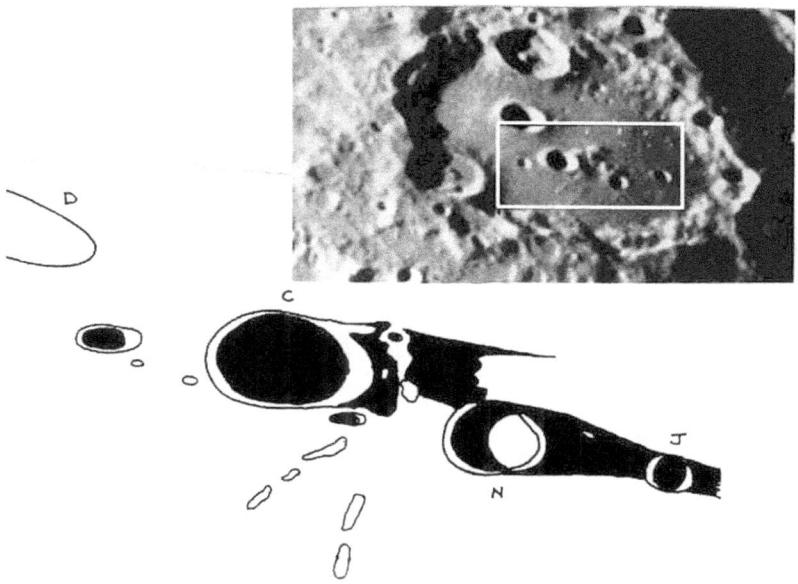

The drawing above was made using this first homemade telescope. The drawing is of the small craters within the rectangle in the photograph of the crater Clavius on the Moon.

The photograph was taken by a large professional telescope.

I was able to see as much detail as is in the photograph.

If that doesn't convince you that a simple, cheap and homemade telescope can perform well - what will?

So, let's browse through my observing log books to see the sort of projects I undertook.

I've selected those comments and observations that reflected my changing addictions to practical astronomy.

It was not until I read through my log books that I realised what my family and friends have had to put up with over the decades.

My early observations were done on scraps of paper and few have survived - a bit like the astronomer Edmond Modeste Lescarbault (1814 - 1895) who recorded his observations on the discarded brown paper bags that his wigs came in or on planks of wood which were later planed off to be recycled.

My earliest surviving observation is dated July 16th 1959.

Drawing The Moon

I had attempted to draw an entire half of the Moon and obviously I had given up after a short time when I realised that it was a pretty stupid thing to attempt.

Patrick Moore's advice, as always, was full of wisdom.

"Concentrate on a tiny part of the Moon and draw it big"

I did and started to draw individual lunar features; mostly craters.

Almost immediately I was collaborating with other BAA Lunar Section members looking for changes on the Moon and also charting the areas near to the Moon's limb.

Remember that when I started observing the Moon nothing was known about the face that is permanently turned away from the Earth.

If someone had predicted that the side of the Moon turned away from us was covered in Flying Saucer refuelling stations nobody could have proven otherwise.

However, the Moon librates; it appears to rock as it passes around the Earth. There is a good simulation of this motion on YouTube [30].

Because of libration it is possible to see 59% of the Moon's surface from Earth. The zone just around the limb which can occasionally be glimpsed was a 'hot' topic of study by BAA members in the 1950s and 1960s.

One aspect was the prediction of craters not visible from Earth by tracing their rays coming over the limb onto the visible side of the Moon.

I will expand on this project in Chapter 9.

These observations reflected my early interest in observing 'useful' things rather than just sweeping around the sky at random.

So many people start off in astronomy buying a large expensive telescope and, within a few months, it is put away in a cupboard or sold at a huge loss because the user has become bored.

Frankly, there are a limited number of things to see in the heavens that can maintain an undirected interest for long.

Stars generally look like stars even in very powerful telescopes unless they are double stars. Galaxies are still fuzzy patches and the planets, whilst offering more variety, can become a little boring after a few dozen glimpses.

When people have asked to see through my telescope I have always been somewhat embarrassed. They expected to see colourful galaxies, intricate details on Jupiter and to see the polar caps on Mars.

Well, Mars usually looks like a tiny pink shimmering disc and nothing like the magnificent pictures returned from the Hubble Space Telescope.

To maintain a lifelong interest in astronomy rather than a fleeting and rather disappointing flirtation, amateur astronomers need a project.

That is what I have done and my various projects are described in the following chapters.

Observing Planets In The 1960s

Most of my observations in the late 1950s and 1960s were of Venus and I was an active member of the BAA Mercury and Venus Section - more of which in Chapter 10.

I did make some drawings of Mars and Jupiter - see below - and Saturn in the 1960s but never seriously concentrated on them.

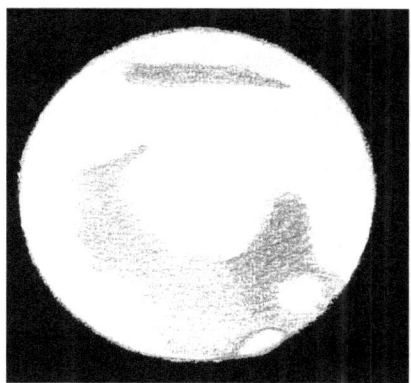

Two Magazines Hit The Newsstands

In the autumn of 1967 a new astronomy magazine was published with Patrick Moore as Editor-in-Chief [31].

It was Patrick Moore's intention in setting up *'Planetarium'* magazine to offer a British alternative to the American *'Sky & Telescope'* which had dominated the world of astronomy magazines since 1941.

I bought all the issues of *'Planetarium'* as they appeared on the newsstands - all nine of them.

From the start the appearance of the magazine was erratic and there was a huge increase in the cover price between the first and the last issue which appeared in September 1969.

Many people, including me, could not understand how an astronomy magazine called *'Planetarium'* could work because the title implied it was about running planetaria (planetariums?) and would contain very little of interest to amateur astronomers.

"What's in a name?"

asks Juliet.

Quite a lot when it comes to naming successful astronomy magazines as Patrick discovered!

The project folded in its original form but was reincarnated as *'Astronomy Today'*.

This appeared twice and vanished without trace leaving subscribers like me wondering what scheme Patrick Moore was next going to cock up - sorry! I meant to say cook up.

It soon appeared in the form of a quirky magazine called *'Astronomy and Space'*.

If *'Planetarium'* had been a bit weird then *'Astronomy and Space'* was off the weirdness scale!

The first two issues started a serialization of the book *'The Planet Mercury'* by E M Antoniadi which had been published in French in 1934 and was reprinted as a translation into English by Patrick Moore in 1974 [32].

The original book was published before it was discovered that all the features mapped on the surface of Mercury were illusions, see Chapter 10 of this book.

Thus, this book about Mercury had a quirky historical interest but the contents were largely irrelevant to readers.

Another article in *'Astronomy and Space'* was a claim that earthquakes were correlated with the lunar tides and, in particular, perigee Spring tides on the Earth.

I puzzled long and hard over the diagrams and charts but could make nothing of it.

Maybe the fact that this paper had been translated from Welsh made it all the more obscure?

Patrick had put out an appeal for articles and I sent him an article about Lunar Occultations.

I received the following letter from Patrick.

```
FROM: PATRICK MOORE.                    FARTHINGS
                                      39 WEST STREET,
                                       SELSEY, SUSSEX.
                                          PO20 9AB
                              SELSEY 3668

        Dear Geoff,

                This is absolutely splendid -

        I am more than gratefu',and it can go

        straight to press.

                I really do appreciate it. More

        anon.

                I am more or less all right

        again, but I had a nasty spell! Damn

        nuisance ...

                See you anon - and once more,

        MANY thanks for this article.

                                Ever

                        Patrick
```

Anyway, this magazine only survived for a few issues, my article was never published and - yet again - subscribers like me were left out of pocket.

2001: A Space Odyssey

In 1968 this film was premiered in the UK on May 15th [33].

I went to see it several times in its full original 70 mm format in 'Cinerama' in a West End London cinema equipped with special projection optics and a deeply curved screen.

My wife and I sat in front of the huge curved screen and were enveloped in both the screen and the all-round sound system.

WOW! What an experience!

It was impossible not to be overwhelmed by the story and its development in such amazing 'special effects' on the screen.

Unlike *'Destination Moon'* which I went to see in the 1950s, '2001' has not aged and I watched it for the umpteenth time only the night before typing these words.

In the mid-1970s I went to see the film in London with a work colleague.

When the Discovery spacecraft first appeared [34] my colleague turned to me and, in a loud whisper asked

"Have you noticed that the spacecraft is exactly the same shape as horse's spermatozoa?"

The magic of the film was briefly lost!

Nobody really understands how to interpret the plot of the film; in particular the last ten minutes.

Stanley Kubrick deliberately left the viewer to interpret the plot in his or her own way; there is no 'right' interpretation.

There are two YouTube videos that provide interpretations of the film's plot. A serious attempt can be seen at [35] and a very funny jokey two-minute interpretation can be viewed at [36].

In the 1980s I painted a huge astronomical mural in my dining room. The centrepiece was my version of the 'Space Child' that appears at the end of the film as seen below.

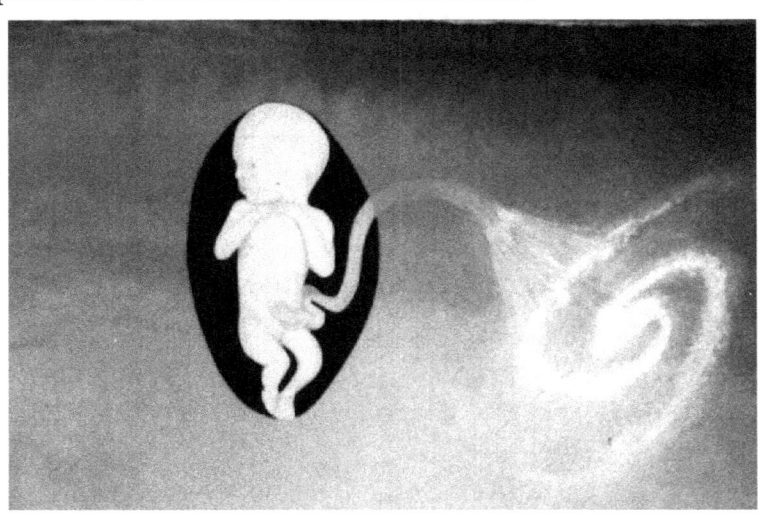

In my interpretation of the 'Space Child' final scene of the film I have the child drawing information from our Galaxy via the umbilical cord. The child is in the process of creating the second great phase in the evolution of humans.

The child is looking away from the Earth because humans will, from this point on, will be partners in a universal civilization in which the Earth will be no more than a sentimentally remembered home, distant in space and time, where humans were cradled.

The full astronomical mural was about two metres high by about six metres long.

Many constellations were represented with stars in their correct positions and named (using Norton's Star Atlas) and the planets included the latest information - such as the rings of Uranus which had just been discovered.

This mural took me about two years to complete.

Because my family got used to it we were surprised when visitors walked into the room and goggled at the wall. To the family, the mural was background but to a newcomer, it was a staggering vista of blue skies, stars, planets and - me piloting a UFO!

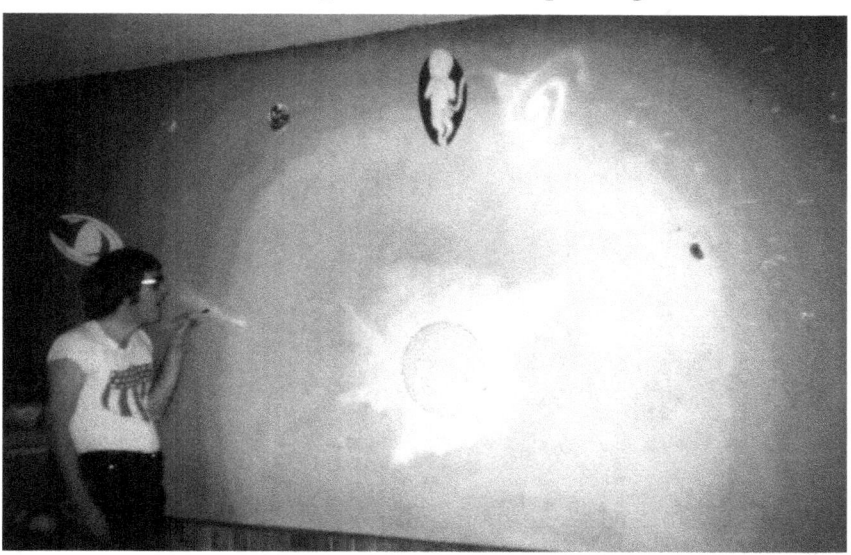

Above I am piloting a UFO with those immortal - but sexist! - words ballooning from my lips

"My mission - to boldly go where no man has gone before..."

The Apollo 11 Moon Landing

July 20th 1969 was an unforgettable date. Men first landed on the Moon [37].

The whole approach and landing was broadcast live. Everyone was listening to the broadcast as the Lunar Lander was flying over the Moon's surface approaching its designated landing place.

I, however, was up the garden with my telescope trained on the Moon.

The neighbours came rushing out believing that I was watching the Lunar Landing Craft flying over the craters and mountains.

I explained that the spacecraft was far too small for me to see with my telescope.

What I was watching for was the flash and explosion of gases if the craft crashed onto the Moon.

The neighbours eyed me suspiciously and shuffled away shaking their heads.

They seemed to think that I was in some way gruesome but, as a scientist, I knew that I stood a chance of seeing the lander craft explode if it crashed.

I was not being gruesome. I was just being a scientist.

Once the Lunar Lander was safely on the Moon I packed up my telescope with mixed feelings.

It would have been exciting to have seen the explosion of unspent fuel in the event of a fatal crash. But, at the same time, I was glad that the astronauts were safe - if only in a peculiarly restricted definition of that word!

Sitting atop a rocket on the Moon knowing that if there was any problem with that rocket the pair would be stranded on the Moon, and very dead, for a very long time indeed.

I stayed up that night to see Armstrong descend the ladder onto the Moon's surface.

At 2 h 58 m UT on 21st July 1969 the first television pictures showed Neil Armstrong starting down the ladder.

Well, in reality, he started to climb UP the ladder because the first pictures broadcast by the BBC were upside down! Once this glitch was corrected I watch the first footprint formed on the

Moon.

Local Astronomy Clubs

I joined the Bournemouth Astronomy Society in 1969.

This was run by a young enthusiast called Colin Pither who deserves to be famous for having bravely ridden on the pillion of Patrick Moore's ancient motorbike which Patrick called 'Vesuvius' because of its habit of exploding and sending out clouds of smoke.

The Bournemouth Astronomy Society was a small thriving club and I gave my first public astronomy talk on 4th March 1971 as shown below.

```
BOURNEMOUTH ASTRONOMICAL SOCIETY


Below will be found the Society's Lecture Programme from
1971 January to May.   I sincerely hope that you find this
varied programme of interest to you, and that we shall look
forward to seeing you at these Meetings;   all of which
commence at 7.45 p.m.

1971  January 7th.      "Cosmological Problems"
                         by A.C. Curtis, F.R.A.S.

       February 4th.     "Manned Trip to Jupiter"
                         by R.E.W. Jansson, Ph.D., B.Sc.

       March 4th.        "New Light on Mars"
                         by G. Kirby, B.Sc.
```

Unfortunately, there was a problem with my maiden talk.

I photographed pictures in books for use in my talk onto slide film and sent the roll of film away to be processed as one had to in those days.

Then the Post Office went on strike so that, as I stood up to give my talk, my slides were languishing in some far away postal sorting office.

I had to *ad lib* my way through the evening.

It was not a success.

The Bournemouth Astronomical Society amalgamated with the Poole Astronomical Club in the early 1970s to form the Wessex Astronomical Society (WAS) which is still a very active and large group of enthusiasts based in Wimborne in Dorset.

In 1972 a coordinated effort was made by the Wessex Astronomical Society to observe the grazing occultation of a star at the limb of the Moon.

The track within which the star should be seen flashing on and off as the edge of the Moon sliced through the beams of starlight reaching our telescopes passed over Hayling Island, a seaside resort in Hampshire.

Members of the Wessex Astronomical Society set up three groups of telescopes across the predicted track and we all saw the star being occulted by the mountains and craters on the edge of the Moon.

More details of this expedition and its successful observation are given in Chapter 12.

It was soon after this event that I was elected Chairman of the Wessex A.S.

I cannot fairly comment on my performance except to say that during my first year in office the committee passed a resolution that the Chairman should not be allowed to stay in office for more than two consecutive years.

Ho Hum...

The years 1972 and 1973 were challenging years for me to be Chairman.

It had been decided to build an observatory for the Society.

Eventually a suitable site was chosen on a consolidated refuse tip at Hengistbury Head; a peninsular of deserted land sticking out into the English Channel. It was a dark site and very suitable.

Discussions with the Dorset County Council Planning Authorities were started and they imposed many difficult conditions; their understanding of what an observatory was like was limited.

Jumble sales were organised to raise money for this worthy project.

I had always been of the opinion that an observatory would not be much used as observers had their own telescopes in their back gardens.

I held a poll of members and it was indeed the overwhelming view that members would not use a society observatory so the whole project was cancelled.

However, today the Wessex Astronomical Society operates a fine observatory at Durlston Country Park near Swanage in Dorset. It is used for open evenings for the public which are hugely successful with large crowds attending events [38].

So my doubts about the viability of a club observatory in 1972 were wrong.

The Durlston Observatory

There were also problems which I had to address which revolved around the creation of a monthly magazine for members.

It was named simply *'Astronomy'*.

Later an American magazine appeared with the same name and we did - briefly - wonder if we would be sued for breach of copyright but we chose our magazine name first and no writ claiming a zillion dollars compensation arrived.

In those days there were no home computers and printers. Every word had to be typed onto stencils and produced on a hand cranked duplicating machine.

Mistakes were careful corrected with a red fluid painted directly onto the stencil sheet. This dried and the stencil was then overtyped - a time consuming and painstaking process.

The early editions of the club magazine had a varied and interesting selection of topics as well as the *'News and Views'* that astronomy club members have come to expect from their magazines.

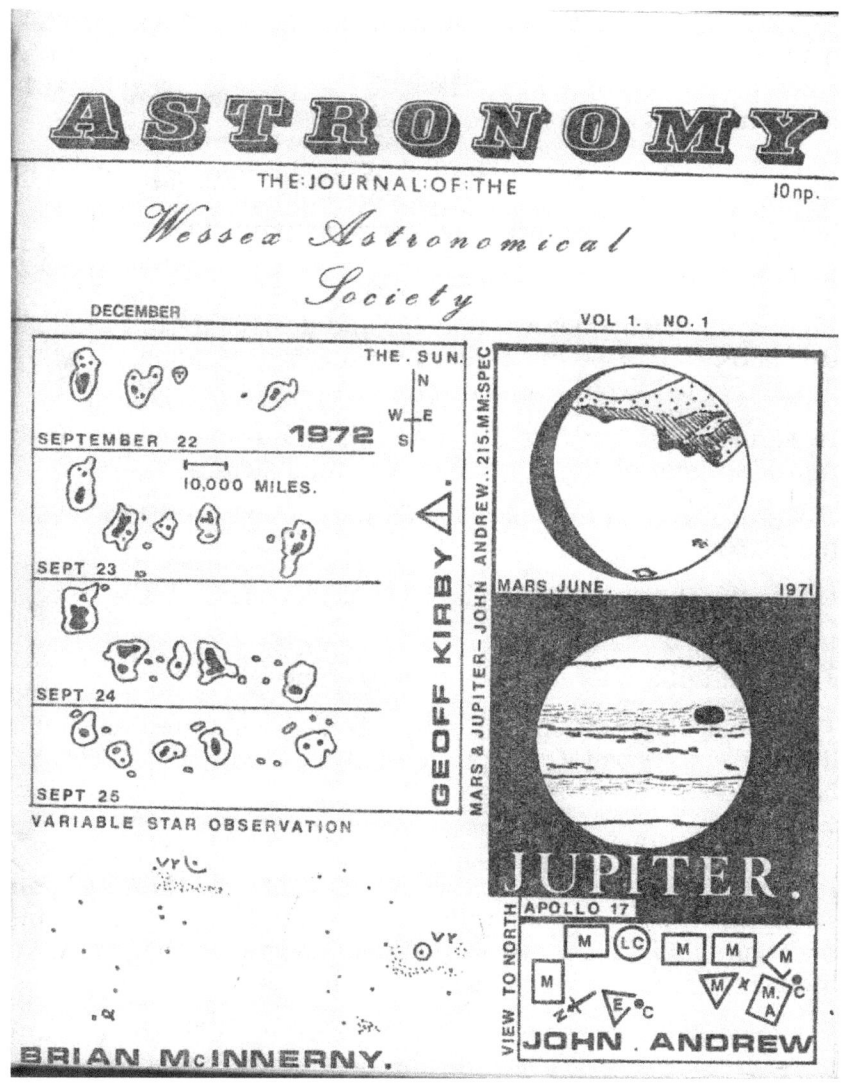

However, there was always the thorny problem of determining copyright of material.

Contributions of original material by members were no problem.

They were only too pleased to see their contributions in print.

However, some sources of information were aggressively

protective of their copyright material and, in the 1970s, the leading US astronomy magazines were threatening to sue the pants off anyone copying and redistributing any of their material; income taking priority over education and inspiration.

Indeed one famous US astronomy magazine did threaten to take one small British astronomy club into court and bankruptcy when it was discovered that the club had copied a page from the magazine and given copies to members.

It was only when the club convinced the magazine that they would never see a penny of the huge amount claimed that the magazine owners gave up their aggressive pursuit.

Breach someone's copyright at your peril!

That is why many pictures that I would dearly love to include in this book cannot be displayed; the copyright issue is either too difficult or too expensive to resolve.

In some cases, such as the BAA, my request to negotiate to reproduce material in this book were simply ignored.

Towards the end of my period as Chairman of the Wessex AS in the mid-1970s an AGM was held.

Unusually, two candidates had been proposed for the post of Chairman to succeed me.

This meant that there had to be an election.

This was a complete surprise and left me unprepared - I should have produced ballot sheets for such an eventuality no matter how unlikely it was.

Then inspiration struck.

I rushed out to the toilet and came back into the meeting room streaming a long length of toilet paper behind me. I went around the room giving each member a sheet and asked them to write the name of their favoured candidate on their sheet.

The toilet paper sheets were then gathered in, votes counted and my successor was elected into the post of Chairman for a two-year period.

What a bum election that was!

Comet Kohoutek - What A Flop!

In December 1973 Comet Kohoutek was widely hyped in the press as to be the 'Comet of the Century!'

As it approached the Sun it gradually brightened and it was forecast in some newspapers as going to be fifty time brighter than Halley's Comet in 1910 which was itself very bright because the Earth passed through its tail. In fact some entrepreneurs made fortunes selling gasmasks in 1910 to protect against the 'poisonous' Halley Comet tail gases!

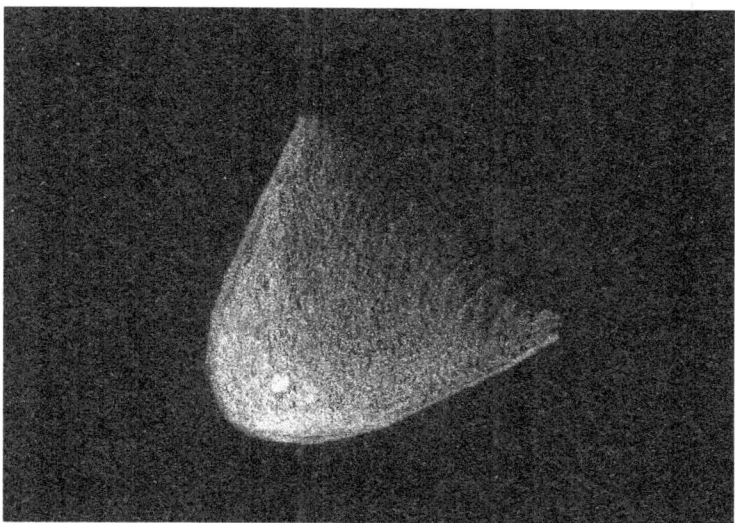

In fact Comet Kohoutek was a flop - I caught a view of it on the morning of 2nd December 1973 at 06.30 UT - but it was rather pathetic after all the media hype of a *"Daylight Comet blazing across the sky!!"* - see my drawing above.

Kohoutek cults arose forecasting that the Earth would be destroyed when the comet 'collided' with the Earth.

For example, David Berg, founder of the *'Children of God'*, predicted that Comet Kohoutek foretold a colossal doomsday event in the United States in January 1974. The cult members distributed Berg's message of doom across the country.

The majority of U.S.-based members then fled in anticipation to existing communes, or formed new ones, around the world.

Older astronomers today measure disappointment in units of 'Kohouteks'.

The failure of Comet Kohoutek to live up to its hype was one 'Kohoutek'.

Chatting up an attractive prospective sexual partner in a bar, believing you are making a good impression and then being given a deliberately false telephone number at the end of the evening with a coy smile inviting you to *"...keep in touch"* - that is five Kohouteks.

Travelling from the United Kingdom to Hawaii in July 1991 to see a total solar eclipse only to spend the day halfway up a mountain being drenched with rain - as happened to a friend of mine - would be rated at about ten on the Kohoutek scale.

Meteor Watching

The Lyrid Meteor Shower is usually active between April 16 and April 25 every year. It tends to peak around April 22 or April 23.

And so, on the night of April 21st 1974 two teams of members of the Wessex Astronomical Society set out to observe these meteors.

I led a group based in Weymouth and a second group set up on Hengistbury Head east of Bournemouth; both sites being in the county of Dorset.

We used stopwatches to record the time of each meteor appearance and we marked the trails on star charts made from laminated copies of pages from Norton's Star Atlas.

There are more details of this activity in Chapter 11 but I will say here that this cooperative venture was very successful with many meteors being observed.

By comparing the tracks of the same meteors from the two locations about 50 km apart it was possible to work out the paths of the meteors through the atmosphere.

This is an ideal group project to get armchair members of an astronomical society out under the stars and enjoying the social aspects of astronomy.

Mercury Occulted By Flimsy Underwear!

One night I was observing Mercury low in the western sky through my telescope when the field became blurred and then Mercury disappeared.

Glancing up I immediately saw the reason; Mercury had slid behind a pair of my neighbour's flimsy lace knickers pegged on her washing line.

I dropped a quickly scribbled note through her letter box, rang her doorbell and rushed back to my telescope.

The note asked the lady if she would remove her knickers so that

"I could see a particularly attractive heavenly object."

Whether or not she obliged will not be revealed here!

My Second Telescope

For two years starting in 1970 I worked on building a new telescope based on a homemade 215 mm (8.5 inch) aperture mirror.

Below it is shown in all its glory!

As usual, I minimized the cost by searching scrapyards, using materials left over from the building of my bungalow and generally buying as little and as cheaply as possible.

The slotted metal mounting was made from scrap bought from a local scrap yard for ten shillings (50 pence).

The moving parts were made from string and Meccano parts costing me £1.

The brick pillar used bricks left behind from the building of my bungalow. The mortar cost me 50 pence.

The sweater was knitted by my mother and the beard I grew for free!

I forget what the mirror blanks and grinding kit cost but probably the whole telescope cost me less than £20.

More details of this telescope are given in Chapter 7.

During the first night of observations of variables stars with my newly completed telescope I saw the variable RV Cyg on October 6th 1974. I recorded that it was the reddest star I had ever seen - even redder than Herschel's famous *'Garnet Star'* - μ Cephei [39].

I describe my decades of observing variable stars in Chapter 14.

October 11th 1974 was a night of destiny for a snail.

I quote from my log book:

"When packing up, I managed to crush a snail between the pages of this logbook which accounts for the yellow stain above"

Ah! The hazards of astronomy - for snails!

In the 1970s I started observing and timing Lunar Occultations - of which much more in Chapter 12.

Spode's Law

Patrick Moore used to quote Spode's Law which referred to the astronomical fact that the more interesting or rare an observation is the less likely it is to be seen.

The reasons could be bad weather, equipment malfunctions, observer illness, etc.

I remember sitting in a bus shelter with a friend in the depths of rural Dorset waiting to time an important occultation of a star by the Moon - see page 320 for more details. It was a perfectly clear sky - not a cloud in sight. The Moon and star were being watched through my telescope; what could go wrong?

An aircraft went across the sky and put a contrail right across the Moon at the exact moment of the occultation so that the star could not be seen.

A minute either way and the contrail would not have obscured the star.

And that, my dear readers, is Spode's Law!

My log book records that on November 2nd 1974 Spode's Law struck again.

I quote:

"Disaster tonight!

I discovered at 22.50 UT that there was a lunar occultation predicted for 23.08 UT!

I rushed out and opened up the observatory but, in swinging the telescope round to the Moon, I sheared off the lower polar axis with the result that the telescope crashed to the ground.

However, I still managed to make a timing of the event with the telescope propped up on bricks!"

Little interfered with my observing passion. Not a lot stopped me from getting out into the observatory.

On December 9th 1974 I recorded

"Having moved house five days ago, these are the first observations from the new garden. I used the 100 mm (4 inch) refractor by balancing it on emptied packing crates."

Observing The Most Southerly Stars From The UK

On January 17th 1975 I made the sort of observation that kept astronomy alive for me. As I have said above - and will say again - an interest in observational astronomy will only thrive if you take on projects.

These can be very simple like the following recording in my log book.

"I searched for the star furthest south that I could see from my garden with binoculars. I managed to pick up π Puppis at about 2.5 degrees altitude due south."

In theory I ought to be able to reach -40° declination (my latitude *minus* 90° *minus* about 1° for atmospheric refraction). It is surprising how many unfamiliar southern constellations are partially visible from the UK.

The table on the next page lists some challenging stars to try and find in binoculars or a small telescope. All rise no more than 5 degrees above the southern horizon from the south of England.

How far north in latitude can these stars be seen?

In the following table the undimmed visual magnitude (Mv) is shown. The apparent brightness could be several magnitudes fainter due to atmospheric extinction. The column 'Dimmed' shows the typical apparent magnitude when viewed from the south of the UK.

Also shown is the date when the star is approximately due south at midnight from the UK. This will give a guide when to look out for a particular star.

Star	Mv	Dimmed	Dec	Culmination
π Puppis	2.8	6.1	-37	January 13th
β Pyxidis	3.9	6.2	-35	February 2nd
α Pyxidis	3.6	5.2	-33	February 4th
κ Velorum	4.6	7.9	-37	February 11th
ε Antiliae	4.5	7.2	-36	February 16th
ι Antiliae	4.6	7.9	-37	March 8th
o Hydrae	4.7	7.0	-35	March 20th
β Hydrae	4.6	6.5	-34	March 23rd
p Centauri	4.9	6.8	-34	April 21st
χ Lupi	4.0	5.9	-34	June 6th
ε Scorpii	2.2	4.1	-34	June 22nd
η Sagittarii	3.1	6.4	-37	July 14th
γ Coronae Australis	5.0	8.3	-37	July 27th
γ Gruis	3.0	6.3	-37	September 7th
β Piscis Austrini	4.2	5.6	-32	September 16th
α Piscis Austrini	1.1	2.2	-30	September 22nd
γ Sculptoris	4.4	5.8	-32	September 28th
β Sculptoris	4.3	8.3	-38	October 2nd
α Sculptoris	4.3	5.3	-29	October 24th
β Fornacis	4.4	5.8	-32	November 5th
α Fornacis	3.9	4.9	-29	November 11th
δ Fornacis	4.9	6.3	-32	November 18th
β Caeli	5.0	8.3	-37	December 4th
α Columbae	2.6	4.5	-34	December 18th
ζ Canis Majoris	3.0	4.1	-30	December 29th

Notice how many constellations are shown which are often thought to be only visible from the southern hemisphere.

Observing Asteroid Eros In 1975

On 17th January 1975 I picked up the asteroid Eros (433) on one of its rare passages close to the Earth.

I found it in binoculars and estimated its visual magnitude (Mv) at 8.1 using comparison stars in the field of the variable star Y Tau.

Some thirty-five minutes later I made it 7.9 and believed that this variation was real even though it was only a small difference.

Even later in the night I made it 6.9

In fact Eros does vary quite a lot in brightness during its rotation.

Unfortunately, I was not able to follow it throughout a complete rotation (5.27 hours) but I got about 80% before clouds came in.

The plot below shows how the brightness appeared to change on 25th January 1975.

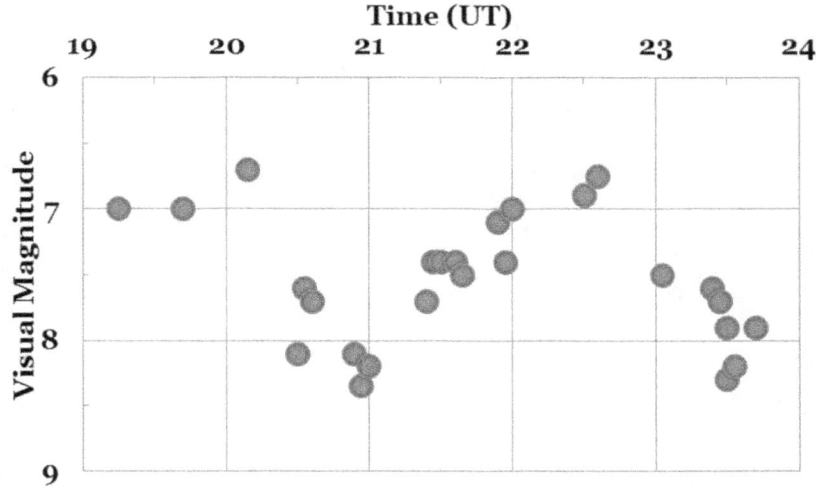

My estimates of the brightness of Eros as it rotated.

I reported these observations to the BAA Minor Planets Coordinator and they created a vigorous debate as to whether they were reliable.

It was justifiably pointed out that the claimed changes in

brightness were large over a small angle of rotation. At that time it was believed that Eros might be only a little irregular in its shape.

In 1975 no asteroid had been seen as other than a point of light.

However, to get such large changes as I had claimed would require the asteroid to be very far from spherical in shape.

It was suggested that perhaps Eros might have been producing flares as gas escaped from it surface as it approached the Sun.

When we look at recent space probe images of Eros as it rotates - see below - we can see that it is very irregular and that the brightness might indeed change considerably during a rotation especially if it were rotating end over end as viewed from Earth.

Furthermore, the light curve is different for each close passage of Eros to the Earth because the angle of illumination by the Sun and the position of the Earth relative to the rotational poles of Eros will be different.

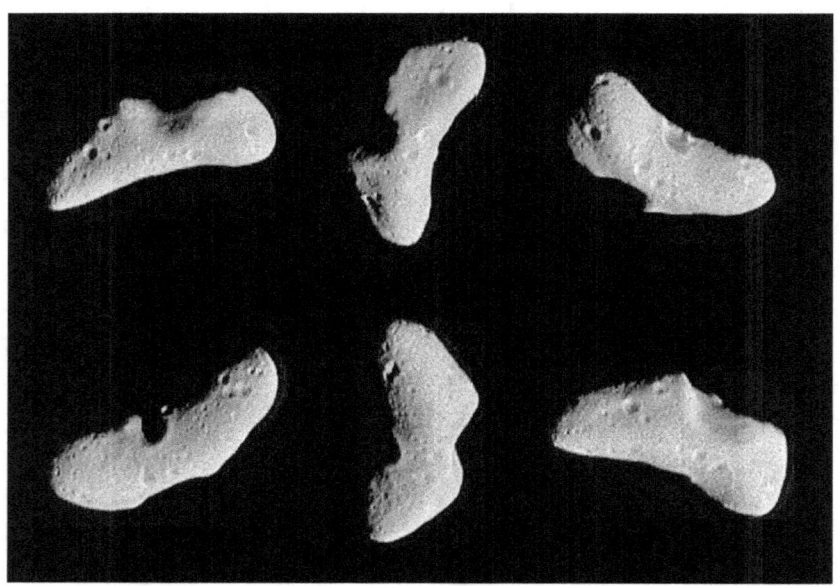

Images of Eros rotating as seen from the NEAR space probe.

Unseasonably Cold Weather

On March 30th 1975 I wrote

"Snow fell and laid over this Easter weekend and it became very cold especially in the fierce northerly wind."

Despite this, I made thirty-six variable star estimates using binoculars whilst stretched out on a sun lounger in the garden.

I made thirty-one observations using binoculars of variable stars a week later which was the coldest April night on record. There was a very severe frost.

On May 4th 1975 I recorded

"A very fine night with no haze. It was breath-taking just to sweep around at random with the binoculars."

I made forty-five variable star estimates - the largest number in one night up to then.

Illness did not prevent me from getting out in the garden. On June 4th 1975 I wrote

"I woke up with a really bad cough in the early hours and decided to do some variable star observing from my sun lounger in the garden with my binoculars until I felt better"

A kill or cure remedy?

A Child's Birth Heralded By A Comet

July 2nd 1975 was a momentous day.

I recorded in my log book

"My second son was born at 2 am BST this morning and I was at the birth. Tonight I am a little tired having had no sleep last night."

Despite having been up all the night tending to my wife in her lengthy labour, I still managed to do many variable star estimates the following night!

The need for sleep is an illusion suffered only by non-astronomically people.

On July 11th 1975 the first of many observations of comet Kobayashi-Berger-Milon (1975h) were made. It was estimated at visual magnitude 5.6 on July 21st so it was quite a good comet in a telescope but not unaided eye brightness.

It was never seen as more than a formless circular blur by me.

A Total Lunar Eclipse

There was an excellent total eclipse of the Moon on November 18th 1975.

I made several dozen crater timings. These are best estimates of time that the edge of the umbra is halfway across a crater. These measurements are used by professional astronomers to work out the diameter of the Earth's shadow. This may sound an odd thing to do but there is an error of about 2.5 % in the observed diameter of the Earth's shadow on the Moon compared with theory. This is believed to have something to do with the Earth's atmospheric conditions.

I made several timings of star occultations (stars disappearing or reappearing from behind the Moon - see Chapter 12) whilst the Moon was in total eclipse.

These observations were very important at the time because star occultations by the Moon could be used to give a very precise fix on the size, shape and orbital position of the Moon.

Such observations cannot usually be made around the time of Full Moon because the starlight is drowned out by the glare of the Moon. However, when the Moon is in eclipse stars can be seen that would otherwise be invisible.

During the total eclipse I could see stars as faint as visual magnitude ten whereas the usual limit is about visual magnitude seven; a factor of sixteen times brighter.

The Moon was brick red.

I measured the brightness of the eclipsed Moon. This is an easy thing to do using my 'eyepiece on a stick' technique. This proved to be a very simple and accurate method for measuring the brightness of the Moon.

The idea is simple. Tie an eyepiece on a long pole. Hold the eyepiece near the direction of the Moon. A tiny image of the Moon is seen which can be compared with a nearby bright star. The distance of the eyepiece is adjusted until the Moon's image and star are equal apparent brightness.

As the Moon's brightness changes during an eclipse, its relative brightness is inversely proportional to the square of the distance of my eye to the eyepiece.

The brightness of the Moon changes a great deal during the course of a total eclipse - sometimes it dims by as much as 100,000 times.

Subjective estimates are made on the Danjon Scale [40] and these correlate, to some extent, with atmospheric clarity which is itself a function, for example, of recent volcanic activity.

The measurements using the 'eyepiece on a stick' technique are calibrated by knowing that the full, unobscured Full Moon has a magnitude of -12.7.

This method is simple and accurate.

An article written by me was published in *'Sky and Telescope'* in the mid-1970s demonstrating my method of measuring the brightness of the eclipsed Moon with an eyepiece on a stick.

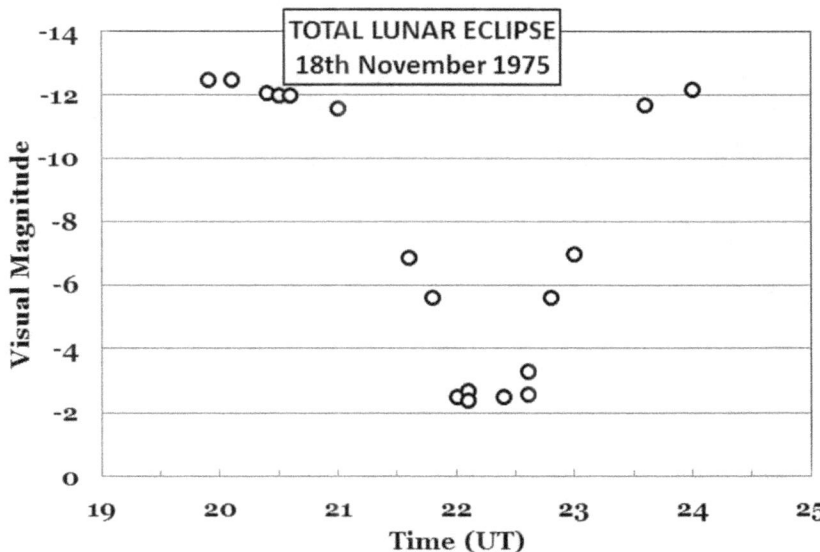

The above diagram shows results obtained by this method by my friend Don Miles and me.

This shows that the Moon was dimmed to about visual magnitude -2.5 at mid-eclipse. This corresponded to a dimming of about a factor of 12,000 times.

Overall in 1975 I made observations on 85 nights (23%) which is very good. I probably did not miss a clear night.

The Unwelcomed Curate And A Brilliant Comet

I set up the telescope to time a lunar occultation on January 20th 1976. However, a new curate called - for some inexplicable reason - and I missed it!

Yet another reason for my resolutely atheistic outlook!

On January 23rd 1976 I made my first eclipsing binary star observation - of U Cep.

Observing eclipsing binary stars became a major part of my observing programme in the late 1970s and throughout the 1980s. Chapter 14 includes information on this topic.

I picked up Comet West at 6 am on March 23rd 1976. Conditions were very hazy but I saw a tail of about $1/5$ degree in length.

This became one of the most spectacular comets of second half of the 20th century.

I didn't see much of it but photographs from more favourable climates and locations showed it in its full glory.

This photograph of Comet West is reproduced with permission [41].

Mars, A Nova And Another Lunar Eclipse

A rare event occurred on April 8th 1976. Mars occulted the star ε Gem in the early hours of the morning.

It was at very low altitude and in the direction of a nearby electricity switching station which obscured part of the horizon.

I set up my refractor looking out of the wide open spare room window with the telescope propped up on cushions with me prone on the bed to observe.

Although I saw Mars and the star until about the time of the occultation, seeing conditions were so bad that the two objects merged into a shimmering mass of light so that I could not actually see the star disappear behind Mars.

The house did get very cold with the window wide open at 2 o'clock in the morning!

My wife had much to complain about - with indisputable justification.

On October 23rd 1976 I recorded in my log book:-

"I had a 'phone call last night telling me that a nova had been discovered - named Nova Vulpecula - and that this had been discovered the previous evening. Today I got details by post and I drew up a chart based on the Smithsonian Astrophysical Observatory Catalogue.

During gaps in the clouds (at one point rain pelted down on my refractor!) I struggled to find the nova.

No sign of it!

I called my informant Jim Muirden only to be told that his position was one hour in Right Ascension in error! Apologies from Jim and a renewed search by me. I did find it and estimated it at visual magnitude 7.1."

A penumbral eclipse of the Moon occurred on November 6th 1976.

It was quite obvious.

Sometimes, penumbral eclipses are not noticeable.

I estimated the Moon's brightness using the 'eyepiece on a stick' method described earlier.

There was a measured 25% drop in brightness.

I Start Hunting For Novae

On March 7th 1977 I took nine pairs of photographs for 'The Astronomer' magazine 'Nova Patrol' survey.

This marked the start of my long series of photographs of our Milky Way Galaxy searching for so-called 'New Stars' or Novae.

More of this in Chapter 13.

Moon And Venus Observations

I noticed on March 28th 1977 that, although the Moon's phase was about 60%, I was clearly able to see Earthshine on the darkened hemisphere.

This is about the closest to Full Moon that I have ever seen the Earthshine.

This would make a very easy 'Unaided Eye' project for beginners in astronomy.

Look for the 'Earthshine' of the darken part of the Moon lit up by light reflected from the Earth.

Note whether it is visible or not. Also note the state of the atmosphere and phase of the Moon.

Is there a correlation?

What is the largest phase at which the Earthshine can be seen?

I discuss the Earthshine more in Chapter 9.

In April 1977 Venus moved from being an evening object to being in the morning sky.

At this time it is between the Sun and the Earth although it usually passes above or below the Sun in the sky.

Mercury and Venus were easily seen with the unaided eye from my bedroom window during a break from decorating on April 1st 1977. This was two days before Venus reached inferior conjunction, i.e., crossed the line joining the Earth and the Sun.

The crescent shape of the illuminated Venus was easily seen in 7x50 binoculars.

On April 3rd 1977 Venus was picked up in my 100 mm refractor just a few hours before inferior conjunction during the daytime.

At 7.00 UT I saw it in binoculars and, with great difficulty, I saw it with the unaided eye at 7.15 UT. At this time Venus was very high above the ecliptic and so could be seen in the evening and morning skies on the same day.

The crescent was horizontal and above the Sun.

WARNING. Observing Venus close to the Sun is **VERY DANGEROUS**. Only a fool would do such a thing. In my case I hid the Sun behind a house roof to see Venus at Inferior Conjunction.

That was still a daft thing to do however as Venus was only a few degrees from the Sun!

The following day I recorded that I saw Venus in the evening sky having seen it in the morning sky the previous day.

A Potpourri Of Observations

On July 19th 1977 I recorded that

"The Ring Nebula in Lyra was very clear and sharp and the core of the globular cluster M13 in Hercules was resolved into a mass of sparkling stars. A very nice night!"

It is nights like these that remind me why I love astronomy so much.

Light Pollution is a problem which astronomers around the world have been grappling with for decades.

Chapter 19 covers this topic but I will briefly illustrate some of the problems and frustrations here.

On September 13th 1977 I took many nova patrol photographs until 22.30 UT when a neighbour's bedroom lights came flood-lighting into the garden.

They *never* closed the curtains.

I got into my pyjamas ready to go to bed but then noticed that the lights had gone out. I rushed out and restarted my photographs up the garden wearing just my pyjamas!

It was a 'brass monkey' [42] night but that didn't deter me.

On November 15th 1977 I observed in my log book that I made my first artificial satellite observation. The BAA supplied the basic elements of the orbit and I computed the track across the sky using the Ministry of Defence computer at my work.

Whoops! Did I really write that!

The first satellite I observed was catalogued as CHINA 6R and it was on its 6161st revolution around the Earth.

This was start of many years tracking and recording artificial satellites of which a lot more in Chapter 15.

On the night of December 12th 1977 I took a long series of photographs for a friend who was to give an astronomy talk.

However, I put the fixer in the developing tank instead of the developer! So - no photographs!

Spode Strikes Again!

By the end of 1977 I had taken a total of 543 nova patrol pairs of photographs.

I was beginning to lose interest and my interest in tracking satellites was growing. At least every time I got a positional fix on an artificial satellite I was contributing to science but, after many years of searching for a nova I had found nothing.

All those sleepless nights and not a single nova discovered - see Chapter 13.

I had also made observations of 48 different variable stars in 1977 so my interests were overlapping to a large extent and were getting more than I could cope with - time to select which to concentrate upon.

Brass Monkeys Were Staying Indoors!

Snow fell on April 10th 1978 - a bit late for that! In fact Weymouth very rarely gets snow at any time and certainly not in April.

I took my sons out on a toboggan.

It was very cold that night (the temperature dropped to -5° C and I was out there on my sun lounger making satellite observations.

On October 6th 1978 I recorded

"Despite violent diarrhoea and vomiting earlier in the day, I am now out observing!"

What devotion!

What stupidity!

In 1978 I had made observations on 80 nights which is about the number of useful clear nights each year in Weymouth on the UK south coast.

On New Year's Day 1979 I recorded

"About 70 mm of snow everywhere in the garden -

temperature about -8ºC and snow on my boots. From my sun lounger I attempted 15 satellites but only got timings on two."

Then on the following evening I wrote

"Snow frozen hard underfoot. Eight satellites timed from my sun lounger"

On January 20th 1979 it was again a fine night but

"...most satellites were missed due to need to wash my three children's hair and get them to bed."

Family came first!

The Moon passed through the Hyades star cluster on December 30th 1979 and I made eight lunar occultation timings. This was really quite exciting as stars were popping in and out of occultation all around the Moon. One of the stars occulted was Aldebaran which was observed with 7x50 binoculars.

In 1979 I secured 154 satellite fixes.

By January 26th 1980 I made my 300th satellite fix.

That sounds a lot but Russell Eberst in Scotland had made over 100,000 by this date. Now there's a devoted observer!

My logbook pages are crammed with satellite observations throughout 1980 with little else recorded.

My last observations of 1980 were made on November 12th and my next logbook entry was in July 1985.

The early 1980s were busy times with three lively children to help bring up.

Evenings were spend providing a taxi service to swimming lessons, cub scouts and scouts, dancing lessons and a whole host of other activities involving school and social activities.

The Arrival Of The Home Computer

The first home computers were coming into use by the end of the 1970s.

I had worked with computers since 1962 when I used a huge mainframe to do calculations that you can now do on a mobile 'phone. It was fed with five hole paper tape which frequently got torn resulting in the whole programming process being repeated.

This monster had a memory made of ferrite rings; each ring

carrying a binary 'one' or 'zero'. A block of 4 kB memory was the size of a four drawer filing cabinet with a built in huge cooling system.

The computer made whizzing noises of different pitches according to how hard the processor was working. I would stand at the massive control panel and follow the program it was running. For example, a typical high-pitched 'whizzing' for several seconds was the computer computing a sine function.

Later monsters were fed by Hollerith punched cards as seen below.

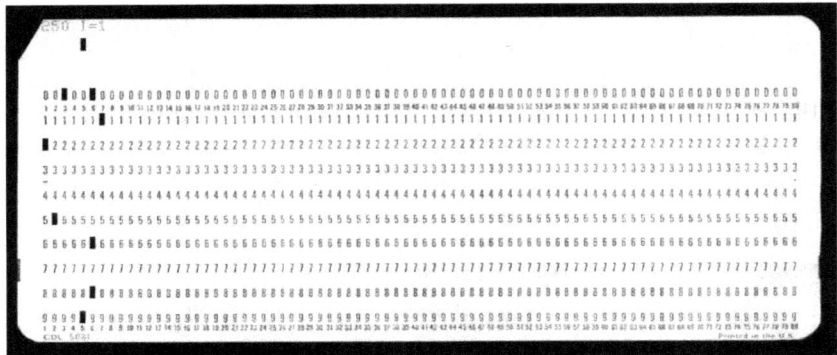

A typical job in the late 1960s might involve feeding over one thousand of these punched cards into the mainframe computer.

These would be carried in steel trays with lids to keep the rain out.

Once I tripped walking to the computer building in the rain. The cards spilt out into a puddle.

Two month's work - almost literally! - went down the drain.

Smaller computers came along in the 1970s such as the PDP 11 series [43] and these were vastly more powerful and were the size of a cupboard.

In the late 1970s a work colleague installed a wardrobe-sized computer in his house and programmed it to play Blackjack.

His friends mocked him.

"Why would anyone ever want a computer in their house?"

We laughed and walked away shaking our heads.

After all, Thomas Watson, president of IBM had once stated [44]

"I think there is a world market for maybe five computers."

As late as 1977 Ken Olsen, founder of Digital Equipment Corporation, stated

"There is no reason anyone would want a computer in their home." [45]

I bought a Sinclair ZX81 computer [46] in 1981 and got very deeply into home computing and gaming so much so that I started writing games for the ZX81 in its machine code language.

This was a challenge with 16 kB of useable memory.

This comprised 1 kB of memory built into the device and a further 15 kB which was added by a very dodgy chip pack plugged into the back of the ZX81.

The BBC computer that I bought in 1984 was a whole new ball game and it allowed me to solve astronomical problems mathematically - all in 32 kB of user memory.

I used the BBC computer to predict artificial satellite orbits seen from my home, analyse variable star data and a lot more.

I moved on to an Amstrad 1614 in 1989 which had 614 kB of memory and was the last computer I owned which had no on-board program storage; all the programs had to be loaded from large 5 ¹/₄ inch diameter floppy discs.

I Embarrass My Son In Front Of Patrick Moore

In 1985 Patrick Moore became involved in one of my most embarrassing events as far as my eldest son was involved.

In that year the Norman Lockyer Observatories outside Sidmouth in Devon were sold by Exeter University to East Devon County Council who planned to demolish the observatory buildings and build a housing estate on the site [47].

This was strongly opposed by the astronomical community.

There was a great history to the site and the instruments. The buildings had huge potential for developing a public observatory.

It was where observations of the Sun by Sir Norman Lockyer showed the existence of a new element which was named Helium after the Greek name Helios for the Sun.

Norman Lockyer was a great scientist and was, for fifty years, the editor of the prestigious science journal *'Nature'* [48].

The observatories had been actively used since 1972 by the

Sidmouth and District Astronomy Society and the society had plans to expand the facilities and open the observatory up for greater public use.

A plan was crafted to obtain financial and material support to get the site fully operational again.

I was invited along as a local keen amateur astronomy and I took my thirteen year old son along to meet Patrick Moore who was to spearhead the appeal to prevent the site being demolished for housing.

It had been very wet before the day of Patrick's visit and Patrick was being interviewed and photographed by the Press in a rather boggy field adjacent to the huge Mond dome.

I was about 3 metres from him when I slipped on a wet patch of grass, waved my arms and legs about in a most spectacular fashion and fell on my back in the mud.

I tried to get up but my feet went again and this time I landed on my front. I wriggled on the grass like a mud wrestler with an invisible but highly skilled opponent,

Patrick stared at my antics unbelievingly and kept on talking to the Press.

As my son and I walked away, me plastered in mud, his shame was overwhelming.

"I just cannot believe you did that in front of Patrick Moore!"

Later that day we finished up in the canteen of Exeter University.

My son and I were on a table of six with Patrick.

We were served lamb which was extremely tough.

My son and I were trying hard not to spit out our indigestible tough meat in front of Patrick.

However, Patrick gave the lead by removing his meat from his mouth and pronouncing

"This lamb must have had a long and happy life! Pity we were inflicted with the corpse!"

I Build A Large Telescope And An Observatory

In the early 1980s I embarked on two major astronomical projects.

One was to build - literally from scratch - a 305 mm (12 inch) aperture Dobsonian telescope.

At about the same time I built a large rotating observatory dome in which to house the telescope.

Chapters 6 and 7 describe the building of this telescope and Chapter 8 describes the building of the observatory.

At last on July 7th 1985 I wrote in my log book

"Back in Business! I have completed the 305 mm aperture Dobsonian telescope and I have calibrated the altitude and azimuth scales....."

At this stage the telescope was standing in the middle of the garden protected only by a plastic sheet.

I spent the next few months building the observatory - see Chapter 8 - and making very few observations.

I first tried to find Halley's Comet on October 9th 1985. The comet was predicted to be about visual magnitude 10 but I did not find it.

Halley's Comet Arrives

I got my first sighting of Halley's Comet on November 10th 1985. It was found in the large telescope but was then seen, by averted vision, in my binoculars.

By December 5th 1985 Halley's Comet was easily visible in my binoculars at visual magnitude 5.5

I took my wife and three children to The Canary Islands to see Halley's Comet in the spring of 1986. The trip is described in Chapter 18.

The comet was rather disappointing but the trip was very enjoyable.

The above photograph shows the comet in the centre of the frame within the star fields of Scutum. I took this picture on the top of Mount Teide with the hand operated *'Braille Barn Door'* camera mount shown below.

This cheap but effective camera mount is described in Chapter 7.

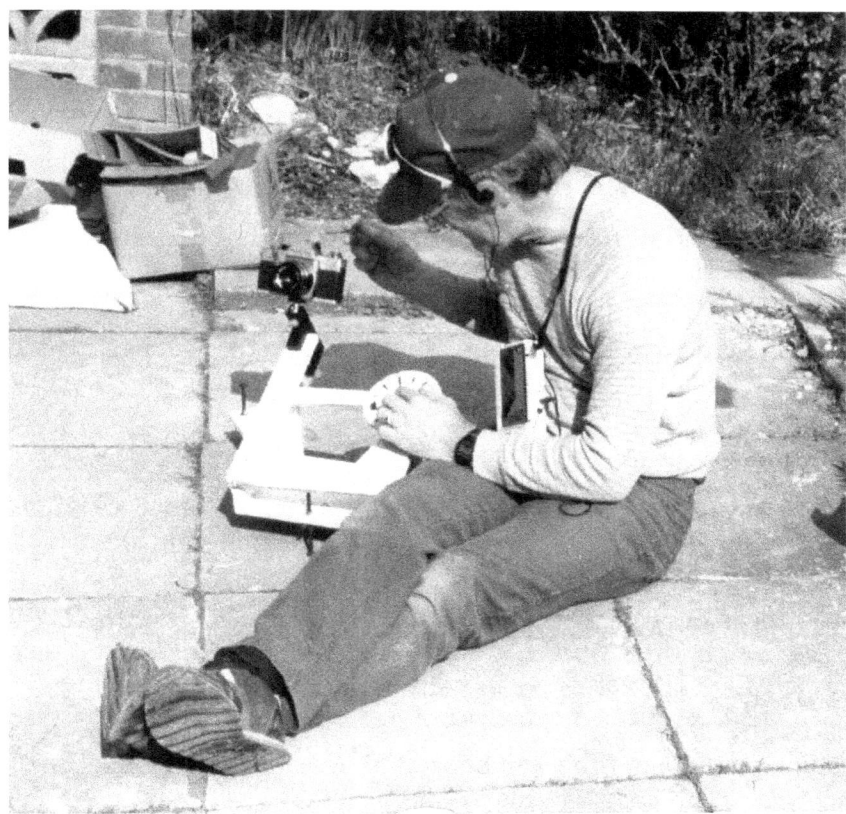

I made a statement in my logbook on March 2nd 1987 that I was now going to concentrate on lunar occultations, eclipsing variables and tracking geosynchronous satellites.

The latter were going to be a challenge as Gordon Taylor (then at the Royal Greenwich Observatory) had told me that the brightest are about visual magnitude 11.5 with most fainter.

However, the challenge of seeing something the size of a large car illuminated by the Sun at a distance of about 40,000 kms is something I had to achieve.

I found my first geosynchronous satellite on March 19th 1987 - and nearly wet myself with excitement!

It was a minute speck of about visual magnitude 11.5 which was near the limit of visibility in my Dobsonian telescope in the urban light polluted skies where I lived.

It was stationary in the field of view of my home-made telescope as the stars drifted past.

I watched it for an hour or so because I was afraid that I would lose it if I moved the telescope.

More information on observing artificial satellites and, in particular, geosynchronous satellites is given in Chapter 15.

BAA Winchester Weekends

Over the period from the early 1970s to mid-1980s, I regularly attended the BAA 'Winchester Weekends' held in the Teacher Training College on the outskirts of Winchester.

These were great fun and a good opportunity to meet likeminded amateurs as well as sinking a few drinks in the evening.

A series of talks were organised over the weekends which were always well attended.

The 'Winchester Weekends' are described in more detail in Chapter 4 along with some of the interesting characters who attended such as Bill and Ethel Granger.

Bill was a large eccentric man who went everywhere with his cat 'Treacle Pudding' on his shoulder and Ethel had a thirteen inch waist; the smallest ever recorded. This waist was the same size as a one litre bottle of wine.

Some pictures of Ethel can be viewed by following the link [49].

They would turn up on a motorbike and sidecar but - unexpectedly - the tiny Ethel would be driving the bike and the huge Bill with cat would be shoe-horned into the side car.

Bill and Ethel's alleged sex life has been published and a strange tale of fetishes, red hot needles and body piercings it turned out to be [50]! Be warned, that document is very sexually explicit!

There are more stories about the BAA and the characters I have encountered in Chapter 4.

I Meet Professor Stephen Hawking

In 1987 I met Professor Stephen Hawking.

I know that is blatant name dropping but it was not as momentous as it sounds.

In the course of my work for the Ministry of Defence I was stuck with a difficult mathematical problem of computing the effect of ocean turbulence on sonar performance.

I contracted Dr Barry Uscinski to work on this problem for me.

Dr Uscinski worked at the Department of Mathematics and Theoretical Physics (DAMTP) in Cambridge University.

I used to enjoy the rarefied academic atmosphere of visits to DAMTP.

One day I was in the staff coffee lounge with Barry and some members of his team when Professor Stephen Hawking came in in his wheelchair accompanied by his team of helpers.

He stopped at the table where we were sitting.

Barry greeted him cheerily and introduced me.

Stephen acknowledged me with a nod.

By this date Stephen Hawking was no longer able to communicate with people like me who did not understand his fragmentary and - to me - incoherent noises.

I sat there dumbstruck to be in the presence of this great genius and all I could do was nod at him.

I think he must have regarded me as an idiot although I guess that he was used to people sitting dumbstruck in his presence.

He moved on to another table where he drank some orange juice that was held to his lips whilst he wore a child's Pelican bib.

It seems incredible that Professor Stephen Hawking is still alive as I write this thirty years later.

Sadly, Dr Barry Uscinski was killed when a replica Spitfire he was flying crashed in Australia in 2010 [51].

Visiting The Canary Islands

I returned to the Canary Islands with two friends in July 1987.

One of these friends had previously been employed at the observatories on Tenerife and La Palma and this gave us open access to the observatories which was wonderful.

Below we see me posing at the highest point on La Palma, above the observatories which gleam brilliant white in the setting sunlight.

The evening blanket of cloud can be seen below the level of the observatories and this blocks the light pollution from the town on the coastline.

In the above view we see the William Herschel dome at left, the Swedish solar tower telescope next to it, the Isaac Newton dome almost hidden and the Dutch Jakobus Kaptyn dome.

This trip is described further in Chapter 18.

Heather Couper Likes My Suggestion

In 1986 the British Post Office invited the public to submit ideas for commemorative sets of stamps to be issued over the next five years.

I wrote to Heather Couper who was then President of the BAA suggesting that a suitable topic for celebration would be the centenary of the BAA which would occur in 1990.

This suggestion was well received and a postcard was sent to me by Heather.

She wrote

1986 May 29

"Dear Geoff,

Thanks very much for the tip-off about British Commemorative Stamps.

Yes, the Centenary is becoming a regular Council item!

I am in the process of writing off to Michael Butcher and will keep you informed of progress.

Sorry not to have seen more of you at Winchester!

Heather (Pres)

The astronomically themed stamps duly appeared in 1990 [52].

The BAA was not specifically mentioned however.

An Astronomical Competition Scam

In 1988 there was a rather curious fiasco in which the BAA got involved.

A leading manufacturer of washing powder held a competition.

On every packet of their product they reproduced a photograph of a part of the heavens. In the original photograph had been an image of Saturn which had been removed leaving just the starry background.

The competition was to invite people to mark on the photograph the position of Saturn at the time that the pictures was taken and the person putting their mark closest to the correct position would win a huge cash prize.

The date and time of the photograph was given.

I don't know how many amateur astronomers entered this competition but there may - just possibly - have been one or two who did not send in an entry.

From the date and time quoted it was easy to work out exactly where Saturn was using information in the BAA Handbook.

I cut out my copy of the photograph from the cardboard box package and using a precision steel ruler marked the exact position where Saturn had been erased. I knew I was accurate to better than a millimetre; I put a lot of effort into getting this right as did countless other astronomers.

The inventor of this competition must have been crapping himself when all these correct solutions came flooding in.

It was just like the people who promoted Hoover products with free flights to the USA which effectively bankrupted the company [53].

A number of Hoover executives were sacked for their parts in the fiasco. They included the Managing Director of Hoover Ltd and president of Hoover Europe, the Hoover vice-president of marketing and the Director of Marketing Services

This management disaster cost the company almost £50 million and the British division of Hoover was sold to the Italian manufacturer Candy.

What was going to happen now that the washing powder executives were overwhelmed with vast numbers of winning entries?

They simply gave the prize to someone whose cross on the photograph was nowhere near the actual position of Saturn!

The BAA got involved by challenging the result with the competition organisers but to no avail.

The organisers defended this debacle by saying that the rules governing this type of competition made it mandatory that no skill must be required to win a prize.

Therefore, the position of Saturn in the photograph was where the competition organisers decided it should be and not where it actually was!

The same rules apply to 'Spot The Ball' competitions like the one shown below where the ball is blanked out and the competitor has to mark where the ball was.

The black disc shows where the ball was deemed to be by the competition organisers and this position, by law, must be chosen at random and not be where the ball had actually been when the photograph was taken! [54]

In fact it is obvious when looking at the typical solution reproduced above that the ball was actually nowhere near where the organisers claimed it was.

In the above example, all the players and the referee are looking to the left and most are running to the left even though the ball is claimed to at an implausible height on the right!

My Telescope And Observatory Destroyed

In February 1989 my first marriage ended after twenty-seven years and I moved to a rented property.

This brought my astronomical observations to an end for a few years.

I destroyed my observatory and telescope but I saved the optics and sold these.

The Aurora Borealis From Dorset 1989

March 13th 1989 was the night of the finest aurora display of the twentieth century seen in the south of England.

I saw it whilst travelling home from an astronomy talk by Patrick Moore given in the glorious Georgian City of Bath.

As I was driving I became aware that the sky was glowing. I stopped the car in a remote road on the Somerset/Dorset border, got out and ...WOW!

The entire sky was glowing with brilliant colours of the Aurora Borealis.

It was breathtakingly awesome with the whole sky a writhing mass of red and green glows and snaking white plumes. In the centre of the display was the Moon which didn't seem to detract from this amazing display at all.

See [55], [56] and [57] for details of this solar magnetic storm associated with this display which caused a massive failure of the electrical power supply in Quebec.

Using Earthshine To Impress My New Lady Friend

One fine evening in February 1992 I was walking my new lady friend home. The crescent Moon was hanging in the western sky displaying a bright Earthshine as in the picture below.

I looked up at it and said

"It's cloudy in North America"

She was so impressed by my ability to tell the weather half a world away by looking at the Moon that she asked me to marry her on February 29th (Leap Year Day) 1992.

I graciously accepted her offer and we were married the following year and have been very happy ever since.

So, how did I pull off this impressive piece of weather lore?

The illumination of the 'dark side of the Moon' comes from the Earth; the Sun shines on the Earth and some of that light is reflected back to the Moon where it lights up the Moon's night sky.

The crescent Moon is highest in the UK sky when the ecliptic is high - as it was when I impressed my lady friend in February 1992.

The light shining on the dark part of the Moon when seen in a UK evening sky in Spring is reflected mainly from North America - predominantly from the USA rather than Canada because the latter sees the Sun lower in the sky and the ground illumination is lower than the USA.

When the Earthshine is dark the USA must be experiencing mainly clear skies because vegetation, roads and housing have a low reflectivity. When the Earthshine is bright it must be cloudy in the USA because clouds reflect light so much more than unclouded land.

Comet Shoemaker - Levy 9 Collides With Jupiter.

On July 22nd 1994 I joined a group of friends to observe with David Strange's large and excellent telescope the impact scars from Shoemaker - Levy 9 on Jupiter.

These black marks were very distinct and spectacular. As I watched, a third scar rounded the limb to join the two already on the disc [58].

This event and visiting David Strange's superb observatory in the Purbeck hills of glorious Dorset inspired me to get back into practical observational astronomy.

I Buy A Professional Dobsonian Telescope

In April 1996 I took delivery of a 305 mm (12 inch) aperture Meade Starfinder telescope.

This was my first professionally built telescope.

I started observing eclipsing binaries but also turned my telescope to anything of interest. This was the start of a renaissance in observing for me.

However, little got recorded in my log book as everything was now typed into my computerised log book the bulk of which was lost when my computer suffered a fatal crash.

I See Patrick Moore Get An Honorary Doctorate

In July 1997 my wife and I were proud to attend the graduation ceremony of my younger son at Portsmouth University.

An additional feature of this event was the presentation of an Honorary Doctorate to Patrick Moore.

After the graduates had been presented with their degree certificates Patrick was invited to take centre stage dressed in all the finery of the university's Doctorial robes.

Patrick gave an excellent speech. It was partly based on his humility for his honour which - characteristically - he said was not deserved.

It was also in part an inspiration to students to go forth and be hard-working worthy members of society.

He then walked ponderously off the stage and down the aisle passing right past me.

I had to tuck my tie inside my jacket as Patrick's huge gravitational field was tugging at it and it was threatening to fly out and point at him as he walked up the aisle.

I have strong negative feelings about the awarding of honorary degrees to those who have not spent four or five years of their lives working the 'proper' way to gain a PhD degree.

The system of awarding honorary doctorates has descended into farce over recent decades [59].

In 2014, 117 universities across the United Kingdom awarded 957 honorary degrees or fellowships to people, many of whom who had not spent a single day in academic study.

As examples, people entitled to call themselves 'Doctor' include singers Ed Sheeran and Danni Minogue, comedians Jo Brand and Billy Connelly, football manager Sir Alex Ferguson (who has eight honorary degrees) and player Ryan Giggs as well as late celebrity Terry Wogan.

Chris McGovern, chairman of the Campaign for Real Education (CRE), has said

"I think when you award honorary degrees to football managers and pop stars, they're being rewarded with what I feel is a bogus notion of achievement.

"It's an insult to the students that work very hard. There are lots of people who are more deserving of degrees than Alex Ferguson."

I agree.

At least one of the above mentioned receivers of an honorary doctorate actually uses the prefix Dr. in everyday life.

Compare a student who makes huge personal sacrifices and studies for years to become a medical doctor only for a pop star to come along using the same prefix of Dr.

I would like to see the whole sordid business of awarding

honorary degrees stopped.

If someone deserves an award then there are many worthy titles that can be given with a Knighthood being the ultimate award of excellence.

Undoubtedly Patrick Moore's Knighthood was well deserved - but a Doctorate? I think not!

Those who know that I do not have a doctor's degree may accuse me of 'sour grapes'.

I obtained a First Class Honours degree in Physics from London University. This automatically qualified me for entering into a PhD research course.

However, the grant for studying for a PhD was half what I could earn by going straight into employment.

With a widowed and elderly mother to support there was no practical option but for me to take employment rather than embark on a PhD course.

Being known as Dr. Kirby could have had some benefits.

I recall travelling with a work colleague on the London to Edinburgh overnight sleeper train in the 1980s. My colleague was booked on the train as Dr Pattison.

During the night as the train roared northwards, he was aroused by the conductor who said that a young lady was feeling unwell on the train and, seeing that my colleague was a doctor, would my colleague please

"...come along and examine the young lady?"

What temptation!

However, my colleague did the honourable thing and admitted that his doctorate was not in medicine but in physics!

I have wandered far from my autobiographical thread but I feel passionately that honorary degrees should never be awarded and are an insult to those who have worked hard for years to obtain their academic qualifications.

Patrick Moore Comes To Weymouth

With the total eclipse of the Sun approaching, Patrick Moore toured the country in 1999 giving lectures to explain to the public what was going to happen and how best to watch the event safely.

My wife and I went along to hear his talk at the Weymouth Pavilion Theatre.

We agreed at the end of the evening that Patrick Moore was 'losing the plot'.

His talk was jumbled and his spoken delivery was somewhat slurred; this being made worse by the breakneck speed of his delivery.

He showed a few slides which were ancient and uninspiring.

The 'show' was divided into two parts.

During the interval my wife and I debated whether or not to go home and abandon the rest of the event. However, we returned to the auditorium for the question and answer session.

I had the misfortune to be sitting directly in front of an 'Independent Thinker' as Patrick has described them; people who hold absurd theories about the universe and everything in it.

I delve into this strange fantasy world of the non-conformists in Chapter 20 - such as the Flat Earthers and those who believe that they can speak Venusian, Plutonian and Kruger. I urge you to watch a man speaking these space languages on YouTube [60].

Anyway, the 'Independent Thinker' right behind me immediately dominated the questioning session by asking Patrick about Flying Saucers, aliens abducting women and raping them, the canals on Mars, ghosts, etc.

Patrick was peering hard in my direction from the brightly lit stage. I had this fear that, being half blinded by the spotlights, he might recognize me from the 'Winchester Weekend' days and wrongly believe that I had turned into some crazy fool.

I slumped as low in my seat as I could without actually sitting on the floor. My wife and I slipped out quickly at the end of Patrick's show.

Total Solar Eclipse 1999

The total eclipse of the Sun in the South-West of England occurred on August 11th 1999; an event I had been eagerly awaiting since I was a small child.

The previous total eclipse in the UK had been in 1927 and had passed over the centre of England.

For a total
eclipse of the BLUES
come to
WEYMOUTH.

The above postcard was one of thousands produced in 1927.

My mother-in-law who had been born in 1923 remembers seeing the eclipse and all the media fuss that was made about it.

The track of totality passed over North Wales in the north-east of England and went all the way to Durham.

The 1927 eclipse had been a large partial event in Weymouth with over 90% of the Sun's disc covered by the Moon.

The name of my home town has clearly been overprinted onto a mass produced postcard.

I first saw a prediction of the 1999 event in about 1950. It seemed (and was) a lifetime away and I really hoped I would live long enough to see it.

I lived in Weymouth, Dorset in 1999 and the track of totality was predicted not to pass over my house but to go 19 kms south of my home.

However, from the southernmost point in Dorset - Portland Bill - the northern limit of totality for the eclipse was predicted to pass just one kilometre south of Portland Bill but that was out at sea

Various plans were made by my astronomically inclined friends.

Some planned to go to France to view totality, others went on a chartered preserved ship named The Shieldhall (sailing out to the totality path in the English Channel from Weymouth) whilst others travelled to Cornwall where the track of totality passed over the city of Plymouth.

The weather forecast was bad and did not improve as 'Eclipse Day' approached. I opted to stay in Dorset but was ready to drive to Plymouth should the weather be promising on the day despite predictions of millions of cars jamming the roads into Devon and Cornwall.

On the day my partner and I opted to cycle the 18 kilometres to Portland Bill from our house. Even if the eclipse was not total, being less than 1,000 metres from the northern limit meant that the sky would get spectacularly dark and it might even be possible to see the 'Baily's Beads' phenomenon [61] as the limb of the Sun passed along the mountains and valleys of the Moon's limb.

Indeed, some observers deliberately place themselves in this very narrow zone where the limb of the Sun flickers down the valleys on the Moon's limb.

On the morning it was cloudy but my partner and I set off on our bikes for the ride to the southernmost point of Dorset. Through the clouds we got glimpses of the crescent Sun - that was when it really got exciting! The eclipse I had been waiting for over six decades to see was underway!

Cars were jammed all over Portland; their drivers desperate to get as far south as possible.

By the time my partner and I arrived at the southern tip of Portland every field was full of cars - never previously had local farmers had such a financial bonanza!

By the time we and two friends arrived it was getting significantly dark. As the time of the eclipse approached the cloud was continuous.

In Devon and Cornwall it was raining!

There is a movie of the scene from Portland at [62].

However, we saw a dark patch on the horizon in the west which was moving towards us. It got larger, the surroundings quickly dimmed to almost perfect darkness and the huge black stain on the clouds moved quickly past us to the south.

It was then so dark that only the immediate ground about us could be seen and the beams of Portland Bill lighthouse shone out.

Cheering broke out despite the obscuration of the clouds and a large group of Hell's Angels revved up their bikes to show their appreciation of this spectacle.

It transpired that few of my friends and family had seen much of the eclipse. My daughter stood in the rain in Plymouth watching a large screen on which was shown the view from a military aircraft high above the clouds.

A Transit Of Venus 2004

The highlight of 2004 was the Transit of Venus.

The planet Venus orbits between the Earth and the Sun and is about the same size at the Earth. On rare occasions it passes between us and the Sun and appears as a black disc moving slowly across the Sun's face.

Transits occur in pairs of years 8 years apart but the pairs are about 120 years apart. The previous time when this happened was in 1882.

Amazingly, the weather in 2004 was perfect and I took many photographs. The picture above shows me using a telescope to show some friends the transit in Dorchester in Dorset.

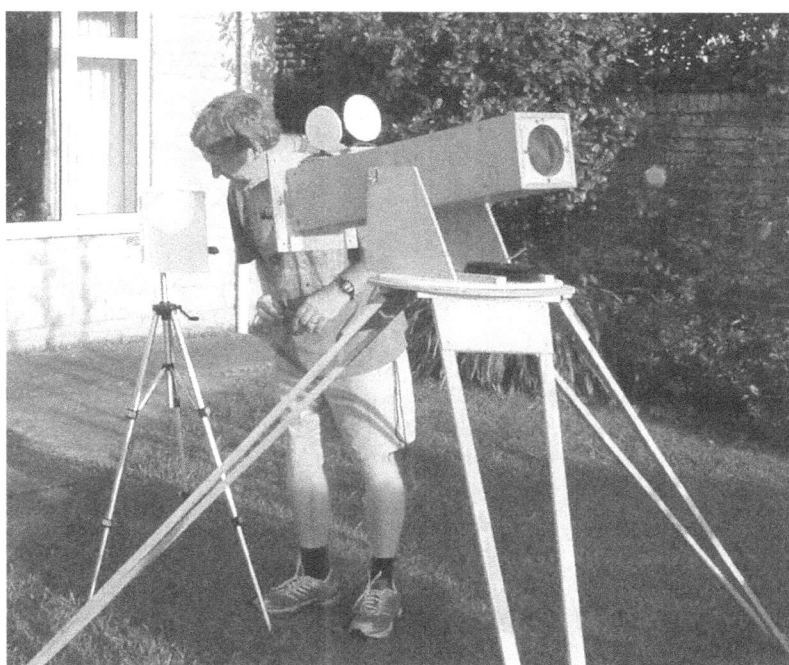

The above picture shows me observing from my front garden by projecting an image of the Sun on a white card. I then photographed the image on the card - a simple procedure which worked well as shown by the example below.

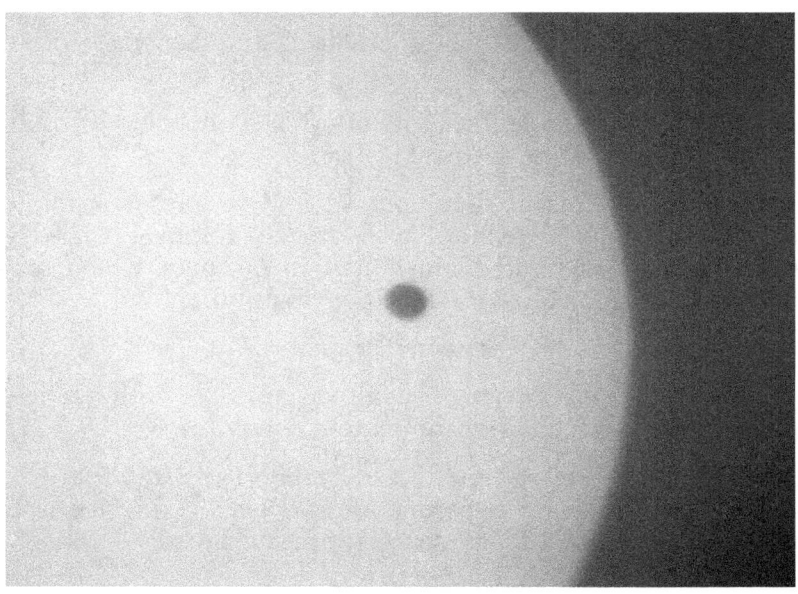

I Break My Neck In Three Places!

In July 2009 I fell head first down a flight of stairs at home and broke my neck in three places.

In addition, I smashed my head at high speed into a wall which broke my nose and split my face open.

I called to my alarmed wife

"Quick! Get the camera!"

This spectacular and potentially fatal accident needed to be recorded for posterity!

Had I 'flown' a small distance to the right I would have gone head first through the glass panel on a door and I would have died a painful bloody death in a large pool of blood.

On arrival at hospital I was immobilized by being encased in plastic blocks and subjected to several MRI, X-Ray and ultrasound scans.

I had no feelings in my arms or feet.

The duty consultant told me I had a 50:50 chance of being paralyzed from the neck down.

However, after a few days the feeling gradually returned to my arms and I was able to sit up in bed wearing a rigid neck brace.

The neck brace came off after many months of wearing it continuously and - very luckily - my neck and face made perfect recoveries.

The story of this accident and its aftermath are told with much gore and humour at my website [63].

Needless to say this accident put an end to any astronomical observing for a while because, even after the removal of the rigid neck brace, I had limited movement in my neck which made observing uncomfortable if not actually painful.

Recent Years

I did quite a lot of observing of faint variable star with my newly purchased large Meade Dobsonian telescope.

However, there came a time in 2010 when I realised that I had not looked through my telescope for two years. I decided to sell it and from that point I have scarcely observed at all.

One reason has been my deteriorating eyesight.

By 2015 I had become very short-sighted. The focal length of my left eye was minus 60 mm and the right eye was not much better. I had to wear spectacles with very thick lenses which interfered with viewing through a telescope eyepiece.

I was unable to observe without my glasses because both eyes were highly astigmatic and these gave very distorted images.

Despite these eyesight problems I decided to try again and get back into limited observational astronomy.

When I started to write this book in the summer of 2015 I was rummaging through boxes containing my astronomical bits and pieces and found the achromatic doublet refractor lens that I had been given in the 1960s.

I had scrapped the rest of the telescope but I had kept the lens.

I set about building a mount for this lens using a drainpipe purchased from a hardware store together with eyepieces also in my 'rummage' box. I had an equatorial mount bought for a few pounds from a friend.

The photograph above shows the refractor lens as reincarnated in 2015 which cost me less than £100 to get up and running again.

A professionally made refractor of this aperture would have cost me about £2,000 to purchase; see the chart on page 148.

4. The British Astronomical Association (BAA)

I Join The BAA In 1959

The BAA was formed in 1890 by a group of amateur astronomers who shared a common interest which was unsatisfied by the professional organisations - mainly the Royal Astronomical Society (R.A.S.) at that time.

In particular, ladies were discouraged from membership of the R.A.S.

The BAA was one of the first scientific organisations in the world to encourage and enrol women as equals to men. The first Council in 1890 included four women: Margaret Huggins, Elizabeth Brown, Agnes Clerke and Agnes Giberne.

The impression gained when I joined in 1959 was that it was dominated by old men. In fact, this was a slight misperception because some Presidents in the 1950s were relative youngsters such as R. L. Waterfield who became President in 1954 at the age of 54 and Alan Hunter in 1956 aged 44.

Never-the-less, the overall impression was of warring cliques of old 'greybeards' as Richard Baum has described them [64].

Despite the early successes of women gaining positions on the

BAA Council in 1890, the first female President was Dr Heather Couper in 1984 at the age of 35.

This evidence of a gentlemen's club run by 'grey bearded' octogenarians was very powerful back in 1959 when I joined. Successive presidents appeared to be selected on the basis of ensuring that every one in the geriatric 'inner circle' got their turn in the presidential office just before they died.

It was curious that in 1959 that I had to get a character witness statement that I was a person of upstanding morality and integrity in the community before I could join the BAA.

This was not as stringent as it seems because I was sponsored by the father of a school friend who had no knowledge of astronomy. A few years previously I would have had to obtain two sponsors to ensure that the BAA did not become, by admitting me to membership, a den of drunken iniquity and the last refuge of riff-raff.

A Subscription Set In Guineas!

When I joined the BAA in 1959 the subscription was 1½ guineas.

GUINEAS!

When did they go out of fashion? Well, for the BAA it was well into the 1960s.

To all readers younger than about sixty I will explain that one guinea was equivalent to £1.05 in modern money.

There was nothing of general or popular interest in the BAA for beginners like me.

The journal had few pictures and the articles were very pedantic and boring.

Reports of meeting were verbatim and extremely tedious and took up many pages.

The BAA Handbook - A Comedy Of Errors!

The BAA Handbook was a joke in those long gone days.

It was full of errors which sometimes were not corrected until a year or two later - long after the currency of the edition containing the errors. For example, a list of no less than thirty-four errors for the BAA Handbook for 1961 was published in 1962 and even more errors in that 1961 volume were reported later in the 1963 Handbook.

By this time the mistakes would have become woven into the fabric of amateur astronomical observation and reporting.

An even greater criticism of the Handbook was heaped upon the pages and pages of positional predictions for comets which never got brighter than about magnitude 20!

I once asked a BAA Council member why so much space was devoted to invisible comets.

"Well..." he mused, *"Someone has to predict them..."*

The BAA Journal

The BAA Journal was tedious to put it mildly.

For example, many pages were devoted to reports of the Solar Section.

When the BAA Council decided to cut down the number of pages detailing every tiny sunspot, the Director of the Solar Section resigned in a huff and was never seen again! [65]

Detailed tabulations of variable star estimates also filled a large proportion of Journal pages. Only a minute fraction of the readers would have used this information and it would have been so much better to publish all these data as a simple graph.

In 1978 a very detailed report on CP Lacerta, a nova of 1938, was published...yawn!

The delay between submission of a technical paper and publication could be more than two years by which time the paper was probably out of date as I found out when one of my submitted papers concerning observations of the planet Venus was still languishing in the Editor's 'IN' tray 28 months after receipt.

The BAA Goes Into A Decline

In the 1960s BAA meetings were always held on Wednesday afternoons in London. Clearly the BAA was not interested in attracting anyone other than retired or independently wealthy Londoners.

The graphs below should have been a serious cause for concern in the affairs of the BAA.

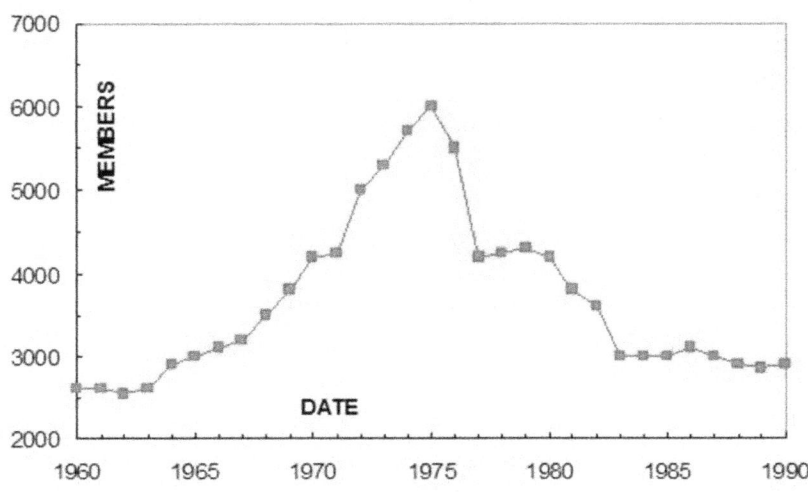

Membership numbers for the BAA.

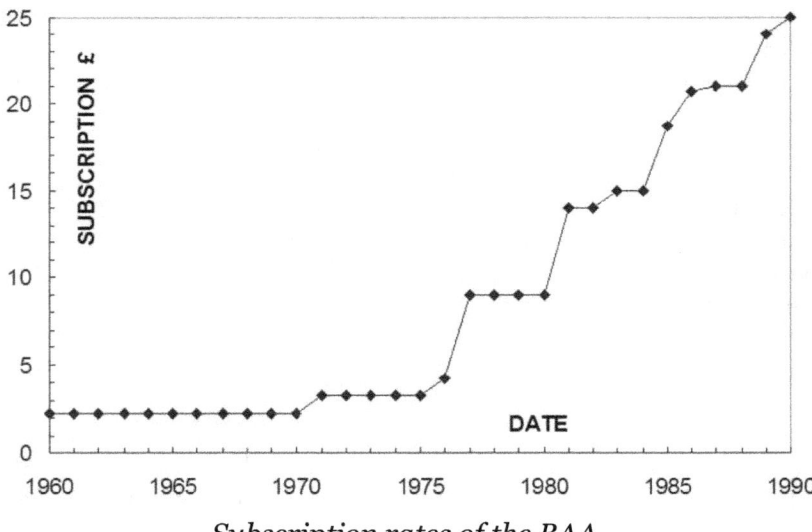

Subscription rates of the BAA

A large increase in annual subscription in 1976 followed by further large rises reaching 1,000% over a twenty year period predictably put an end to the flow of new members and the loss of many existing members.

Between 1970 and 1990 the increase allowing for inflation amounted to 190%.

Membership numbers almost halved from the peak in 1975.

The membership dwindled at a spectacular rate - much to the relief of the President who wrote in his annual report that

"The drop in membership was very welcome as it relieved pressure on the central administrative staff"

Can you believe that the President of a society whose aims were to encourage people into enjoying astronomy was pleased that the number of members was falling fast?

The facts were that some of the clique of old men running the BAA generally did not welcome new younger members and actively sought to deter them with punishingly large fees for joining.

There was a huge and very acrimonious schism in the BAA Council in the mid-1980s. The procession of 'greybeards' into the Presidential post was halted when Patrick Moore, Heather Couper and their (generally) young and enthusiastic successors held the Presidential position.

It can be seen from the chart on page 102 that the catastrophic decline in membership halted in the mid-1980s and, whilst not immediately growing again, at least stabilized at around three thousand members.

The number of active members in 2015 was 3,095 which equates to about the number in 1991. The membership has remained static for the past thirty years; not an achievement to be proud of over a period of enormous growth of interest in astronomy and space.

The BAA's Acrimonious Vendettas

Blatant acrimony was ever present in the pages of the BAA Journal. Indeed, one of the main reasons for me getting the Journal was to see who was being slagged off in the latest issue rather than to enjoy the astronomical content.

Book reviews were one of the means whereby the various warring factions of the BAA hierarchy attacked each other and settled scores.

For example, Alfred Curtis had criticised an astronomical book.

The author responded in the pages of the BAA Journal -

"Mr Curtis is well known for his own disputatious (often surprisingly naïve) ideas ..."

A devastating review of 'Lifecloud' by Fred Hoyle appeared.

"To correct and comment on all the errors and inaccuracies in this book would require another book of almost equal length!"

A scathing review of V A Firsoff's book *'At the Crossroads of Knowledge'* appeared.

"This book is vintage Firsoff. Those readers of a more conservative turn of mind will shake their heads and either smile or grind their teeth"

Another scathing review of a Firsoff book *"The New Face of Mars"* was published. This was followed up by Firsoff who wrote, at the end of a long counter-criticism

"I would also remind the critic that insults and innuendoes are no arguments."

Incidentally, V A Firsoff was one of the BAA's more colourful characters. He was a ski instructor for the British Olympic Ski Team in the 1950s [66].

He was always provocative and frequently wrong in his views, For example, he continued to insist that the rotation period of Mercury on its axis was 88 days long after the evidence proved it was 58.6 days; more of this in Chapter 10.

Right up until his death in 1982 he refused to accept that the universe was expanding [67]. The same was true for Professor Fred Hoyle, a scientist famous for being wrong about many things [68] and right about a few.

But those correct ideas!

Genius!

Firsoff speculated in his books that

"Mind seems to be an entity of the same order as energy and matter"

whatever that meant!

He believed fervently that the lunar craters were of volcanic rather than impact origin but he was wrong. Mind you, Patrick Moore was, for many decades, also a believer that the craters on the Moon were volcanic (of which more in Chapter 9) so Firsoff was wrong in good company.

Firsoff reputedly slept with a loaded gun under his pillow because he believed the Scottish Communist Party was planning to assassinate him! [69]

Here is another random example from an exchange in the BAA Journal concerning a book review:-

Kurt Lambeck: *"If Lyttleton really considers the volume in question to be worthless I hope he will have the courtesy to return the complimentary review copy to the Cambridge University Press. I am sure they will donate it to a library where it will be more appreciated."*

Lyttleton: *"...As for his proposed final destination for my complimentary copy...it will continue to be used as a doorstop"*

All very childish but typical of the BAA geriatric hierarchy's bickering and name calling.

The BAA Journal Letters Section in 1966 carried a vitriolic exchange of letters between Firsoff and others arguing about who invented a system for inserting coloured filters into a telescope optical system.

You might think that this was a trivial matter but to Firsoff and his antagonists it was an opportunity to settle old scores using very strong words and threats of legal action.[70]

I was the victim of a series of letters in the BAA Journal in the 1960s after I had published a letter regarding the Venus Phase Anomaly - see Chapter 10. One of the short letters criticising me included no less than six exclamation marks.

Competent writers don't need six exclamation marks to get their points across!!!!!

Overt criticism and revelations of the sorry state of some of the observing sections sometimes spilt out into the BAA Journal pages.

For example, from the President's Address to the AGM in 1981 we read:-

"Commander L. M. Dougherty has accomplished much in his first session as Director in restoring the Solar Section to a well-organised and co-ordinated body of observers following a protracted period of difficulty and disruption."

What had actually happened was that, following a very angry - almost violent - argument about the conduct and Directorship of the BAA Solar Section, a rival Solar Section had been set up outside the BAA [71].

Members - like me - who had been sending their solar observations to the official Director got letters from the rival Director which pleaded with me not to send my observations to

"...those other people"

I was so fed up with the childish bickering that I simply gave up solar observing at that time.

Even the mild-mannered Hedley Robinson could not hide his concern at the state of the BAA Mercury and Venus Section that he inherited from Dr Jackson who had stated quite clearly his goal of turning the Section into an educational group and to move away from observing.

Hedley wrote of Jackson

"His departure for Canada not only involves the appointment of a new Director, but offers an opportunity of reviewing the activity of the Section..."

Hedley was greatly understating the problems. Members of the BAA Mercury and Venus Section knew exactly what Hedley meant!

The Unfortunate Case Of John Glasby

John Glasby was an internationally respected and very experienced observer of variable stars.

He wrote many excellent books and scientific papers on his area of expertise.

His book *'A Variable Observer's Handbook'* was a classic of great value even now; forty years after it was published [72].

He was also a prolific writer of novels and probably produced over 300 under various pseudonyms in various genres.

However, it was as a renowned variable star expert that he was known to fellow astronomers.

In 1970 John Glasby was prevailed upon by Patrick Moore to write a book on the subject of 'Dwarf Novae' which John did [73].

However, the book attracted criticism because John Glasby reputedly reported observations that he claimed to have made but which were impossible or very difficult to have achieved.

It was widely muttered within the BAA's 'inner circle' that John Glasby had faked data for use in his book.

Howard Miles who was BAA President (1974 - 1976) set up a committee to investigate these claims.

The outcome was published in the BAA Journal for all members to read. It concluded that John Glasby had done the following:

1. Claimed to have observed a variable star when it was invisible due to being very close to the Sun.

2. Claimed to have taken observations of five variable stars from the BAA Variable Star archives which were not actually in the archives.

3. Claimed to have observed a particularly faint variable star when its altitude was scarcely 2 degrees above the horizon.

Faced with such a public accusation of fraud John Glasby left the BAA and Variable Star studies.

He died in 2011.

I was much troubled by this attack on John Glasby when I read the accusations in the BAA Journal.

Was it not possible that these apparent falsehoods could be simply typographical errors or transcription mistakes?

After all, Glasby's book was large and complex and it would have been easy for innocent errors to slip through.

Heaven knows, some of Patrick Moore's books were strewn with errors. I had a copy of one of his books where the orbits of the four inner planets where drawn in a diagram but the names started with Mars closest to the Sun and Mercury furthest moving in the orbit of Mars!

Even the mighty can make absurd mistakes.

As far as I am aware no other astronomical work by John Glasby was ever shown to be of dubious quality

I thought then that John Glasby may have been treated harshly for what could have been simply poor proof reading.

Was this another 'Night Of The Long Knives' executed by the BAA Council?

Patrick Moore and Heather Couper For President!

I won't go on any further about the BAA and its internecine factions because, by the 1980s, the BAA was rapidly being modernised.

The rot was checked around the time when Patrick Moore became President in 1982 and Heather Couper became President in 1984 - the first female President in ninety-four years. They and a few other forward thinking friends dragged the BAA belatedly into the twentieth century.

Heather Couper in the mid-1980s at Greenwich Observatory

My picture above of Heather Couper was taken during a guided tour of the Royal Greenwich Observatory around the time that Heather was BAA President.

What surprised me was that Patrick Moore joined the BAA in 1934 and yet he was not elected President until 48 years later.

I wonder why - arguably - the most famous British amateur astronomer was not given the honour of the Presidential post very much earlier.

I think, in part, that Patrick did not relish the administrative burden of being President and was happy to be on the BAA Council as a Director of a Section.

Heather Couper in the mid-1980s at Greenwich Observatory

BAA 'Winchester Weekends

The popularly named 'Winchester Weekends' were usually held around Easter and consisted of a residential weekend at the King Alfred's Teacher Training College at Winchester.

They are now held at Sparsholt Agricultural College, Sparsholt, Hampshire.

The first event took place in 1967.

I cannot remember when I first attended but it was in the early 1970s. Certainly I appear in the group photograph taken at the 1973 weekend which is reproduced by Martin Mobberley [74].

The Winchester Weekends that I first attended were organised by Alfred Curtis.

Alf died in 1976 and an obituary can be read at [75].

Under Alf's guidance the number of attendees rose from 56 in 1967 to 242 in 1976.

Alf was an interesting and provocative character who was prone to making statements that started with

"In my experience..."

and he would then go on the put forward views that were sometimes wildly at variance with most other amateur astronomer's experience or were just plain wrong.

He once stated from the stage at a Winchester Weekend that

"A refracting telescope performs equally to a reflector of double the aperture."

In fact there very little difference between a good refractor and a good reflector of similar apertures - see page 152.

Alf did a very good job of organising the 'Winchester Weekends'.

It was a complex business booking accommodation, arranging speakers and keeping the peace between the warring factions in the BAA, etc.

If Alf had a fault it was that he took an authoritarian attitude to the organisation and conduct of these weekends which should have been rather more relaxed and informal.

One problem was that nothing was organised for Saturday afternoons. Some attendees sat around chatting but most wandered off into the delightful City of Winchester.

That is what I would usually do with a group of friends.

However, after several years of wandering around Winchester on a Saturday afternoon my interest in the sights faded.

How many times could I see King Cnut's tomb in Winchester Cathedral before wanting something else to do [76]?

It was put to Alf Curtis that the Saturday afternoon might be opened up to firms selling and displaying their astronomical equipment and books but he was resolutely opposed to any commercialisation of 'his' Winchester Weekend.

However, when Charles Wise took over as organiser the Saturday afternoons were opened up to astronomy equipment and book sellers. This was a very popular innovation.

Every 'Winchester Weekend' now has the Alfred Curtis Memorial Lecture and this is a welcome and appropriate memorial to the man who organised so many successful events.

Alf Curtis would probably have disapproved of any sexual activities being indulged in during his Winchester Weekend. It is my recollection that there were only small student rooms containing a single narrow bed.

However, a narrow bed is no bar to sex as I can vouch for although options may be a little limited for those with little imagination or chronic back problems.

Alf was probably under the impression that celibacy was being maintained even on cloudy nights.

Meeting Heather Couper And Nigel Henbest

I met Heather Couper and Nigel Henbest for the first time at an astronomical meeting at Dartington in Devon in the mid-1980s.

I had been invited to speak and gave two presentations; one on 'Variable Star Observations' and the other on 'Observing Lunar Occultations'.

Heather claimed to have met me previously at a BAA Winchester Weekend when she was, in her own words,

"...a spotty teenager."

I have always had a considerable admiration for Heather.

Her cheery and lively character made her very popular especially because she went to such efforts to communicate her enthusiasm for the universe and encourage newcomers to the subject. In the 1980s and 1990s I travelled all over the south of England to be at her public talks.

Nigel gave a talk at the Dartington meeting for which he bought along a large ancient brass slide projector and with this he showed a series of old large format, handmade and exquisitely painted animated slides.

They were superb!

One of these showed Uranus with six moons. The moons could be made to rotate around the planet by turning a tiny crank on the edge of the chunky slide.

The fact that Uranus was shown with six moons dated the slides to the period between 1794 and 1851 during which period it was believed that William Herschel had discovered six moons although four were proved to be spurious by William Lassell in 1851.

After the meeting Heather, Nigel and I ate in a small Italian restaurant with the other guest speakers.

It was the first time I had seen a wine expert in action.

Heather sniffed the cork, sniffed the wine, held a sample to the light, discussed the smell and colour at length with Nigel - I began to wonder if the point was to drink it or not. I almost expected Heather to whip out a portable chromatographic laboratory and come up with a complete analysis of its chemical constitution!

But no, the wine was accepted and sipped delicately throughout the excellent meal.

Meanwhile I was pouring my wine down my throat so fast that it hardly touched the sides. I drink for the C_2H_5OH, not the taste!

At about this time Heather had allegedly shocked Patrick Moore (and others of the BAA's elderly Council members) with an interview with the 'Mail on Sunday' in 1985 in which she was quoted as saying

"It's much more fun to have sexual involvement outside the relationship"

and being pictured in a bath with an inflated model of the Space Shuttle [77].

Until this incident Patrick and Heather had been on good terms.

Indeed, Patrick had proposed that Heather follow him as BAA President. However, after the picture of her in the bath was published the relationship soured somewhat; at least from Patrick's point of view. Heather has never been heard saying anything critical of Patrick.

Heather Couper and Nigel Henbest received a rather barbed mention in a cartoon published in New Scientist in 1994.

There was a cartoon series featuring a nerdy chicken and a fat laidback tabby cat.

Each week they discoursed on things scientific.

One week the cartoon featured 'Boffin Breaks'; holidays for nerds [78].

Cartoonist Kate Charlesworth has the streetwise tabby asking

"Worrabout this? Boffin Makeovers. You arrive like Nigel Henbest and leave like Heather Couper"

"Blimey!" exclaims the nerdy chicken.

Blimey indeed!

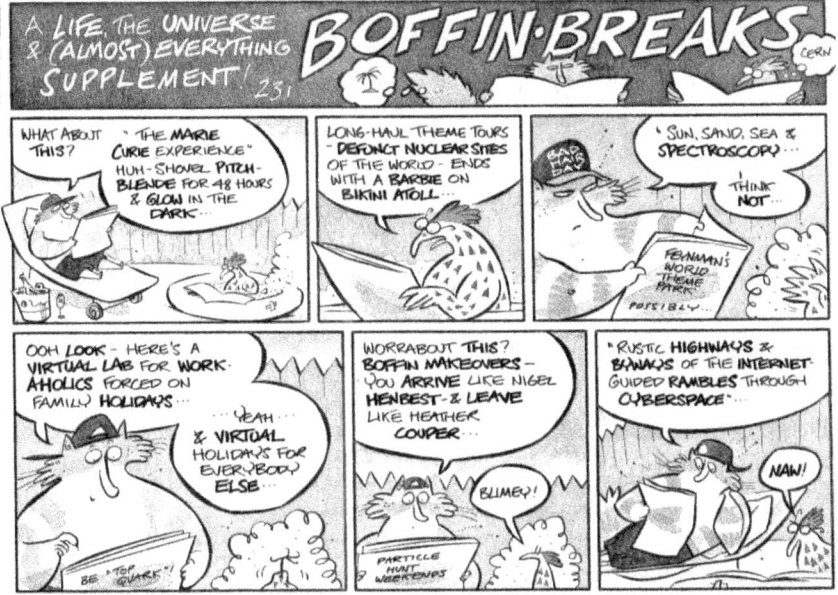

Copyright Kate Charlesworth. Reproduced with her kind permission.

In 1992 I took my new lady friend along to Astrofest. This was about six months after we had first met.

This was to show her what astronomers were like and to make it clear that, if she should be mad enough to stay with me, she was lumbered with an obsessive irrational person who would have windows and doors wide open on the coldest nights and often prefer to be out with a telescope rather than snuggling up in bed.

Heather was at that Astrofest and, when she saw me, she came bouncing up to greet me.

This left my lady friend open mouthed and awestruck because Heather was a very well-known and vivacious personality on television.

Heather's public talks were certainly lively! Her language and content sometimes took the breath away.

At the 1992 London Astrofest I attended she was in full flow with a typically very interesting and engaging talk. She made a comment about a Victorian astronomer, referring to him as a

"Raging woofter"

Patrick Moore then got up and ostentatiously walked out in front of the huge audience. His departure may have been a coincidence and the response to a full bladder but it seemed like a deliberate protest and the audience could not help but notice it.

Do Sex And Astronomy Mix?

Not if the experience of Astronomer Royal George Biddell Airy, is considered. He was widely blamed for the German Gottfried Galle beating the English to seeing and recognising the planet Neptune.

In fact Neptune had been seen several times previously - most famously by Galileo in 1613 [79], but Galle was the first to realise it was a planet.

There were many reasons for the delay in Airy instigating a visual search based on the calculations of John Couch Adams.

Amongst these were the advanced state of Mrs Airy being pregnant with her ninth child when she was in middle age; a dangerous state to be experiencing in the 1840s.

Also, Airy was severely distracted by one of his best assistants - William Richardson - being faced with criminal charges for making his own daughter pregnant and then being charged with killing the baby! [80], [81]

Richardson was eventually found to be innocent after a long trial.

It is curious to speculate that a combination of legal and allegedly illegal sex might have been the main reason for the English not discovering Neptune?

Returning to the question of who was sleeping with whom at the 'Winchester Weekends' I can offer no further speculation.

However, I did gather some Intelligence as to Patrick Moore's sleeping habits whilst attending a 'Winchester Weekend'.

One morning Patrick, Hedley Robinson and I all happened to emerge from our rooms at the same time to go to breakfast.

"How did you sleep, Patrick?" asked Hedley.

"Horizontally, Hedley, Horizontally!" replied Patrick.

I have not yet answered the question above *"Do Sex And Astronomy Mix?"* so let's get down to the nitty gritty.

The 'National Surveys Of Sexual Attitudes And Lifestyles' (NATSAL) [82], [83] have collected large amounts of data on British sexual behaviour. From these data I have calculated the average number of nights in a year when there will be a clash between the attractions of sex and the telescope [84].

For the upper quartile of sexual performers in the 25 - 34 age range and assuming 80 useable clear nights a year (the average for southern England), there will be a clash on about eighteen nights in the year.

For the median sexual performers of amateur astronomers in the 25 - 34 year age range the average number of annual clashes drops to eleven whilst for the lower quartile the clashes drop to one each year.

For an older age group of 65 - 74 years the average number of annual clashes is six for the most prolific (upper quartile) and one for the median.

Thus, for all but the most sexually active young amateur astronomers (male or female) the number of clashes is less than one every month.

Of course, the enormous advances in technology available to amateur astronomers now make it feasible to set a robotic telescope observing whilst the astronomer is enjoying recreational physical activities indoors.

More Of The Winchester Weekends

I loved those 'Winchester Weekends'!

It was a chance to meet up with old friends, make new friends and enjoy the relaxed atmosphere of the bar.

Also, if the nights were clear there were dozens of astronomers with their telescopes anxious to show off their equipment and the sights that could be seen in the sky.

I learned a huge amount at these weekends and certainly, when it came to building my next telescope, I saw plenty of design faults as well as design successes.

For example, I recall one homemade telescope based on a 'yoke mounting' rather like the 5 metre aperture (200 inch) Palomar telescope.

The problem with this type of mounting was that it was heavy, bulky and the whole thing had to be lifted and rotated by 180 degrees when changing the view between the east and west of the meridian line.

Lifting and rotating took two burly astronomers and I could not see the task being untaken by a single observer unless that person had a masochistic wish to get a hernia.

For the 1987 'Winchester Weekend' I was honoured to be invited to give the opening lecture of the Friday evening on March 27th 1987.

This was a difficult spot as attendees had often travelled huge distances on the Friday so were tired and also there was the lure of the bar and meeting up with old friends.

I spoke about Lunar Occultations and I included, at Heather Couper's request, a chart showing my experimental investigation into the effects of alcohol on personal reaction time as shown below

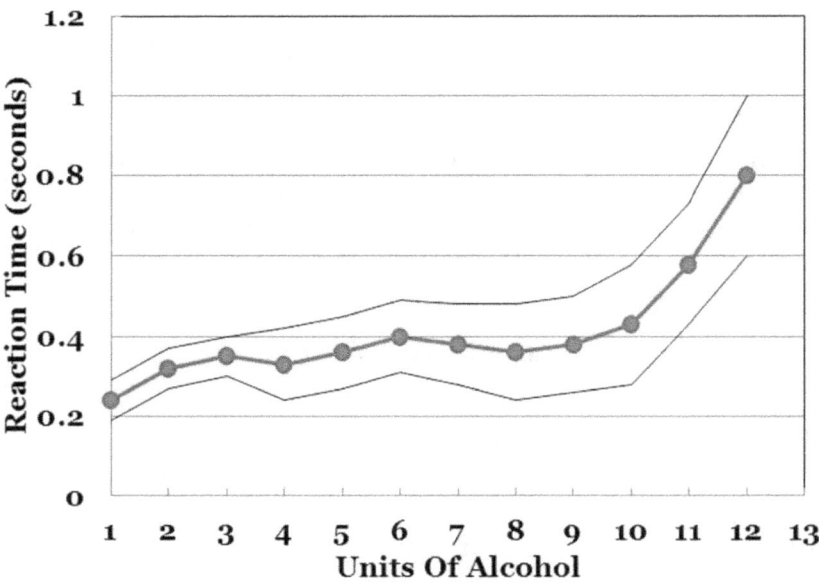

This shows the effect of successively drinking a unit of wine, waiting ten minutes and then making ten measurements of reaction time.

The upper and lower curves show the spread in the measurements.

This process was repeated until I was no longer able to see the stopwatch.

Of interest is that my reaction time is slowed down significantly by only two units of wine and thereafter remains fairly constant until around ten units have been consumed.

My talk seemed to go down well and I was invited on the Saturday evening to a party attended by a small selection of the prominent attendees. These included Heather Couper, Nigel Henbest, John Mason, Le Forbes, Ron Arbour, Alan Dowdell, Hazel McGee and Norman Fisher (who sadly died a few years later.)

A good boozy time was had by all as we partied the night away.

Of course, had the sky been clear, we would not have been partying but outside in the freezing dark looking through telescopes...

Yeah! Right!!

There was one regular attendee and speaker at the 'Winchester Weekends' who greatly affected my fashion choice.

He dressed entirely in black and resembled the 'Dairy Milk Tray Man' who appeared in a long series of TV adverts from 1968 to 2003 - watch the advert here [85].

The 'Dairy Milk Tray Man' was a tough James Bond type of figure dressed entirely in black and usually carrying a coil of rope as well as the box of chocolates. He would undertake daring 'raids' to secretly deliver the chocolates to a lady's bedroom.

The original tagline was

"And all because the lady loves Milk Tray".

I was so entranced with this idea of dressing entirely in black that I spent several years wearing black outfits; black roll neck sweaters, black trousers and black shoes.

The Dark Side Of Patrick Moore

Patrick Moore was a regular attendee at the 'Winchester Weekend' meetings until the mid-1980s when he became a rare visitor due mainly to his growing commitments to his monthly 'Sky At Night' and other TV appearances.

This was a shame as Patrick was undoubtedly a huge attraction and many people went to the 'Winchester Weekends' to meet Patrick.

Martin Mobberley has written a detailed and highly readable biography of Patrick Moore in two volumes totalling slightly over one thousand pages [86], [87].

These volumes are a very welcome and essential guide to the real Patrick Moore rather than the inaccurate and heavily biased autobiography that Patrick published in 2003 [88].

Martin Mobberley has written about Patrick 'warts and all' and has exposed the falsities put out by Patrick to inflate his importance and to score points off his enemies.

In the 1960s and 1970s Patrick Moore fell out with a large number of people.

He classified people as 'Dear Friends' or 'Serpents'.

The latter group - his *'Kingdom Of Serpents'* - became very large at times and, at one time or other, included everyone associated with the Junior Astronomical Society, the entire membership of the Croydon Astronomical Society, everyone who believed lunar craters were caused by impacting meteorites, as well as many individuals such as Heather Couper, Nigel Henbest, James Muirden, Ian Ridpath, Storm Dunlop [89] and everyone associated with the magazines *'Astronomy Now'* and *'The Casual Astronomer'*; the latter becoming *'The Astronomer'* [90].

I often overheard Patrick referring to other participants at 'Winchester Weekends' as 'snakes'.

This dark side to his character was exposed on television occasionally.

For example, on one Saturday morning children's TV programme he was appearing with an aging rocker who was wearing jet black glasses and a typical 1980s rocker's outfit.

Patrick made little effort to hide his contempt and when the rocker was given a card to read, he - the rocker of course - couldn't read it. Patrick leaned over and sneered

"Can I help you with the long words?"

When appearing on a TV quiz show with Graig Charles of *'Red Dwarf'* fame and now a famous DJ, Patrick scowled at Graig and said

"Oh shut up you idiot!"

I never got to know Patrick Moore well enough to be classified by him as serpent or friend although I met him many times.

In fact Patrick came over as an unpleasant person in some of his dealings with those who had the temerity to disagree with him or who failed to give his books rave reviews.

My partner Sandra and I have a friend who lived at Selsey, Patrick's home town. One day whilst queuing at the local Post Office, Patrick came thundering in and walked straight to the front of the queue.

Our friend, being outspoken and outraged, told Patrick to

"Get to the end of the queue or come back later!"

After a scowl in her direction he huffed and puffed, asked her if she knew who he was and left the Post Office unserved.

She told us that Patrick always went to the front of queues and he was widely disliked in Selsey because of his rudeness.

On the other hand, his generosity with his money and time was enormous when it came to encouraging people in their enjoyment of astronomy.

Indeed, his financial generosity was to leave him in a desperate situation when, in old age and infirmity, he was unable to fund his care costs until Dr. Brian May stepped in to help [91].

I will give just three examples of Patrick Moore's lapses of accuracy when it as it relates to his account of his life. These are as recorded by Martin Mobberley in his biography of Patrick Moore.

First, he claimed that he lied about his age so that he could join the Royal Air Force. He claimed that he joined up aged 16.

However, Martin Mobberley's painstaking research shows that Patrick Moore was actually 18 years and 10 months old when he enlisted in the Royal Air Force Volunteer Reserves [92].

Patrick also claimed on several occasions that he was parachuted into Nazi occupied Demark during World War 2.

However, he never elaborated on this dangerous mission probably because he was in Canada at the time claimed as Martin Mobberley has shown [93].

Another fiction was the invention of a fiancée who was killed in

World War 2. Martin Mobberley devotes five pages in his biography of Patrick Moore [94] to this imaginary lady.

I do not intend to repeat the evidence against Patrick Moore ever having had a fiancée who was killed in World War 2; Martin Mobberley has made a very strong case in favour of her non-existence.

It is likely that Patrick made up this story to deflect speculation because he was a life-long bachelor with no known romantic interest in either men or women whilst living with his mother with whom he had a very close relationship until her death.

The dead 'fiancée' does not get a mention until Patrick's mother was very frail and not in a condition to 'spill the beans' on Patrick's invention.

In reality, Patrick did fall passionately in love in the 1950s with the daughter of one of his best friends; a very attractive young lady named Eileen Wilkins [95].

Sadly Eileen did not return Patrick's affection and Patrick was distraught when she married another suiter.

I have a theory, based on no evidence whatsoever, that Patrick's invention of a dead fiancée may have been driven by an increasing pressure on him by the media to explain his bachelor lifestyle living with his mother, his perceived lack of sexual interest in females and his enthusiasm for being involved in organisations involving young people.

As an example of the latter trait from my personal experience, in the mid-1980s Patrick came to my home town of Weymouth in Dorset to open an extension to the local Boys' Club. I was invited to entertain Patrick by the local organiser because he knew of my astronomical interests.

Patrick was particularly generous to all young people but this could have been wilfully misinterpreted by a newspaper hack seeking some sleazy story to peddle.

Of course, Patrick's love for Eileen Wilkins in the 1950s would have squashed most sexual innuendo but Patrick never mentioned that event.

Why should he?

Going back to Patrick's sexuality and the apparent invention of 'Lorna', a dead fiancée, I believe that his attitude and fears could have been influenced by the experience of Gilbert Harding [96].

Gilbert was perhaps the original 'TV celebrity' who had no particular talent but earned a living for being a 'personality'.

Unfortunately, Gilbert, who was widely mocked in the media as 'Filbert Farthing', had only the talent of being grumpy.

His main income was from appearing on the TV quiz *'What's My Line'*. This was very boring as entertainment - a sample of this programme has survived on YouTube [97].

However, the majority of viewers - my mother and I included - only watched in case Gilbert Harding lost his temper and started to insult and berate members of the public or his fellow team members on live TV.

Gilbert agreed to be interviewed in depth by John Freeman for Freeman's series of personal interviews *'Face to Face'*.

Gilbert burst into tears when asked about his unmarried life and his closeness to his mother.

I recall my mother and I were shocked by this person, who we only watched every week in the hope that he lost his temper and was rude, falling apart on live television.

It was revealed that Gilbert Harding was a painfully sad celibate homosexual.

He was great friends with Barbara Kelly [98] and her husband Bernard Braden [99]. Barbara was a long term team member of *'What's My Line'* and Gilbert was a close friend of them both.

He interceded and strongly advised the couple not to send their son to boarding school because Gilbert was fearful of what could happen to young boys based on his own experiences.

Having myself been through a boy's boarding school in the early 1950s I would have given the same advice as Gilbert.

My experiences have been published in my book *"My Lost Childhood"* [100].

The media tore into Gilbert for his homosexuality even though being a celibate homosexual was not illegal in the 1950s.

However, it closed doors and gave rise to public repugnance.

He died in 1960 soon after this revelation of his homosexuality.

Had Patrick Moore fallen foul of the media in his early years on TV and been falsely presented as a homosexual his career would have suffered a serious setback. That's the way it was in the

1950s and early 1960s.

Patrick Moore would have known what happened to Gilbert Harding and I believe this is why he may have invented 'Lorna' [101] although he could not reveal 'her' to the world until his mother was in no state to expose the truth.

Another factor that must have later played heavily on Patrick's thoughts was the outrageous attack on his long-time friend Arthur C Clarke by the Sunday Mirror in 1998 when Arthur was accused of being a paedophile [102].

Patrick Moore was unsympathetic to gay men saying

"Homosexuals are mainly responsible for the spreading of AIDS; the Garden of Eden was the home of Adam and Eve, not Adam and Steve!" [103]

However his friendship with Arthur C Clarke appears to have not wavered over the decades despite his friend apparently being gay - perhaps because Arthur claimed that he had been celibate for decades [104].

There is little to suggest that Patrick Moore was anything other than a celibate and sexually 'straight' person who devoted his enormous talents and most of his income to helping others, enjoying his astronomy and being a devoted son to his mother.

I recall an edition of the 'Wogan' chat show when Patrick appeared alongside the very explicit and sexually outspoken sex therapist Ruth Westheimer popularly known as 'Dr Ruth' [105].

She was, as usual, extolling the wonders of uninhibited sex.

The late Terry Wogan turned to Patrick and asked

"Do you have any trouble with your sex life, Patrick?"

Without a second's pause he quipped back

"Do I look stupid?"

Brilliant!

He gave nothing away and left Wogan lost for words [106].

Was Patrick Moore A Racist?

Patrick Moore was very vocal about immigrants saying that

"...they should be returned to their own country."

This view is still widespread but is not realistic because people cannot just be 'sent back where they came from'.

Many countries of origin - especially those from which illegal immigrants mostly flee - refuse to take them back.

What is the point of putting someone on a flight 'home' when the foreign immigration authorities will put that person on the next flight back to the UK? The airline will be heavily fined for transporting a passenger without an entry visa and the UK taxpayers will pick up the cost of the two-way flights.

Patrick Moore disliked career-orientated women, non-white skinned people, the police, long-haired sponging students, the European Union and all Germans - except Werner von Braun who was, in my opinion, one of the worst Germans to survive World War 2 [107].

I travelled to the beautiful Georgian City of Bath in the mid-1980s to hear Patrick give a public lecture.

I forget what topic Patrick was booked to talk about but that didn't matter because he had picked up the wrong box of slides as he left home so we had a talk on a totally unrelated topic!

As he entered and mounted the stage he looked around the audience and muttered in a loud stage whisper

"Good! No blacks in here tonight!"

Patrick was outspokenly racist about non-white people. To their faces he would be coolly polite but, when out of earshot, he would make comments such as

"Blacks, Wogs and Coons, what can you do with them?" [108]

"Living in a mud hut in the Congo Basin one minute, Head of Security the next: what do you expect?" [109]

Martin Mobberley reports in his excellent biography of Patrick Moore that he made jokes which severely embarrassed his friends and inflamed his enemies.

For example Patrick publically 'joked' in 2005 [110]

"Do you know what you should throw a Pakistani if he were drowning? His wife and children!"

I wonder what Patrick's reaction was to the 1969 pop song *'To Be Young, Gifted and Black'*. Had he been aware of this song he would no doubt have expressed an unprintable opinion!

Patrick Moore was an enthusiastic supporter of Apartheid in

South Africa and refused to visit the country after it became 'The Rainbow Nation' under Nelson Mandela.

Ironically it was Winnie Mandela who led the drive to reinstate astronomical research and education in post-Apartheid South Africa.

In Chapter 20 of his autobiography Patrick rants against black people, immigration policies and the police [111].

He cites two criminal cases reported in the press in 2001.

One man was given three months' imprisonment for collecting lost golf balls from lakes whilst another was given a conditional discharge after assaulting a milkman and putting him in hospital.

Patrick Moore wrote

"It may or may not be significant that the first culprit was a white Englishman, while the second rejoiced in the name of Shahid Akram."

Patrick Moore claimed that the press is dominated with the same message that

"Black is Beautiful, White is Wicked."[112]

Well, I suppose it depends on what material one reads.

If your reading material of choice is the newsletter of the British National Party then gathering material of the type quoted by Patrick is quite easy.

However, there are also plenty of sources that provide an opposing view.

For example, in my home county of Dorset a black person is seventeen times more likely to be stopped and searched by the police than a white person [113].

Patrick Moore refers in his autobiography to *"The Stephen Lawrence Industry"*

This is an appalling slur on a hard-working intelligent black youth who was stabbed to death by a gang of white racially motivated thugs [114].

Police corruption and racism wove its insidious evil throughout the investigation into Stephen's brutal murder.

And yet Patrick Moore dismissed all this as an anti-white

'industry'.

Appalling!

According to Martin Mobberley's biography of Patrick Moore, he had vigorously opposed the recruitment of a young and very capable British black lady to the BAA Secretarial office in 1988 [115].

Moore objected

"...does everyone around this table want to learn to speak Urdu?"

He once said that his politics were

"Just to the right of Attila The Hun" [116].

so perhaps we should not be surprised by his objectionable and unacceptable racist views?

In 1948 the ship 'Empire Windrush' docked and nearly five hundred Afro-Caribbean immigrants arrived in in the UK. Many more immigrants arrived over the next decade. It is estimated that the number of people in Britain who had been born in the West Indies grew from 15,000 in 1951 to 172,000 in 1961 [117].

This prompted aggression in the predominantly white inner cities culminating in Enoch Powell's infamous *'Rivers Of Blood'* speech in 1968 [118].

Patrick Moore was 25 when the Empire Windrush docked at Tilbury Docks and 45 when Enoch Powell gave his inflammatory speech.

He was in the thick of an aggressively racist period.

People a generation younger than Patrick Moore have not experienced first-hand the racism of the 1950s and early 1960s.

When my mother put her west London suburban house on the market in 1960 her neighbours warned her that there would be 'consequences' if she sold to a non-white family.

Maybe Patrick should not be judged too harshly for his views. He was, after all, of the generation that watched and enjoyed the *'Black and White Minstrels Show'* on television [119].

This TV show had a male cast who used exaggerated shiny black face makeup and sang popular songs in the style of early traditional *Black-Faced Minstrels'* [120].

They danced, sang and flirted with white girls.

Al Johnson famously did the same in the 1920s [121].

By 1964, the show was achieving viewing figures of 21 million - half the UK population.

The Minstrels also had a theatrical show which ran for 6,477 performances from 1962 to 1972 and is in 'The Guinness Book of Records' as the stage show seen by the largest number of people.

The TV show ran until 1978 at which time Patrick Moore was aged 55. This means that Patrick Moore was very much a man of those times and attitudes.

It was not until the mid-1970s that a backlash against this ludicrous style of entertainment gathered pace with its offensive manner of depicting black people.

Similarly, it is only in very recent years that viewers of the 'Dam Busters' film have started to be concerned that Wing Commander Guy Gibson's dog was named 'Nigger' [122].

Many black dogs were so named until a couple of decades ago, not because it was an insult to black people, but because the Latin for 'black' is 'nigrum'.

Despite this, in a version of the film produced for the US market the dog was renamed 'Trigger'.

Legislation has, quite rightly, suppressed the open expression of degrading or critical comments about non-white people.

However, the thoughts and feelings are still there in private conversations. Bernard Manning's legacy lives on in clubs and pubs [123], [124].

It is my experience that the most vehement insults against non-white people are often generated by people the most removed from non-white communities.

Certainty, middle class white residents of my county of Dorset, where the non-white population is less than 2%, seem to propagate more racist views than dwellers in inner cities.

Patrick Moore was also aggressively anti-German.

"The only good Kraut is a dead Kraut!"

he would opined to the horror of everyone within earshot.

Because Patrick's voice was always at eleven on the volume dial when ranting this meant that his views were heard by a very large number of people.

His view of Germans is one still held by a large number of mostly elderly people.

No other nation caused more pain and slaughter in the first half of the 20th century.

Whilst in the Royal Air Force Patrick flew missions over Germany in World War 2 so there was never any doubt about his patriotism. However, patriotism often still contains a strongly anti-German flavour amongst those born in or before 1939.

I was born in 1939 and endured the full horrors of World War 2 from my heavily bombed community in North London.

I had an uncle who was killed by German poison gas in the trenches at Ypres in World War 1 and a brother-in-law who was killed in World War 2 when his Lancaster bomber was shot down.

What happened at the end of the war to the millions who had cheered Hitler at his rallies in the 1930s?

Where were the Luftwaffe airmen who had machine-gunned children like me in British school playgrounds?

In the mid-1970s I worked as a military scientist and made several visits to Germany.

I met engineers and scientists who had given all they could to develop and manufacture Hitler's armaments which were used to commit such appalling carnage against civilians.

At one of my meetings in the mid-1970s was a German submarine commander.

I promise that this is not a bad joke - he really did have a monocle and facial scar!

He placed on the conference table a wooden frame which held four large superbly carved wooden smoking pipes.

Frequently during the meeting he would pick a pipe, stuff it with very strong tobacco and belch acrid fumes over the rest of us.

At one stage he stated

"At present we Germans are not allowed to take our navy out of the North Sea. But - our time will come!"

I felt a cold shiver run down my spine.

I fancied I saw his right arm twitching as if it were trying to snap up into a Nazi salute - as kept happening to Dr. Strangelove so brilliantly portrayed by Peter Sellers [125].

In contrast, when my business took me to the US Naval Research Establishment in San Diego I would meet a Jewish physicist who had his concentration camp identity number as a fading tattoo on his arm; one of the few to survive Hitler's death camps.

I do not apologize for feeling no warmth to the German nation if only because my family home was destroyed by a V1 'Doodlebug'. I and my parents were trapped by the rubble in our Anderson shelter in the back garden.

In the mid-1970s I was on a business trip the Netherlands with three colleagues and we took time off to tour the tulip growing area.

We called at several hotels but all were full.

Eventually we arrived at a hotel on the outskirts of a town.

The manager apologized and said that his hotel was full.

Then he asked

"Are you Germans?"

When we told him we were British he beamed at us and ushered us in where staff rooms were quickly made available for us.

The Dutch have a long memory of what the Germans did to their country.

One Dutch engineer I worked with was the same age as me; born in 1939.

As the Nazis retreated from The Netherlands they stripped the fields and warehouses leaving the Dutch starving.

My Dutch colleague remembered British bombers flying over his town dropping food. He was overwhelmed that British airmen should risk their lives to fly over Nazi occupied Holland to deliver food, a commodity heavily rationed in Britain.

The few survivors who so enthusiastically fought for Hitler and his Aryan 'Master Race' are now aged over ninety and those who were their victims are, like me, in their seventies and older.

When my generation are all dead then perhaps the animosity towards Germans will die along with those who hold on to that animosity.

So, do not be upset by Patrick Moore's ranting against Germans.

It is a generational characteristic and, like the millions who watched the 'Black and White' minstrels TV shows, they should not be judged too harshly.

The way through the remaining antagonism towards Germans is through humour.

Do watch the hilarious advertisements for Spitfire Ale [126] and the Carling Black Label advertisement mocking the Germans on YouTube [127].

Laugh at Basil Fawlty's encounter with German guests at his hotel [128].

"Don't mention the war. I mentioned it once but I think I got away with it. It's all forgotten now so let's hear no more about it!

said Basil Fawlty.

Patrick Moore would have typified the class of rural middleclass bigots and racists.

He should not - perhaps - be condemned too strongly for expressing views that were endemic of his era and social class.

Patrick engendered much devotion and admiration from his many fans. It was as if his racism was like a huge ugly wart on the end of his nose that everyone could see but chose to ignore because of his generosity and enthusiasm towards the many friends and strangers - albeit predominantly white and British - that he helped.

Patrick Moore's Car 'The Ark'

Patrick owned a Ford Prefect which he named 'The Ark'.

He drove it for over three-quarters of a million miles before giving it up.

Martin Mobberley describes it as having a 1,172 cubic centimetre four cylinder side-valve engine which gave a top speed of about 60 miles per hour, a fuel consumption of 33 miles to the gallon and an acceleration rate of 23 seconds to 50 miles per hour.

The reason I have wandered off the astronomical topic to

mention Patrick's car is that I had a very similar model which I bought for £110 in 1962 after passing my driving test. Such was my state of finances that I only had £50 to pay for the car and I borrowed the balance from a loan shark.

I bought the car from a bombsite car dealer in Southampton and set off home.

When I got to Bournemouth I was stopped at traffic lights - the first time on the journey that I had needed to come to a halt.

I had been taught in a driving school car which had four forward gears and reverse. My Ford had three forward gears and reverse. The reverse was in the same position as the first gear on all other cars - top left.

First gear was down on the left.

When the traffic lights turned green I instinctively stuck the car into gear, up with the clutch and - backwards I shot into the car behind! Luckily not much damage was caused.

I never made that mistake again.

There was no heater so I bought my wife a sheepskin foot muff to keep her feet warm in the winter.

The windscreen wipers ran off the carburettor vacuum system. This meant that they stopped working if the car was going up a hill. I had to keep taking my foot off the throttle pedal to get the wipers sweeping for a second or two.

Various regulations came into force which my car had to conform with and this caused some problems.

The turning indicators were small 'semaphore' arms which popped out from the side of the car.

Flashing indicators became a requirement for cars so I fitted flashing 'bunny ears' on the roof. These came as a kit and required me to drill holes into the bodywork to fit the flashers.

The day after I fitted the flashing orange lights the car refused to start; the battery was dead.

A friend 'jump started' the car and I put a current meter inline with the battery. The meter needle flicked off the scale regularly in synchronicity with the new flashing indicator lights.

Yes - I had fitted the new flashers the wrong side of the ignition key circuit so that the flasher unit was ticking away all the time even though the flashers were not flashing.

Then windscreen washers became a legal requirement. I drilled a hole through the body of the car above the centre of the windscreen and passed a plastic pipe through.

This went around the inside of the windscreen to a squeezy bottle under the dashboard. To wash the windscreen I pressed the bottle and water squirted out over the screen. (Before that I used to open the window, reach around and squirt washing up liquid directly onto the windscreen.

Amazingly, the car passed the MOT test requirement - my plastic tube washed the windscreen so the police could not argue!

I had a lot of problems with that car. The engine kept coming loose and had to be screwed back tight every few months. Then, at about 50,000 miles, the engine developed a loud knocking noise. A worn big end was diagnosed and I put the car into a garage for a reconditioned engine to be fitted.

This was done for £50. Unfortunately, to get my hands on that sort of money I only had a Post Office savings account and the money had to be applied for by post. The Post Office went on strike and I had to wait three weeks for my cheque to arrive. The garage refused to release the car so I was without transport for nearly a month.

Anyway, I eventually got the car back and my wife and I set off touring Wales in it. It chugged up the hills - very slowly - and whizzed down the valleys at nearly 60 m.p.h. but the engine was making a knocking noise - not unlike the old engine.

When I got it home I took it to a local garage where it was stripped it down and it was discovered that the crankshaft had been put in wrong. It should have been supported by three bearings but one bearing was missing so one end of the crankshaft had been waving around unsupported.

Patrick Moore also had problems with his car but I think mine were worse because I sold the car for a pittance with only 60,000 miles on the clock; less than ten times the distance Patrick drove his car.

Patrick Moore Featured In 'VIZ' Comic

Patrick Moore's fame was enhanced ever so slightly when he was parodied in VIZ comic [129]. Under the headline 'EYE AT NIGHT' and a very unflattering photograph the following horrific tale unfolded.

An Aberdeen woman may be forced to sell her house - because she claims TV astrologer (sic) Patrick Moore has been using his telescope to observe heavenly bodies - through her bedroom window!

Glenda McBride, 58, says she has been forced to dress and undress with her curtains closed since peeping Patrick had a new extra powerful lens fitted to his telescope at his observatory in Selsey, Sussex.

"He ought to keep his boggly eyes fixed firmly on the stars and not on my tits" said Glenda yesterday.

Bill And Ethel Granger

Regular visitors to the BAA Weekends at Winchester were Bill and Ethel Granger.

We have already met them briefly on page 80.

What a strange couple!

Delightfully weird and undoubtedly unique.

Ethel was - and still is despite being dead for three decades - in the Guinness Book of Records for having the smallest ever verified waist at 13 inches - the same circumference as a one litre bottle of wine.

Please see - and be amazed by - a photograph of her belt shown at [130].

She would regularly appear at BAA 'Winchester Weekends'.

She was so tiny at the waist that it seemed impossible that her spine and other vital passageways could get through such a small gap. I felt physically sick to see her because she looked as though she would snap at the waist and her top half would fall off.

Please go to [131] to see many amazing pictures of Ethel Granger.

On a cruise to view a total solar eclipse she won the fancy dress prize dressed as a wasp. I think she had an unfair advantage! [132]

Bill was a portly chap accompanied everywhere with his cat 'Treacle Pudding'. It would sit on his shoulders during car journeys, whilst observing the night sky, at meal times and at astronomical meetings.

They really were a delightful but eccentric couple.

Ethel wore large jewellery in her pierced nose years before it became popular. In fact, their biography - see [133] - shows that Bill and Ethel were very much into body piercing and did it on many parts of their own bodies with red hot needles.

Furthermore, it has been reported that it was Bill who originally was the corset fetishist having wore restrictive women's clothing and high heeled shoes since early teenage years and it was him who persuaded - some commentators say that he bullied - Ethel into her extreme corsetry, body piercing and fetish activities.

Bill Smith

Another personality at the 'Winchester Weekend' meetings and at other BAA 'Out Of Town' meetings was Bill Smith.

Dear Bill!

A man who had tenacious courage in the face of old age and growing fragility. He was always kind, always helpful and a really nice person.

I first met Bill at an early BAA Winchester Weekend.

One morning I was sitting at breakfast and Bill sat next to me. He had a goatee beard, a stooping posture and a lively line in conversation.

"My telescope mounting is so stable that I can do THAT and it won't budge!"

he shouted as his fist crashed down on the table.

The plates and cutlery jumped and danced around. Everybody looked - and, on recognising Bill, they got back to their meal, smiling.

He told me about his observatory and home-made telescope. He was a bit of a dabbler in electrics and electronics. Eventually we discovered that we both lived in Weymouth and we became friends until his death about twenty years later at the age of 89.

Bill made audio recordings of the lectures at the BAA 'Winchester Weekends' on spools of magnetic tape. He would spread cables all over the floor and the lectern. These would be connected together by tobacco tins sprouting coaxial connectors like some alien monster's web. He had a 'portable' tape recorder the size of a suitcase. The lecturers all knew and liked Bill and tolerated his eccentric ways.

However, sometimes they forgot he was there.

Bill monitored the recordings with headphones. During one lecture Hedley Robinson thumped the lectern to emphasise a point. Bill's microphone amplified this thump into a thunderclap in his ears and he shot up onto his feet, tearing the headphones from his ears and waving his fist in a mock gesture of anger.

This resulted in a tumult of laughter and applause.

Bill drove a decrepit Austin A30 van.

His eyesight seriously degraded but he kept on driving.

One evening he drove me to observe a grazing occultation of Mars by the Moon (see Chapter 12) - a rare event - which was predicted to be visible over the army ranges at Bovington Camp in Dorset.

On the way home in the dark it started to rain.

Bill was peering closer and closer to the windscreen as the tiny, slow wipers scrapped ineffectually at the glass.

"Geoff old lad, I can't see a damn thing on my side. Can you shout when the road goes left or right please?"

He meant it!

What a terrible journey that turned out to be.

He gave up driving soon afterwards and the world became a much safer place.

I took Bill to the Wessex Astronomical Society meetings for many years.

He grew feeble and his sight deteriorated until he could not recognize me from about two metres distance.

His hearing went and he wore a hearing aid.

This would be turned up so high that it would whistle with intense feedback. He would sit at meetings, his hearing aid screeching out so that all but Bill could hear it.

His legs gave in so that he could only move around with me supporting his weight. He shuffled slowly into the lecture room. I had to take him to the toilet and it became increasingly personal as he got feebler.

Each month I would wonder if he would come to the next meeting - he couldn't hear the lecture, he couldn't see the slides and he couldn't walk around in the interval. He panicked if I went out of sight.

However, he once told me that the monthly meetings at the Wessex Astronomical Society were his life. If he ever stopped getting to meetings he knew that death was near.

Bill was a very kind and emotional person.

The Wessex Astronomical Society made him its only life member. He openly wept and was so emotional that he could not say anything by way of thanks.

In his youth he tested fighter aircraft in the pre-World War I era. He flew open cockpit biplanes and was one of the first members of the Royal Flying Corps (RFC). I went to his house one evening to collect him for a meeting and he was watching a television programme about the RFC and the exploits of the young men that had been his companions.

He was sitting sobbing and I felt so embarrassed.

Bill had a den.

It was the spare bedroom and it was piled high with what, to the untrained eye, was junk. I mean literally piled at eye level with only a small space to move around. He had spent his life accumulating instruments, lenses, machines, tools, anything at all, at jumble sales, etc.

It was an Aladdin's Cave of wonders. There was a one metre diameter parabola made of Perspex which he used for wildlife recording, several telescope mirrors, old cameras - you name anything and he had two of them.

I loved browsing through that junk room.

Bill was an acoustics engineer by profession and had worked for the Ministry of Defence at Portland in Dorset

He retired in 1959, the year before I graduated from university. He drew his pension for nearly thirty years - almost as long as my working life!

He was always inventing and building things.

One project was a photographically recording photometer for measuring star brightness. He rigged up a microscope objective to scan a photographic negative of a star field.

The light transmitted through the film was concentrated on a light-sensitive diode and the current was read on a dial. It was simple but not very accurate or repeatable.

Bill was over eighty when he built this device and he continued to bubble with ideas right up until his death.

Science attracts its 'Independent Thinkers' [134].

Einstein's Theory of Relativity always attracts a host of challengers.

Bill was one who took on Einstein. Bill did have enough scientific knowledge to put up a case - albeit a wrong one.

Bill was convinced that the observed precession of the perihelion of Mercury's orbit, a keystone proof of General Relativity, was due to a simple oversight by the mathematicians.

It was because they were computing Mercury's orbit using the centre of the Sun as the basis of their calculations rather than the centre of mass (barycentre) of the Solar System.

Now, of course, Bill was wrong. No mathematician would be so silly but it was impossible to convince Bill.

Every journey to the Wessex Astronomical Society meetings at Wimborne would involve him going through this 'mighty flaw' in Einstein's reasoning.

My counter-arguments were forgotten.

Bill was not a trained scientist or mathematician; he was a very competent acoustics technician from the 'valve age' of electronics, so it was impossible to argue mathematics with him.

Bill's notions were not all so off course however.

One experiment he did was to clamp his 200 mm aperture Newtonian reflector so that a radar dish about 10 kilometres away on the highest point of the Isle of Portland was on a crosswire at high magnification.

He observed at various times of the day and noted that the radar dish moved up and down in his field of view by about two metres.

He concluded from this that the Isle of Portland was moving up and down by about two metres in tune with the tides. The argument was quite plausible - but incorrect.

He argued thus.

His observatory at Hillcrest Road, Wyke Regis looked towards Portland over Portland Harbour. As the tide came in, the extra weight of water pressed the harbour floor down and both Wyke Regis and Portland 'bent inwards'.

When he let me examine his evidence I found that he had only made three observations over the course of a few hours so he could not possibly say that the tide was a factor.

I reckoned that the effect was due to varying atmospheric refraction caused by the heating of the air as the day warmed up.

I consulted some books on surveying and they confirmed that effects similar to Bill's observations could happen and by about the observed amount.

This showed that Bill was a good and imaginative observer but lacked the scientific skills to interpret what he was seeing.

Bill was a keen photographer and his photographs of Comet Arend-Roland in 1957 were shown by Patrick Moore on his 'Sky at Night' programme.

The postcard below was from Patrick Moore and shows how highly Patrick thought of dear old Bill.

From PATRICK MOORE, F.R.A.S., Glencathara, Worsted Lane, East Grinstead, Sussex.

Thanks a lot. I was so sorry to have to dash off. Mother is better now, but it was VERY nasty.

The photos are lovely - I hope to show one on The Sky at Night, though I'm not sure how much time we'll have - may I also use one in ASTRONOMY TODAY? You photos are vastly better than mine!

Best wishes and thanks,

PM

Bill used an observatory belonging to a friend who lived across the road. The telescope had optics on loan from the BAA and the mount was of Bill's construction. He described it as 'The Plank' because, instead of a tube, it had the optics at each end supported on a simple plank of wood. You will see my 'plank' telescope in Chapter 7 which was inspired by Bill's telescope.

Bill himself was inspired to build this simple, strong and light mount by Henry Hatfield's telescope which used the same design.

Whilst browsing through his papers after his death I found the following postcard.

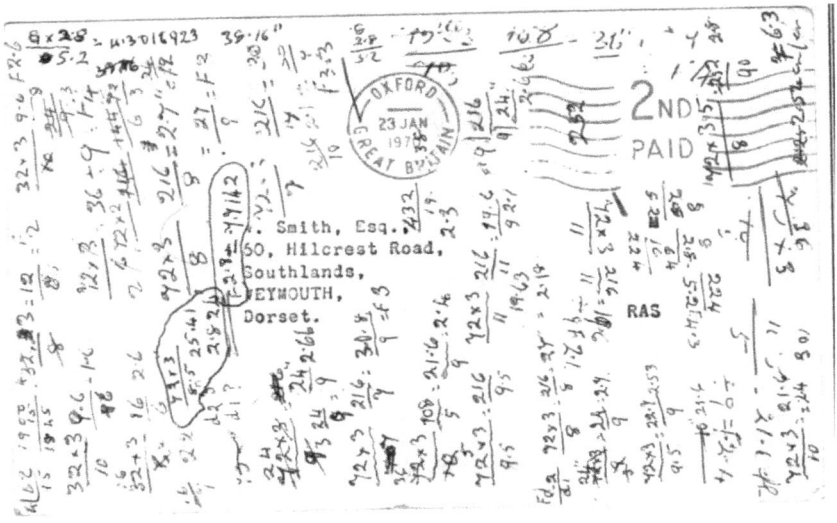

It shows the enormous capability and enthusiasm that Bill had for making simple calculations. What these numbers refer to I do not know but this card sums Bill Smith up in many ways!

Bill eventually became so feeble that he had to go into a care home and his wife Muriel was no longer able to look after him.

He hated it and was persistently trying to 'escape' and go home.

He only lived about a further four weeks. I visited him frequently in the care home and took his wife to see him.

Bill had many relatives in Australia and always wanted to emigrate to be with them. Even at the age of 88 he was planning his new life in the antipodes but it was not to be.

I Part Company With The BAA

During my many years in the BAA and attending its 'Out Of Town' meetings, especially the 'Winchester Weekends', I had met many interesting people.

The refusal of the BAA to publish a paper I had written especially for the BAA Journal upset me - of which more in

Chapter 15.

I reviewed whether I really needed to continue belonging to the BAA especially as its annual subscription was now, in my view, very high.

I was about to leave work on March 31st 1992 under a redundancy scheme and I had been divorced in that same year and had given up my share of the family house and most of my savings to my ex-wife.

I was out of work and living in a rented property without any immediate prospects of a significant income.

I therefore resigned from the BAA having been a member since way back in 1959.

Alan Dowdell picked up on my resignation and asked me if I would share my reasons with him because I had been an active member for so long.

My thoughts and reasons for my resignation were summarised in the letter I sent to Alan, reproduced below.

Since then I have felt no reason to consider re-joining the BAA.

My Letter Of Resignation From The BAA

8th April 1992

To Mr A Dowdell

Dear Alan,

I was flattered that my resignation from the BAA reached the attention of Council and I think I owe you and the BAA a more detailed reason than I was able to give you last evening.

It was not an easy decision because I have been a member continuously since 1959 and have been active as an observer and contributor to the Journal and the BAA Sectional programmes for much of that time.

However, it comes down to economics and value for money.

I was made redundant at the end of March and my income has been approximately halved. Although I am far from pleading poverty this has made me review my finances and the BAA was one of many expenses which I have given up.

For personal reasons I have not been able to observe for three years and, as a reluctant armchair astronomer, I find the BAA now offers little that is not covered more cheaply by others.

I was never able to attend the BAA meetings. Being held on a Wednesday afternoon in London, they were clearly only attractive to the self-employed or the independently wealthy who either lived in London or could afford to travel there.

Although I am no longer in full-time employment, I now cannot afford the train fare to London so BAA meetings do not count as a lure to persuade me to stay a member.

The recent BAA questionnaire showed that 75% of members rarely or never attend meetings and the current policy on meetings does nothing to persuade this 'missing mass' to join meetings.

Thus, my contact with the BAA is through its Journals.

Dividing my annual subscription by the number of journals received each year gives a nominal cover price of £4.75 which is very expensive when compared with 'Astronomy Now' at £1.50 (subscription rate), 'Popular Astronomy' at £2.50 and 'Sky and Telescope' at about £1.80.

I have decided to concentrate only on 'Astronomy Now' as this, together with my subscription to 'New Scientist' which I read for more general interest, provides me with what I am seeking in the astronomical literature.

Looking at the February 1992 issue of the BAA Journal I found that I read 15 pages out of the 68 published.

For 'Astronomy Now' May 1992 issue I read 45 pages out of 64 published. The latter gives me news items in more detail, in colour and quicker than the BAA Journal and this is important to me as a reluctant armchair astronomer.

And so I left the BAA.

5. Do You *Really* Need A Telescope?

Unaided Eye Observing

Newcomers to the wonderful hobby of astronomy will quickly decide that they need a telescope.

My advice - which Patrick Moore would have endorsed - is to start simple and wait before obtaining a telescope.

There is much good observing that can be done with the unaided eye or with very simple aids - including just the inside tube from a toilet roll.

More of this later.

Start your hobby with the unaided eye.

Go out and look at the stars.

Learn the shapes and names of the constellations. Identify the brightest stars in the sky and look up on Wikipedia or from free astronomy software such as Stellarium [135] their characteristics; distance, size, colour, lifespan, etc.

Find and watch the planets as they move through the background star patterns.

There are plenty of projects to keep you busy using only the unaided eye as described on page 539.

If this is not enough then do not rush out with your credit card to buy a telescope.

"Binoculars are far better than a small telescope."

Sir Patrick Moore [136]

Observing With Binoculars

The next stage, in line with my progression as an amateur astronomer, is to buy a pair of binoculars or a monocular.

A monocular is basically half of a pair of binoculars. These are cheaper than binoculars (but not usually cheaper by half!) and work just as well.

Have you ever thought why binoculars are so popular when a cheaper single telescope would work just as well? Using both eyes offers no additional resolution or light gathering power.

Never the less, binoculars are popular and can be used for terrestrial purposes such as bird watching, geological field trips, scanning the countryside, etc.

I bought a pair of binoculars in the 1960s and have used these extensively for astronomy (mostly observing variable stars and artificial satellites) even though I also had a large aperture telescope to use.

It is amazing how much astronomy can be done and enjoyed using binoculars apart from the sheer pleasure of sweeping around the night sky looking at star fields.

I am not going to spend a lot of space here encouraging you to buy binoculars and enjoy using them for astronomy because all you need to know is available on Stephen Tonkin's excellent 'Binocular Sky' website [137] and in his book [138].

Before you buy your binoculars you must read the section in Stephen's website about the types and sizes of binoculars [139].

My binoculars were rated 7x50. This means that they magnify seven times and the apertures of the main lenses are 50 mm.

Recently I bought a pair of 10x50 binoculars which are much more suitable for astronomical use. The greater magnification allows more detail to be seen.

However, going to higher magnifications increases the 'wobbling' of the view due to the greater magnification of hand and arm tremors and reduces the area of sky that you can see at any time.

There is another important reason for going to a magnification greater than the seven times of my old binoculars.

What I am about to say applies to all types of optical instrument.

If you hold your eye away from the eyepiece when the binoculars or telescope are pointed to a bright daylight sky (NOT pointing near to the Sun!) you will see a bright disc.

This is called the Ramsden Disc and is a virtual image of the object lens for a refractor or the mirror for a reflector.

The small white disc in the above photograph taken looking into the eyepiece of a telescope is the 'Ramsden Disc'. In this case it is an image of the main mirror with the obscuration by the secondary mirror and its support in the centre.

For a refracting telescope including binoculars, this would be an image of the main lens.

The diameter of the Ramsden Disc is given by the aperture of the main lens or mirror divided by the telescope magnification.

The diameter of the Ramsden Disc can be measured using a very simple device called the Berthon Dynameter which is shown below.

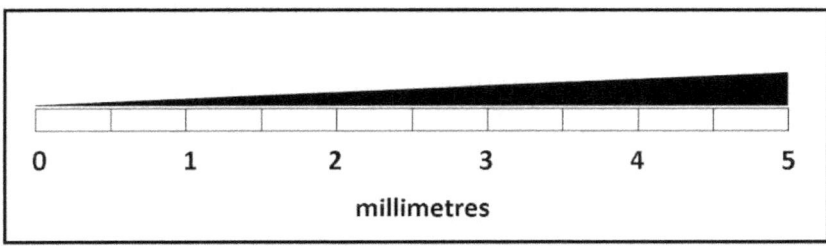

The device consists of two straight-edged pieces of stiff card glued together to leave a tapering gap shown by the dark wedge above.

The scale shows the width of the gap at any point along the wedge-shaped gap.

With such a device the diameter of the Ramsden Disc can be measured to within one-tenth of a millimetre.

For my old 7x50 binoculars the diameter of the Ramsden Disc is $50/7 = 7.1$ mm.

All the light passing through the binoculars or telescope goes through the Ramsden Disc which means that the diameter of the disc must be smaller than the diameter of the eye's pupil.

The diameter of the eye's pupil in dark conditions can be between 4 mm and 8 mm [140].

This means that my binoculars which have a Ramsden Disc diameter of 7 mm will only work at full effective diameter if my pupil is 7 mm or greater.

Anyone with a pupil diameter of 4 mm will only receive a fraction of the incoming light equal to $(4/7)^2$ which is 32% of the incoming light.

For a pair of 10x50 binoculars the diameter of the Ramsden Disc is $50/10 = 5$ mm and this is far more likely to suit dark-adapted pupils of 5 mm or larger.

Thus, my 7x50 binoculars are not ideal and I have found that images through my new 10x50 binoculars are much brighter even though the apertures are the same.

One disadvantage of using binoculars is that they have to be held up and this causes neck strain and arm ache. I am particularly keen to avoid neck strain having broken my neck in three places in 2009 and having been held in a rigid neck brace for months, see [141].

Unsurprisingly I still have neck problems.

There are many ways of mounting binoculars to avoid these problems described on Stephen Tonkin's website [142] where there are many excellent ideas for holding binoculars. My solutions are further discussed in Chapter 7.

Moving On From Binoculars

You may become so addicted to observing the wonders of the night sky through binoculars that you will not feel a need to move on to a telescope. I probably have spent as much time using binoculars as using my telescopes over the five decades during which I have owned both a pair of binoculars and a large telescope.

There are three options open to you when you think you are ready to move on to owning a telescope.

The first option is actually not to buy a telescope but to join a local astronomy club. Many clubs have telescopes for loan to members, have members who are happy to let people view through their telescope or the club may have an observatory.

The second option is to make your own telescope.

This is what I have mainly done over many decades and it has given me enormous satisfaction to look through a telescope that I have built. I will spend Chapters 6, 7 and 8 describing how I ground and polished my mirrors, mounted the mirrors in a telescope and then built observatories to house the instruments.

The third option is to grab your credit card and buy a telescope. This can be an expensive and disappointing option and this is why the first two options should never be lightly dismissed.

If you decide to buy a telescope then look for a second-hand telescope.

Unless they have been rather badly misused, telescopes do not lose their performance with age. There are telescopes out there over a century old which perform as well as when they were made.

OK! So they will not have computerised 'GoTo' mountings but - how do you think astronomers managed before computers were used to guide telescopes?

If you have learned your way around the heavens with the unaided eye and then with binoculars you will not need computerised mountings.

I have always pointed my telescopes by eye because I know where all the interesting objects are having become familiar with the skies early in my astronomical life.

There are vast numbers of telescopes languishing in attics whose owners rushed out to buy a totally unsuitable model - far too big and complicated - because they had just seen a picture of some magnificent galaxy imaged by the Hubble Space telescope and expected to pop out with their new and very expensive acquisition and see a similar sight.

These poor deluded people will happily part with their scarcely used telescopes for a fraction of what they paid just to eradicate the embarrassing memory of what they had done and the money they had wasted.

Building Your Own Telescope

Around 1957 I found a little book *'Making and Using a Telescope'* by Percy Wilkins and Patrick Moore [143] describing how to make a simple reflecting telescope.

It showed readers how to grind and polish a mirror and build a simple mount. It was just what I wanted and I set to work immediately.

My telescopes have always been cheap and simple but actually performed rather well and have given me enormous pleasure.

The photograph above shows my first telescope from 1957.

If you think you are unable to build a useful instrument have a good look at this picture.

The mounting was made from an orange box given to me by a local greengrocer enhanced with free scrap wood. To move the telescope horizontally (in azimuth) I nudged the box with my foot.

To raise and lower the telescope (in altitude) I would turn a small Meccano handle which wound string in or out which was attached to the telescope tube.

The eyepiece holder was the cardboard tube from inside a toilet roll and the single lens eyepiece was ground by me from a piece of window glass.

The only items I bought were a mirror making kit for £5 and the elliptical secondary mirror which cost £2; both from Brunnings of High Holborn [144].

Thus, I ended up with a significant sized and very useful telescope for less than £8 in 1959 money.

Of course I had to grind and polish the mirror which was time consuming but required only a limited amount of skill.

I will describe the process of making a telescope mirror in more detail in Chapter 6.

The home-made mirror, whilst not perfect, resulted in observations which were published by the BAA Journal.

Don't pass over the next three chapters muttering

"I could never make a telescope!"

You can - it's easy!

The one Golden Rule of telescope making is that there are no Golden Rules.

I suspect that more utter rubbish has been written about the techniques for making mirrors than any other aspect of astronomy.

One rather ancient book stated that the polishing of mirrors must be conducted in a basement room, north facing with no more than one square foot of window area so that perfect temperature control can be maintained.

I ground and polished my most recent mirror in the garden, during hot summer days and produced a mirror which tested accurate to one-twentieth of a wavelength of light (about 25 nanometres or one millionth of an inch) - which is excellent!

For a good example of how to make a really cheap but usable telescope seek out a little book by Reg Spry - see [145].

It is a wonderful book which pokes an unrepentant finger in the eye of those who say that you cannot have a worthwhile telescope without going over your credit card limit.

I have already quoted from Reg Spry's list of components for his homemade telescope on page 5.

Remember that this telescope cost £21.25 in 1976 prices which inflates to about £160 in 2016 value. This is about half the cost of a modern commercially built telescope of the same aperture.

What I am going to do in the next three chapters is to tell you what I did to provide myself with a series of telescopes starting with my first crude homemade instrument in 1959 leading up to my commercially manufactured Meade 305 mm (12 inch) aperture Dobsonian telescope which I purchased in 1997.

In the next three chapters I will lead you on an adventure involving abrasive grinding powders, inhaling finely ground airborne glass, dowsing burning boiling pitch on my mother's cooker and razorblades moving back and forth over my eyelashes.

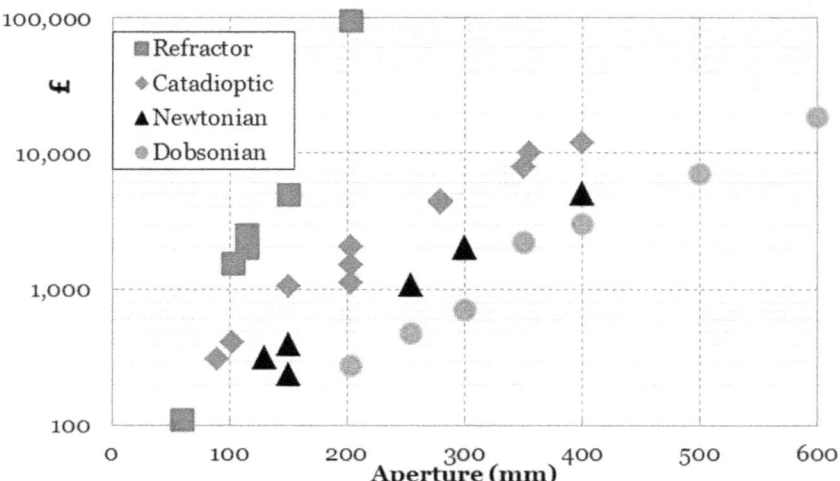

Before that let's pause to look at the chart above.

This shows the typical prices you would have to pay for various types and apertures of telescopes available for sale in the UK at the time of writing (2016).

It is remarkable that the data points are not more scattered on this chart. This is probably more due to manufacturers and retailers setting prices relative to their competitors than relative to the true costs of manufacturing the instruments.

Broadly speaking, telescope costs increase exponentially with aperture such that, for reflecting telescopes, increasing the aperture by 60 mm doubles the cost whatever the size and type of telescope.

For refractors the increase is much more rapid.

This is obviously a very broad rule of thumb because the cost of a telescope will depend on the accessories such as computer or manual control, auto-guiding features for photography, etc.

However, even with these varying options the costs do rise broadly and consistently very rapidly.

A typical Dobsonian telescope will cost about 50% of the price of an equatorially mounted Newtonian of the same aperture.

A Newtonian telescope will cost about 40% the price of a Catadioptic telescope of the same aperture and all will be hugely cheaper than a substantial but similar aperture refracting telescope.

So, is it worth spending a large amount of money on a small refractor rather than a large Dobsonian?

Choosing Your Type Of Telescope

Of course it all depends on what you want to do with your telescope.

If you are intending to observe the Sun using narrowband filters then it's a refractor for you.

However, for the overwhelming majority of applications the refractor loses out to reflecting telescopes when balancing performance against cost.

Dobsonians are cheap and perform well but are very inconvenient when it comes to pointing at faint objects because of the lack of equatorial setting circles.

Some observers mount their Dobsonian telescopes on a contraption called an 'Equatorial Platform' [146]. This device will move the whole Dobsonian telescope in an equatorial manner for a limited period of time.

I have never had experience of such a device so cannot comment further here.

Maybe it works fine.

Maybe it is a pain lifting your large Dobsonian on and off the platform.

I really couldn't say.

The useful light gathered by a reflector and a similar aperture refractor is much the same. There is some loss of transmitted light in a refracting telescope due to absorption of light passing through the glass objective lens.

There is also a loss in the Newtonian system due to the obscuration by the secondary mirror and reflection loss on both primary and secondary mirror surfaces.

Refracting telescopes will always suffer from some chromatic aberration which is the fault that light of different wavelengths fails to focus at the same point resulting in coloured fringes and flares around images.

Reflecting telescopes do not suffer from this defect.

It is claimed that well-made refracting telescopes have negligible chromatic aberration.

I had the opportunity in the 1970s of viewing through Norman Lockyer's 150 mm (6 inch) aperture refractor with which he discovered Helium in the Sun.

The view of Sirius was frightening. The field of view was splashed with sparkling colours like you would never see through a reflector.

About forty years ago I had the opportunity to observe through the United States Naval Observatory 26 inch (660 mm) Washington refractor - see Chapter 17 - and being dismayed at the psychedelic residual colour flickering around bright stars.

This was not what I had expected from the famous telescope that was used by Asaph Hall to discover the moons of Mars in 1877.

This is a natural consequence of chromatic dispersion in the lenses of refractors.

In terms of resolution of fine detail there is little to choose between refracting and reflecting telescopes.

I know that many readers will be rolling their eyes and grimacing at this statement!

There has been a long held popular 'wisdom' in the telescope-using fraternity that refracting telescopes are far superior to reflectors when comparing instruments with the same aperture.

Alf Curtis, the organiser of the early BAA 'Winchester Weekends', once stated

"A refracting telescope performs equally to a reflector of double the aperture."

Clearly this cannot be true of light gathering ability because the useful area of a mirror, even with some loss by the obscuration of a secondary mirror, must produce a brighter image than that of a lens half the diameter.

Of course there are differences in performance of a refractor and reflector. This must be the case because of the different construction of the two types of telescope.

Refractors do not have a secondary mirror with its necessary supporting mechanism to cut out some of the light entering the telescope and also to diffract some light away from the optical axis.

On the other hand, refractors will always have some residual chromatic aberrations that colour and scatter light from an otherwise perfectly focused image.

In an excellent article on telescopic resolution [147] we find the following quote:-

"The resolving power of a reflector is some 5% greater than that of a refractor of the same aperture..."

In fact, having used both refractors and reflectors of various apertures, I agree that there is very little difference between refractors and reflectors of the same aperture.

The resolving power of amateur sized telescopes was investigated in a paper published in the BAA Journal in 1982 [148].

This took a large number of telescopes and their resolving powers were determined experimentally. The apertures ranged from 40 mm to 320 mm and various magnifications were used.

A conclusion drawn from this experiment was

"Refractors show a better resolution than reflectors of equal aperture"

This conclusion was not supported by the data and was based on an inappropriate mathematical method of analysing the data.

How such a paper got through the referees and appeared in the BAA Journal is a bad reflection on the Journal and all associated with it at that time.

In fact, Nigel Henbest picked up this point in a scathing criticism of the paper [149].

Firstly Nigel referred to the bias towards refractors previously stated by one of the authors in a 1981 paper published in the BAA Journal.

He then pointed out that, although the telescopes tested ranged in aperture from 40 mm to 320 mm, the largest aperture refractor was about 110 mm and the smallest reflector was about 90 mm.

The stated aim of the study was to compare the resolution of reflectors and refractors *of the same aperture*. However, there were very few cases where the two types of telescope could be compared directly at similar apertures.

Nigel showed that the mean resolution where telescopes of both types could be sensibly compared showed that the resolutions of both types of telescope were statistically identical!

Nigel went on to write

"I hope it is only a rare lapse that the name of the British Astronomical Association could be seen to lie behind an invalid assertion which will undoubtedly be quoted widely amongst amateur astronomers and telescope - makers in the long-standing 'reflectors versus refractors' debate".

Well said Nigel!

An extensive testing of telescopes by a US team compared different apertures, different mirror accuracies and compared refractors with reflectors [150].

I quote from the summary

"I was amazed to discover that a well-made 6-inch reflector can approach the performance of a 7-inch refractor. The conventional wisdom about refractors being better than

reflectors has definitely been over-stated...the difference between an excellent refractor and an excellent reflector is very subtle."

Opinions have varied enormously on the relative merits of reflectors and refractors.

It is my experience that the chart on page 148 gives a very good explanation why the owners of refractors believe their small telescopes greatly out-perform a large reflector.

A commercially built refractor of about 120 mm aperture will cost in the region of £2,000. For that same amount of money you could have bought a Dobsonian with an aperture of about 350 mm (14 inches) aperture. This latter would give ten times more light gathering performance and undoubtedly a much better resolution and contrast when observing the Moon and planets.

In fact, the 305 mm Dobsonian that I built shown on page 77 cost me only about £120 in the 1980s equivalent to £380 in 2016 monetary value. Paying out six times that amount for a 120 mm aperture refractor would have been ridiculous for what I intended to observe.

Let me repeat here that I am not talking about specialist refractors such as those for observing and photographing the Sun. If that's what turns you on then expect to spend a lot of money for a small telescope.

Let's now look at what performance you can realistically expect to get from a telescope; particularly in urban light polluted observing locations.

This will help you to choose a telescope.

What Limiting Magnitude Will You Achieve?

The aperture determines the amount of light collected and the resolution of detail.

An old rule-of-thumb is that the limiting magnitude (LM) in perfectly dark conditions with perfect eyesight and a perfect telescope is related to telescope lens or mirror diameter D (mm) by

$$LM = 2.7 + 5 \log_{10}(D)$$

Well, on the top of La Palma in the Canary Islands or in the middle of the Atacama Desert this might be true.

In practice, many amateur astronomers - like me - have to observe from urban backyards plagued by street lights and beams from neighbour's unnecessary security lights.

Double the aperture of your mirror and you will collect four times as much light from stars and galaxies - and four times as much light pollution! It is by no means obvious that you will see significantly fainter stars with a larger aperture mirror or, indeed, glimpse those fainter galaxies and nebulae.

In fact, increasing the magnification of a telescope keeps the apparent brightness of a star much the same but spreads out and dims the stray background light.

Thus, magnification is important in determining the limiting magnitude; more so than aperture in light polluted skies.

As magnification is increased the image of a star eventually ceases to be a point source of light. It takes on the appearance of a disc of light due to atmospheric turbulence, the effects of diffraction and the unwanted scattering produced by the optical components of your telescope.

At high magnifications you will be trying to glimpse the star as a shimmering disk against the background of a light-polluted sky. If aperture is now increased the relative surface brightness of the star disc and the background sky glow will stay more-or-less the same thereby putting a halt to further improvements in limiting magnitude.

So, if you expect to be observing from a light polluted site don't go for a large aperture mirror in the expectation that you are going to see very faint stars.

There is an excellent website [151] which allows the limiting magnitude of stars to be estimated in poor seeing and light polluted environments.

It is user-interactive and requires parameters to be entered such as the limiting magnitude of stars viewed with the unaided eye (to assess background sky brightness and visual sensitivity), the angle of the star above the horizon and type of telescope.

The website has an excellent explanatory page making it very easy to use.

The diagram below has been produced from predictions of limiting magnitude (LM) using the website under typical urban conditions of my observing site in my garden.

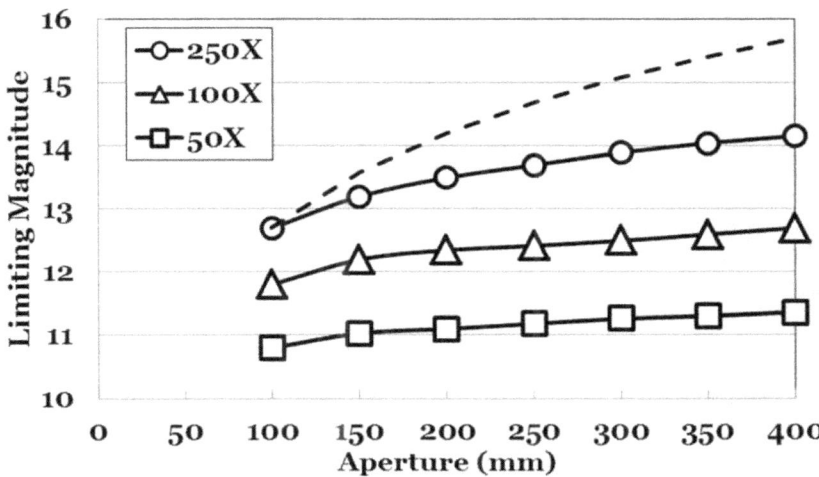

In this chart the dashed curve is the theoretical limiting magnitude in perfect observing conditions, namely

$LM = 2.7 + 5 \log_{10}(D)$

where D is the diameter of the telescope lens or mirror (mm).

The messages here are clear.

1. Acquiring a large aperture and expensive telescope to see faint stars in urban environments will lead to disappointment.

2. The faintest star visible in your telescope in light polluted skies depends far more critically on magnification than aperture.

Doubling the telescope aperture from 200 mm (8 inches) to 400 mm (16 inches) will increase the limiting magnitude by a mere 0.6 magnitudes at most. The theoretical increase is about 1.5 magnitudes in perfectly dark skies.

The difference in the cost of commercial Dobsonian telescopes in the United Kingdom of apertures 200 mm (8 inch) and 400 mm (16 inch) is currently about £2,800.

Would you really pay out all that extra money for such a small gain in limiting magnitude?

I don't think so.

This proves that there is limited advantage in using a large aperture to capture glimpses of faint stars in urban light polluted skies.

The same advice is broadly true for extended diffuse objects where the relative surface brightness of the object and the background light polluted sky is unchanged (more or less) by aperture and magnification changes.

We should by now understand that the detection of faint stars, nebulae and galaxies is a complicated function of many factors of which aperture and magnification are just two.

Another factor is the sensitivity of the retina.

Fainter stars can be seen easier by averted vision than by looking directly at the star image and this averted sensitivity also varies with the position of the image on the retina as well as the colour of the star.

There maybe gains to be achieved by using narrowband filters.

These are available to block out the largely monochromatic light from mercury and older types of sodium vapour street lights. However, their use is less effective for modern sodium lights where the emitted light is spread over a wider part of the spectrum.

The new LED street lights have a full 'white light' spectrum and cannot be significantly mitigated by filters.

However, I am dealing here with the broad principles of limiting magnitude; there is plenty of information available online if you want to go into this subject in more detail.

Resolving power

Another popular reason for acquiring a large aperture telescope is to get a finer image resolution so that very close double stars can be split and fine surface details on the planets and the Moon can be seen.

There are four main factors that limit the finest detail that can be distinguished by eye on the Moon and planetary surfaces and also in your ability to see separately close double stars.

These are

1. Your visual acuity.

2. The quality of your telescope.

3. The diffraction of light due to the aperture of your telescope.

4. Atmospheric turbulence.

It should be made clear here that I am talking about looking through the eyepiece of a telescope; that antique method of viewing celestial objects used by people who do not wish to spend vast sums of money just to sit indoors at a laptop whilst a webcam takes thousands of images outside at a robotic telescope which will later be stacked and turned into a psychedelic work of art by specialised computer software.

That's fine if you have lots of spare cash and no inclination to be close to nature.

I, on the other hand, have sympathy with the Scottish amateur astronomer whose cries of pain bought his wife rushing to the observatory to find her husband stuck to his instrument when his eyelashes froze to the eyepiece.

That's real astronomy!

Your Visual Acuity.

If your eyesight is poor - like mine - then there is not much you can do about it.

I am highly myopic (short sighted) such that without my spectacles everything more distant than 60 mm is blurred.

In addition I have considerable astigmatism which distorts the image on my retina.

My spectacles correct these defects so, in theory, if I observe through a telescope whilst wearing my spectacles I should get a clear image.

However, my spectacle lenses prevent me from getting my eye close enough to the eyepiece to have an adequately wide field of vision.

If I observe without my spectacles I can get my eye close enough to the eyepiece to enjoy the full field of view but the astigmatism in my eye distorts the image. Stars appear as fuzzy ellipses and planetary and lunar detail is blurred.

Add to these problems my developing cataracts which dim and scatter light and you can see that, for me at least, observing perfect images through a telescope is simply no longer possible.

OK! I know that eyepieces with prescription corrections for astigmatism can be obtained at an eye wateringly high price but my astigmatic correction changes between annual eye tests.

It is possible to buy eyepieces with large 'eye relief' which allow spectacle wearers to get the 'Ramsden Disc' close to their pupil but these eyepieces are often significantly more expensive than those with small eye relief where the user places her eye against the eyepiece rim.

The Quality Of Your Telescope

Back in the 1960s telescope prices depended very much on the accuracy with which mirrors and lenses were manufactured.

Mirrors would be quoted as 'one-half wave', 'one-quarter wave' or 'one tenth wave'.

This specified the maximum error in the shape of the surface of a mirror or lens as a function of the wavelength of light in the middle of the visual region; usually taken to be about 550 nanometres.

Today the accuracy is more often quoted as the PV value; which is the maximum Peak to Valley error on the surface.

When I was starting out in the 1960s the quality of the mirror had a large effect on the price of the telescope and I'm sorry to say that some very poor commercial telescopes were sold with badly made mirrors which did not live up to the claimed specification.

This was because telescope mirrors were made by hand - at least in the final stages of polishing the surface to the required parabolic shape. This final stage was so skilled and time consuming that bad mirrors were turned out quickly to make a fast profit.

In recent decades, commercial mirror making has been mechanised and everyone now buying a telescope has the right to expect a mirror or lens that will work close to perfection.

It was not just the manufacture of object lens and mirrors that was poor in the 1960s. Eyepieces were unreliable in their description.

For example, I paid a lot of money for an eyepiece in 1963 which was inscribed 'one-third inch focal length' (8 mm). When I tested it I found it had a focal length very similar to another very much cheaper eyepiece of about one inch (25 mm) focal length.

When I complained to the dealer I was fobbed off by the throwaway comment that

'British manufacturers always exaggerate the power of their eyepieces. That's life!'

Of course, various consumer laws have limited - but perhaps not entirely stopped - these outrageous practices and modern telescope buyers should be confident that they are buying a 'Fit for Purpose' instrument that is accurately described and performs well.

The Diffraction Limitation Of Your Telescope.

Because light behaves as a wave as it goes through the optical system of a telescope, a process called diffraction spreads out images. This blurs out detail on the Moon and planetary surfaces and turns star images into small discs, called the Airy Disc.

There are also rings of light around the Airy Disc but these are faint.

For light in the middle of the visible spectrum - conventionally taken at around 550 nanometres - the diameter of the Airy disc is about 1.4 arcseconds for a telescope of 200 mm (8 inch) aperture.

The diameter of the Airy disc is inversely proportional to telescope aperture so that a 400 mm (16 inch) aperture telescope will produce an Airy disc about 0.7 arcseconds in diameter under perfect atmospheric conditions.

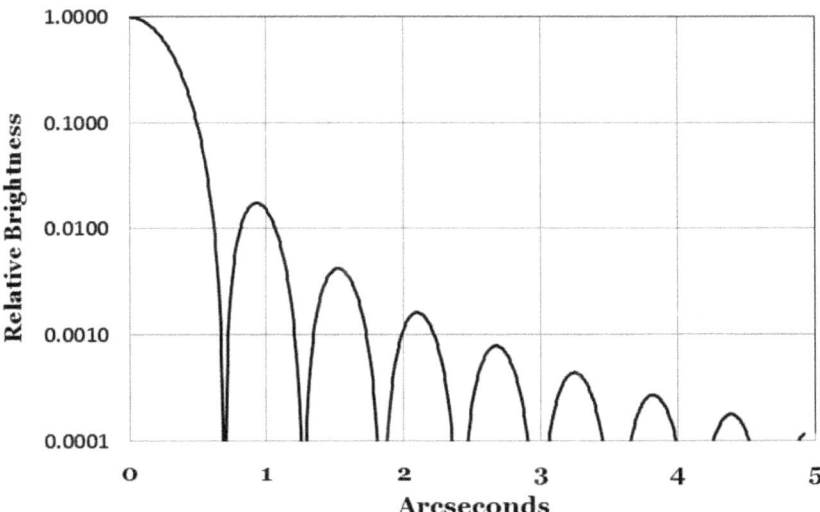

The above diagram shows my calculations for the diffraction pattern of a 200 mm aperture refracting telescope with perfect optics under perfect atmospheric conditions viewing a star. The star is being viewed with a narrow band optical filter centred on 550 nm which is yellow light.

It can be seen that the peak brightness of the diffraction rings falls rapidly from the centre of the image. The first diffraction ring has a brightness of 1.7% relative to the centre of the Airy Disc.

Thereafter the ring brightness falls to 0.4%, 0.16%, 0.08%, etc.

Thus, only the first few rings will be seen.

This diffraction pattern sets the ultimate limit for amateur visual resolution of close double stars, fine detail on the Moon and planets, etc.

As Scottie might have said whilst looking through a telescope at a star

'Ye cannae change the laws of physics!'

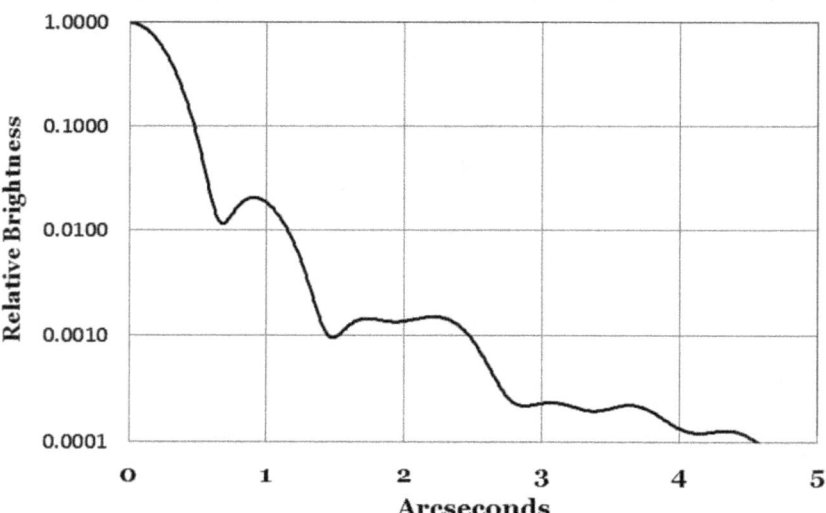

The above figure shows my calculations for the diffraction pattern of a 200 mm aperture reflecting telescope with perfect optics under perfect atmospheric conditions viewing a star. In this case the telescope has a central obstruction (secondary mirror) 40 mm in diameter.

The star is being viewed visually over the full range of colour of a normal eye which is approximately from 400 nm to 700 nm.

The star has a surface temperature of 5,000 kelvin making it similar to our Sun with a peak luminous intensity at about 550 nm wavelength.

We now see that the first diffraction ring is about the same relative brightness as before (2.1%) but the diffraction pattern is thereafter confused as each optical wavelength produces its own diffraction pattern and these interfere with each other.

Obviously, if you want to be able to separate visually two stars which are very close together or want to see the finest detail on the Moon or planets then you need a large aperture.

However, atmospheric turbulence smears out the light and this can - and often does - wipe out any advantage of using a large aperture telescope to see fine detail.

If you expect to be using your telescope visually in an area with poor seeing then there may well be no resolution advantage in obtaining a large aperture telescope.

Atmospheric Turbulence ('Seeing').

If you use your telescope in superbly steady atmospheric conditions such as a mountain top on an island (for example La Palma in the Canary Islands) the image you see visually may be sharp much of the time.

However, in suburban backyards the atmosphere will be turbulent due to heat from the neighbourhood buildings and from local atmospheric disturbances such as the mixing of air masses of different temperatures.

This routinely occurs at my coastal observing site where the difference in temperature between the nearby sea and the land disturbs my viewing.

I once checked the resolving power of a 305 mm (12 inch) aperture Newtonian telescope by trying to separate the components of the stars σ1 and σ2 Lyrae (both about 2.3 arcseconds separation). I used masks with circular holes over the main mirror to change the aperture.

The clearest and sharpest separation of the star images usually occurred at about 100 mm (4 inches) aperture. This is consistent with the fact that visual resolution is nearly always limited by atmospheric turbulence in urban environments.

Turbulence in the atmosphere occurs in a layer of random cells of differing refractive index which are about 100 - 200 mm (4 - 8 inch) in size - see [152] for a good article on the topic of atmospheric turbulence and how it affects the view through telescopes.

In that article we read

'Seeing is one of the biggest problems for Earth-based astronomy: while the big telescopes have theoretically milli-arcsecond resolution, the real image will never be better than the average seeing disc during the observation. This can easily mean a factor of 100 between the potential and practical resolution.'

There is an excellent animation of the effects of atmospheric turbulence on the view of a star at the website at [153].

From my backyard observing site I have found that only on about 2 - 3 nights each year is the image in my telescope rock steady and limited only by the optics.

The website at [154] uses a popular means of assessing the 'seeing' of a telescope viewing a star. This is reproduced below.

Category	Description	Diameter of star image (arcseconds)
I	Boiling image without any sign of diffraction pattern	> 4.0
II	Important eddy streams in the central disc. Missing or partly missing diffraction rings	3.0 - 4.0
III	Central disc deformations. Broken diffraction rings	1.0 - 2.0
IV	Light undulations across diffraction rings	0.4 - 1.0
V	Perfect motionless diffraction pattern	< 0.4

The chart below shows the relationship between the diameter of the diffraction-limited Airy Disc and aperture - the full curve.

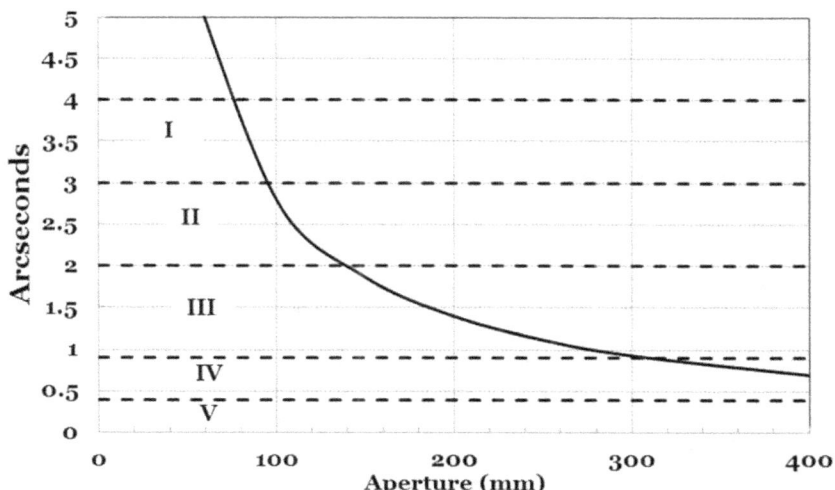

Also shown are the diameters of star images due to atmospheric turbulence using the scale in the chart above.

The effects of the atmosphere dominate the visual 'seeing' on most nights and it is rare to achieve better than 1 - 2 arcseconds in any telescope - Class III in the above chart.

Thus, on many nights, telescope aperture has little effect on resolution.

Even professional observatories suffer from atmospheric turbulence which turns star images into discs of shimmering light.

On the majority of useful nights, the 5 metre (200 inch) Palomar telescope has a visual resolving power of about 1 - 2 arcseconds - the same as a telescope of about 75 mm under ideal conditions [155].

All of this applies to 'old fashioned' amateur visual observing.

The modern technique of taking movies at the eyepiece and using computer technology to stack and merge the best of hundreds or even thousands of frames largely circumvents the problems of atmospheric turbulence.

However, this book is about low tech, cheap and easily accessible amateur astronomy.

Professional astronomers now have additional means to overcome the effects of atmospheric turbulence by using such mechanical techniques as Adaptive Optics and processing techniques like Speckle Interferometry [156].

Professionals can also send their telescopes into space which is not within the means of amateur astronomers; especially those trying, like me, the keep the cost of our hobby affordable.

So, do not commit yourself to a huge aperture telescope just because you expect to see faint stars, faint galaxies or resolve very close double stars.

On the majority of nights in an urban or semi-urban site your telescope may perform no better than one of 100 mm aperture on both angular resolution, detection of subtle contrast differences and star limiting magnitude.

Do Not Be Misled At Star Parties

A warning about visiting gatherings of amateur astronomers.

Do not get swept along by the large telescopes that some enthusiasts set up.

In the United States it is particularly popular to set up one or two huge 'portable' Dobsonian telescopes, with apertures as large as 400 mm (16 inches), in a town or city so that members of the public can see the wonders of the heavens for themselves and - hopefully - take up astronomy as a hobby.

That's all very well and a laudable aim.

However, how many of those novices get put off by believing that nothing worthwhile can be seen in the heavens with a telescope smaller than the leviathans bought out to star gazing parties?

I have tried to show above that large aperture telescopes often do not perform better from urban environments than telescopes half the size - or even smaller - on the criteria of viewing faint stars and galaxies and resolving detail on the Moon or planets.

To be fair, gatherings of amateur astronomers usually have a wide range of telescopes available to look through.

The novice should look through as many telescopes as possible and decide for herself whether it is worth buying a telescope that will break the bank or settle for something just as good but a lot smaller and cheaper.

A Warning From Two Poor Telescopes

I am now going to tell you about two telescopes to dissuade you from falling into the trap of buying telescopes whose performances were disappointingly poor and which were sold at unjustifiably high prices.

The first is a cheap and small refracting telescope known popularly in astronomy circles as a 'Supermarket Telescope' which I came across in a friend's home.

These typically are piled up near the shop entrance in a gaudy box on which we may read such absurd claims as

"Up to 750 times magnification!"

The mounting was more wobbly than a blancmange.

The 'telescope' had a single plastic lens of about 35 mm (1.4 inches) aperture.

I looked down the tube from the object lens end knowing full well what to expect. Behind the lens was a disc with a hole in it. The purpose of this disc was to prevent the outer area of the main lens being used.

This is because with single plastic lenses, the image formed is horrible; blurred, highly coloured and useless. By blocking out this outer zone the image is improved - but not by much.

The clear aperture of the telescope was reduced by this blocking disc to about 25 mm (1 inch).

The magnification claimed on the box was ludicrous. The eyepiece supplied actually gave a magnification of 2X - that's not a typographic error. It really could only magnify by a factor of two.

Using this mockery of a 'telescope' with its wobbly mount, disgracefully reduced aperture using a 'hidden' zonal shield and ridiculously weak eyepiece, the user would see nothing more than could be seen with the unaided eye; possibly even less!

How better to put a young person off astronomy by having these sorts of devices on the market?

My second example of a bad design is a Newtonian reflector mounted on an equatorial tripod as seen above.

I bought this second-hand.

Luckily I did not spend much on it and I reused the equatorial mounting for a homemade refractor; of which more later. The rest of the telescope went into a rubbish skip.

It had a main mirror 110 mm in diameter and so it should have had a reasonable performance; especially for a beginner's telescope.

My initial views of the Moon were not good. Wherever I put my eye near to the eyepiece I had a black blurry obscuration in my vision.

My experience of looking through many telescopes told me that something was badly wrong and I soon found out what the problem was.

The secondary mirror that reflected the light from the main mirror into the eyepiece was grossly undersized.

The effect of this was that only about one-third of the light hitting the main mirror was reaching the eyepiece.

Notice carefully that

1. The main tube is short compared with the diameter.

2. The eyepiece is a long way from the centre of the main tube.

These two facts are a warning that there may be a problem with the telescope.

I am not saying that there is a problem with this general configuration of telescope; just that there might be a problem as, indeed, there was with my telescope.

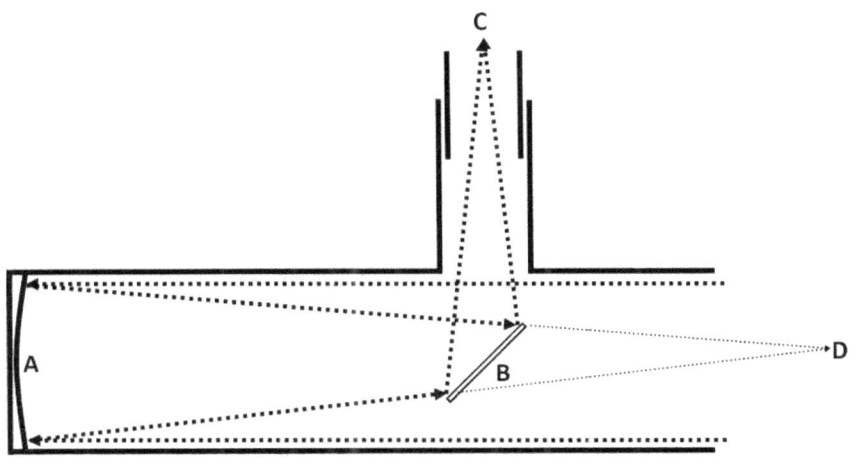

The diagram above shows the light path inside a Newtonian telescope.

The light from the main mirror (A) is reflected as a converging cone and hits the secondary mirror (B). The light then travels to the eyepiece (C).

The focal length of the main mirror is the distance from A to D which is also the combined distance from A to C via B.

If the secondary mirror is too small - as it was in this telescope - the light from the outer zone of the main mirror misses the secondary mirror and returns back into space from whence it came.

It is easy to work out by simple geometry the minimum size of secondary mirror that will intercept all the light reflected from the main mirror.

By making the simple measurements shown in the diagram above I worked out that only 35% of the light hitting the main mirror was arriving at the eyepiece.

The rest of the light was missing the secondary elliptical mirror and was heading straight back into space.

Had the secondary mirror been 50% larger then 85% of the incident light would have reached the eyepiece.

A much simpler way to test if the secondary mirror in a reflecting telescope is too small is by measuring the diameter of the Ramsden Disc using the cardboard Berthon Dynameter as described on page 144.

I did this and found the diameter of the Ramsden Disc for this telescope to be 2.9 mm.

The diameter should have been 5.5 mm (equal to the aperture divided by the magnification) proving that the secondary mirror was grossly undersized.

So, although the telescope looked good with a 110 mm aperture mirror, the light gathering ability of this telescope was equivalent to a lens 55 mm in diameter. This was pretty much equivalent to one half of a pair of binoculars.

What the designer of this telescope should have done was to make the main tube longer and move the eyepiece and secondary mirror about 100 mm away from the main mirror. The secondary mirror would have then been a good match.

This would mean the eyepiece being closer to the tube. It was already a ridiculous long distance from the tube in the original design as seen on page 166.

The mounting also had a gross error in that the latitude setting scale was printed in reverse. The rotational axis of an equatorially mounted telescope must point at the north pole of the celestial sphere. This angle above the horizon is equal to the telescope's latitude.

I have set up the mounting in the picture above for a latitude of 60 degrees and that is the angle at which the rotational axis is set. However, the angular scale is reversed so that the reading is 30 degrees.

I wonder how many novices with this dreadful model of telescope have not noticed this manufacturing blunder and wonder why their telescope does not track stars and planets properly?

If you are a novice in the business of buying a telescope for astronomy I strongly advise you to get help from members of your local astronomy club to avoid the above problems.

I have tried in this chapter to warn you of a few of the pitfalls of acquiring a telescope.

If you have taken on-board my advice based on seven decades of experience you will not rush into buying a huge commercial telescope before you have experienced the wonders of the heavens with the unaided eye and with binoculars.

Only then should you think carefully about acquiring a telescope.

Decide whether you intend to bring the cost down substantially by making your own telescope or whether you will take your credit card to its limit with a professionally made telescope.

Alternatively look out for a used telescope.

In the next three chapters I describe my own experiences of making telescopes and observatories.

I hope you will understand that I am not telling you what to do. Some telescope makers will say that what I have done is all wrong.

All I will say to you is that this is what I have done and, for me, it worked.

6. Mirror Making Is Such A Grind!

"Grind Away, Jolly Boys, Grind Away!

Such Work Is But Play.

Smooth And Polish Carefully,

And Sing All So Cheerfully.

Grind Away, Jolly Boys, Grind Away!"

William Kitchener 1825

How I Made My Mirrors

I have to stress - as I will do throughout this book - that this is not a manual of advice on astronomy; especially not on procuring and using a telescope.

There must be hundreds of books giving advice on obtaining or making a telescope.

Some of them are valuable and some are not.

All I will do in this chapter is to tell you what I did.

Grinding a mirror is easy - polishing is the more difficult part.

The process of turning a disc of glass into a near-perfect parabolic shape is sometimes treated like a religious ritual with complicated ceremonials; rules which MUST be followed even though these are worthless.

Let me illustrate the difference between my cheap, simple and pragmatic approach to mirror making and the ritualistic approach I have encountered in books.

One fault common to amateur mirrors is the 'turned edge'. This is a narrow zone around the edge of the mirror which is out of true. Every book or article on mirror making that I have read says that a mirror with a turned edge must be taken back to the grinding and polishing phase.

This would take many hours, days - even possibly weeks.

One of my mirrors had a turned edge so I painted the offending zone with black paint.

This took about fifteen minutes and the problem was solved as seen above.

This solution would not be acceptable for a professional mirror but the effect of the black paint zone is purely cosmetic.

Even professional telescope mirrors are sold with turned edges.

I once looked through a large telescope in a public observatory in Arizona. The view was poor and I suspected that the mirror had a turned edge.

I did a quick test for a turned edge and confirmed the defect.

This was a disgrace but professional mirrors often have this defect and users do not know how to test for it.

This is what you do.

Focus the telescope on a star.

Move the eyepiece out to turn the star image into a disc.

Move the eyepiece in through the focus point to make another disc.

If the two out of focus discs are sharp edged and the same in appearance the mirror has no turned edge.

If the discs have a fuzzy, spikey edge on one side of focus and a sharp edge the other side of focus the mirror probably has a turned edge. Other mirror defects can produce a similar result but a turned edge is the most probably cause.

So, my view is that if a mirror has a narrow black edge painted on it and a few scratches on the surface it's fine. It may not look good but, if it works perfectly well, use and enjoy it.

Grinding Your Mirror

You will need a mirror blank and a glass tool of the same aperture.

The mirror needs to be a material not much affected by temperature changes.

The tool can be any old material as it will be thrown away when the mirror is finished.

So, when buying a mirror making kit, do not be scammed into buying an expensive disc for the tool.

Plate glass was the common material sold for the mirror when I started out making my own telescopes.

Pyrex came along in the 1970s and has been popular to the present day.

However, Pyrex was sometimes sold unannealed.

I took the Pyrex blank for one of my mirrors to Eric Mobsby, a wonderful mirror maker from Shillingstone in Dorset who is, alas, now dead and sorely missed. I have written an appreciation of Eric's work at the end of this chapter.

He put the Pyrex blank between crossed polarised screens and recoiled at the multi-coloured psychedelic effects. It was hopelessly strained due to having been cast and cooled too quickly.

"Throw it away!"

he implored.

I could not afford to throw it away but he warned me that unannealed Pyrex will often yield over a few years and turn a perfect parabola into a misshapen surface reminiscent of the profile of the Alps.

Be warned!

If you buy Pyrex, make sure it is annealed! It may seem good to be getting a Pyrex (or similar) mirror but you may be better off in the long run making do with a plate glass disc or paying out for a modern material such as Zerodur, Corning Incorporated's Ultra Low Expansion Glass (ULE), Cervit, etc. [157]

These low expansion materials must still be checked for strain with crossed polarising screens.

Do not be tempted to use discarded porthole glass as recommended in some old telescope making books. These are usually armour plated. This means that once the surface skin of tensioned glass is penetrated, the whole disc may explode in your face.

The focal ratio is the focal length divided by the mirror diameter.

The choice of mirror diameter and focal ratio determines the quantity of glass to be ground away.

The chart below shows the volume of glass to be removed if the edge of the mirror is not reduced.

In reality, some of the mirror edge is removed but this is a relatively small amount compared with the volume removed from the centre of the mirror.

The point of the chart below is to show how the glass volume to be removed increases very rapidly as the mirror diameter increases for a given focal ratio.

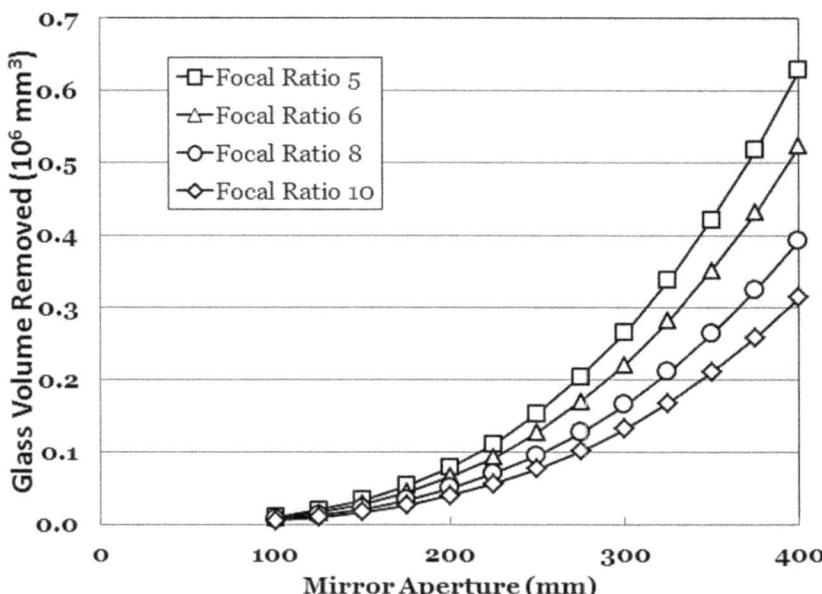

This chart is deceptive however.

In my experience the time take to grind a mirror does not depend much on diameter.

This is, in part, because the grinding action on the mirror depends on its weight pressing down on the grinding tool. This increases approximately as the cube of diameter.

So for example, if we are making a mirror with a focal ratio of 5 then the volume of glass removed would be about eight times greater for a 400 mm mirror than for a 200 mm diameter mirror.

However, the time taken, assuming that the grinding time depends inversely on the weight of the mirror pressing down on the grinding tool, will reduce this time by a factor of $(400/200)^3$ which is a factor of eight.

This reduces the grinding time for the 400 mm diameter mirror down to about the same as the 200 mm diameter mirror.

Of course the actual time taken is a more complicated process but this simple calculation explains why, in my experience, the time taken to grind a mirror blank is not much dependent on mirror diameter.

For my 305 mm mirror I did not have a blank disc the correct size so I screwed together two wooden discs and stuck 25 mm squares of plain window glass over the surface with Araldite adhesive.

According to the experts this cannot produce a satisfactory mirror because the wooden discs will flex and cause distortions in the grinding surface.

My wooden-based tool worked perfectly and produced a mirror of better than 20 nanometres accuracy.

You need Carborundum grinding powder.

This can be bought in mirror making kits from telescope equipment suppliers. However, I got mine more cheaply at a local craft shop where it was sold for polishing pebbles (lapidary).

Be careful not to get course grit mixed in with the fine grit or your mirror will be scratched.

I once ordered a mirror kit by post and the grit came in plastic bags tied with string. Loose grit was everywhere. I sent it back!

Mind you, if you do get a scratch on the mirror, so what! It is only a cosmetic blemish and will not affect the performance of the mirror in the slightest.

In the USA, an observatory night assistant once fired six bullets at a professional telescope mirror and caused six large craters in the surface. There was no detectable change in the mirror's performance - see [158]

Don't worry about scratches - even through some books say that you must go back to the start.

Poppycock!

A scratch-free mirror is purely aesthetic.

Grinding is easy but tedious and the principles are well described in books so I won't go through it here in detail. The book I used in the 1980s and which I still recommend is a classic textbook by Neale Howard [159].

There is one tip I have not seen in print.

Having decided on a focal length you will need to grind until you have the right depth of curve. How do you know when you have got the correct depth at the mirror's centre?

The books I have seen say that you should slosh water over the coarse ground mirror to make it reflective and form an image of the Sun onto a sheet of paper. Measure the distance from the mirror to the image - that is the approximate focal length.

Well, when the Sun shines on the wet mirror it dries it out quickly and the image is very indistinct - not a good technique. Adding a drop of washing up liquid helps. Better still is to cover the rough ground mirror with light oil which does not dry out in sunshine.

My technique was to place a precision steel ruler across the mirror and measure the depth of the curve with a set of car feeler gauges.

Do you remember those?

These are a set of precision thickness steel strips and were used to set the gaps on contact breakers in the distributers of pre-computerized cars.

These are still readily available online.

The depth of the mirror at its centre (d) is given by

$$d = D^2/16F$$

where D is the mirror diameter and F is the focal length.

The error in the measurement of focal length ΔF is given by

$$\Delta F = (F/d)\,\Delta d$$

Let's illustrate this with an example.

Assume the mirror is 300 mm in diameter (D) and the focal length (F) is 1600 mm which is similar to a recent mirror I made. The depth of the curve on the mirror (d) works out at 3.52 mm using the above equation.

Feeler gauges can measure to an accuracy of 0.03 mm (Δd) so that the uncertainty in focal length is 14 mm which is better than 1%. This is much more accurate than shining a very fuzzy image of the Sun from a wet mirror onto a sheet of paper!

With careful measurement across the mirror it is even possible to check whether the shape is spherical or has distortions. The picture below shows one of my mirrors 2½ hours into rough grinding marked out with the depth of curve at various points (in inches) across the mirror as measured with the feeler gauges.

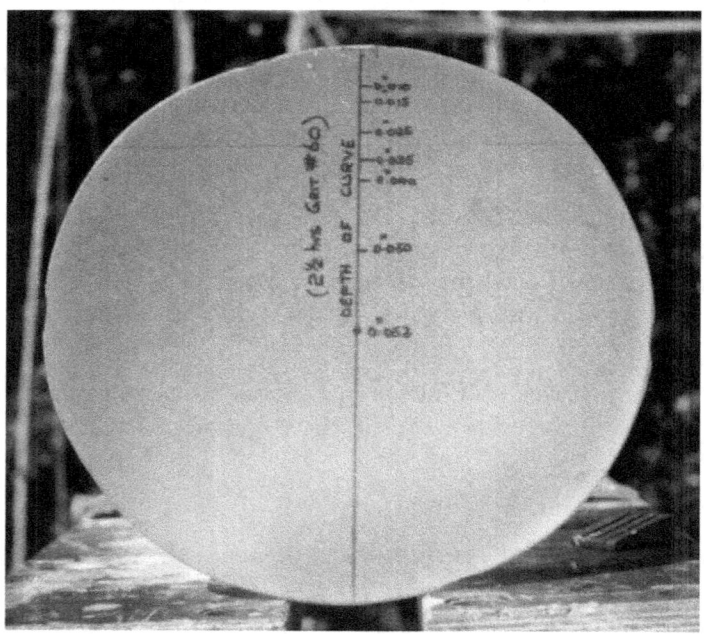

Once the required depth has almost been reached you continue grinding with increasingly fine Carborundum grit being very careful to clean up the mirror, tool and work surface at every stage.

As you move to finer grades of grit the surface of the mirror becomes very smooth and translucent.

The photographs below show me making my 305 mm mirror in the 1980s breaking all the rules. I'm outside in the sunshine, I'm not using a barrel filled with concrete, I'm not working under scrupulously clean conditions and I'm using a wooden-based tool.

Polishing Your Mirror

Polishing is done using a pitch surface and polishing powder. The melted pitch is poured onto a disc to form a layer about 5 - 10 mm thick. Channels are cut to allow the used powder and fine glass dust to drain away.

Some text books I have read tell you to use only the finest Swedish pine pitch to which the exact amount of pure beeswax and turpentine must be added to get the correct consistency for polishing.

For my mirror I asked the late and great Eric Mobsby's advice on where I could get my finest Swedish Pine Pitch. He looked at me with incredulity.

"Use the pitch they put on road surfaces" he said.

"Hang on Eric" I protested *"What about the beeswax, the pure turpentine and the lumps?"*

"If you are really worried, strain the hot pitch through a silk stocking" advised Eric.

Unluckily, I'd just laddered my last silk stocking climbing over a farm gate during the annual South Dorset Transvestites' Cross Country Run so I could not strain the molten pitch as Eric recommended.

Even so, it worked fine!

The smallest quantity of pitch I could get was a 25 kg drum which cost £12. I only needed about half a kilogram so the rest eventually got thrown away.

Pitch is dreadful stuff.

I melted it in a washed out food can on the cooker.

For my first mirror back in the late 1950s I did this on my mother's gas cooker whilst she was at work. I had the gas turned up too high and the pitch caught fire. I opened the window and threw the blazing mass into the garden.

Large quantities had boiled over onto the cooker's enamel surface. It quickly set hard and would not budge. Eventually, a great deal of rubbing with white spirit removed all trace but the smell lingered for days.

My mother could not understand why she could smell hot tar when there were no roads being repaired nearby. I agreed that it was a puzzle!

I travelled to London and called on a very famous supplier of telescopes and telescope accessories - Dudley Fuller - to buy my polishing powder.

Dudley sucked his teeth, whistled quietly and shook his head.

'Polishing powder? Just can't get it I'm afraid!'

I was gobsmacked.

If the premier supplier in the UK could not get polishing powder then how were mirrors being made?

There was something very fishy going on but I was not prepared to argue. For some reason he did not want to sell me any polishing powder. I left his showroom dejected and caught the train home - a substantial rail fare wasted!

I wondered what to do next.

I called on Eric Mobsby and he gave me a small jar of cerium oxide which was ideal.

Some older books tell you to use jeweller's rouge. This is bright red and stains permanently crimson everything that the pitch has not already stained stinking black.

It is awful stuff.

Cerium oxide, in contrast, is non-staining and just as good as rouge - many people say it is better. It is readily available online for lapidary.

Polishing is done in much the same way as grinding. The mirror is rubbed back and forth with random (short) strokes on the pitch lap. I did it all by muscle power.

However, if you plan to make lots of mirrors and are mechanically minded you can build a polishing machine.

The picture below is of a home made polishing machine made by Eric Mobsby.

One of the most exciting times in mirror making is when the opalescent mirror starts to show a shine. This happens very quickly but it takes a lot of polishing to get a full shine. When I saw a shine start on my first mirror I set it up at night, propped against a house brick, and picked up a star with a simple eyepiece. First light on my first telescope - unfinished mirror, brick and eyepiece!

I was thrilled at how bright that star was. I was also thrilled that those photons had been travelling for hundreds of years solely to bounce off my mirror and enter my eye. That sense of wonder and excitement often grabs me when I'm out with the night sky.

Couch potatoes never know what it is like.

The Foucault Mirror Test

There are several ways of testing the accuracy of a mirror's shape.

The Foucault test consists of shining light from a very small source, bouncing it off the mirror and observing the way the light comes back to form an image of the source [160].

For a spherical mirror, a point can be found where the light from the source reflects off the mirror and all returns to another point very close to the source. The distance from the source or image to the centre of the mirror is called the radius of curvature - I call it R.

Now, if the mirror is not spherical, the light reflecting off the mirror comes back to different points depending on the shape of the mirror. In the diagram below, the top ray diagram is for a spherical mirror, the lower one is for some other shape.

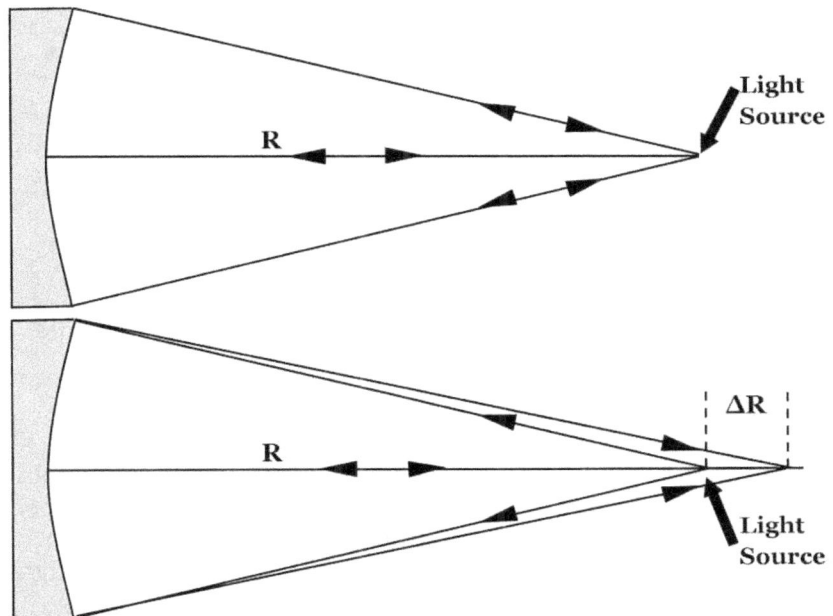

By measuring the change in position where the rays return to the optical axis as a function of zone radius on the mirror we can deduce the shape of the mirror.

For a (perfect) parabolic mirror, the change in distance ΔR is equal to the depth of the mirror's curve. For my 305 mm aperture mirror with a focal length of 1,500 mm the value of ΔR came to about 3.9 mm - an easy distance to measure.

The deviation of the mirror surface from a true parabola can be measured to an accuracy of about one-tenth of a wavelength of light if ΔR can be measured to a precision of about 0.1 mm.

This would result in a good quality mirror.

So how do we measure the changes in R on a mirror?

We use a Foucault mirror tester.

The Foucault Mirror Tester

The principle is to set up a minute point of bright light at a distance close to the radius of curvature distant from the mirror. We then catch the returning rays of light from the mirror as they converge close to the light source [161].

We intercept this converging cone of light with a razor blade.

The diagram below shows what happens in three situations.

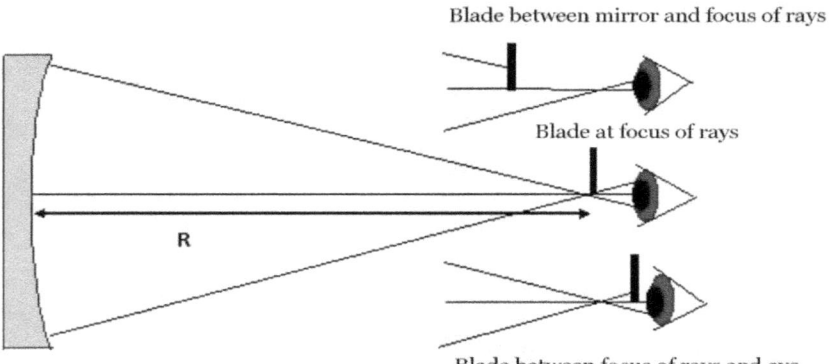

Blade between mirror and focus of rays

Blade at focus of rays

R

Blade between focus of rays and eye

Imagine the razor blade to be between the convergence point and the mirror, intercepting half the cone and the eye is placed close to the blade to intercept the partially obscured cone of light. The mirror will appear to have zones which are half light and half dark.

If the blade is moved out to beyond the image, the light and dark zones of the mirror are interchanged.

Exactly at the ray intersection point the mirror will appear to darken uniformly.

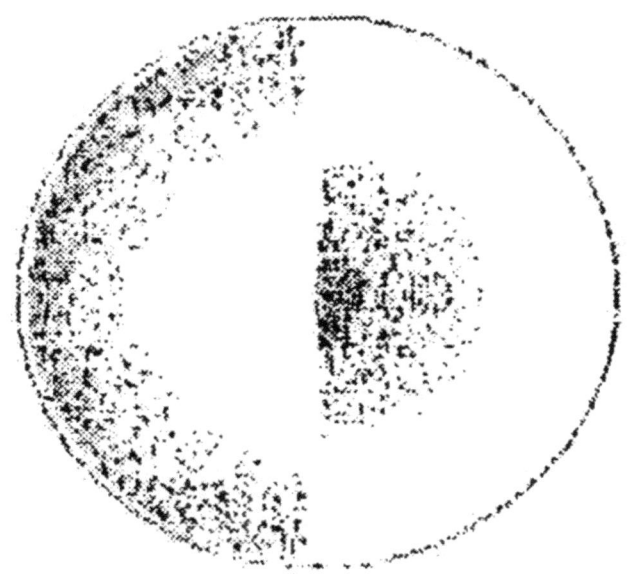

The above sketch typically shows what a parabolic mirror looks like as the knife edge crosses the optical axis.

It can be seen that the dark halves of the mirror reverse at the zone which returns rays converging on the blade edge. By moving the blade to and from the mirror we can measure the value of the radius of curvature as a function of zone radius 'y'.

So, all we need is a means of moving a razor blade towards and away from the mirror, a means of measuring this movement, a means of moving the blade left and right to cut the converging light rays and a light source.

Books tell you how to make such a Foucault tester and these are generally over complicated and require skill in the use of lathes and micrometers.

My Foucault tester was neither complicated nor expensive because I made it largely out of scrap found in my garage and cost me less than one pound.

The tester shown above is a professional device costing a huge amount of money.

The Foucault tester shown above and below was built by me in an afternoon from pieces of wood, an elastic band, a razor blade, a discarded baked bean tin and a bicycle cycle lamp.

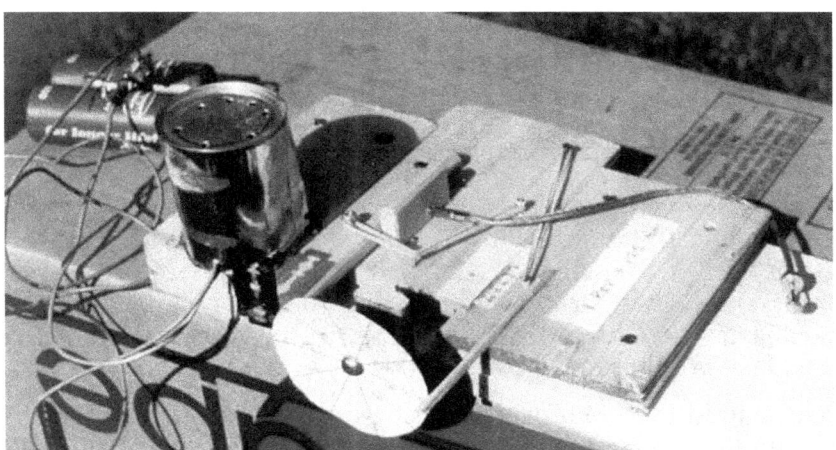

It was accurate enough to produce a mirror with a surface better than one-tenth of a wavelength of light which was forty - yes FORTY times - more accurate than the mirror of the Hubble Space telescope when it was first put into orbit around the Earth [162]!

I have dwelt on this because it is important that inexperienced readers know that astronomy does not have to be expensive. It is easy to get by with simple cheap equipment and even to make your own telescopes with only a very little skill.

These are the components of my Foucault Mirror Tester.

Light Source. We need a very tiny point of light. I took about twenty layers of aluminium cooking foil and pushed a sharp new fine sewing needle into it. I separated the foil layers and found the layer that was just penetrated, the next one being only dented.

This should be a very small and round hole. The foil is glued over a small hole cut in the side of a baked beans tin. (Save the lid!)

It is best to wash out any remaining beans and sauce but this is not essential. Inside the tin place a torch bulb connected to a battery and punch a few holes in the base, which is now on top, to let out the heat. Any beans left will warm up and smell pretty bad.

Old-Fashioned Razor Blade. Mount this on a piece of wood which is a snug fit between two other pieces and capable of rotating on a bolt. Thus, the razor blade can slide freely to and fro.

For my first tester in about 1959 I broke up my mother's best wooden baking board and used this to make a sliding mount for the razor blade. The sliding part of the board was sprung against a screw head by means of a rubber band. The razor blade was moved from side to side by a camera shutter release cable.

The Micrometer. This is normally a precision piece of equipment. In my tester it was a standard bolt from a hardware store with a fine thread. The bolt was wound into a slightly undersized hole drilled into soft wood. This totally avoids backlash - well worth remembering in any application where backlash is a threat.

I measured the pitch of the bolt by turning it twenty times in a nut and measuring the distance moved. My bolt moved 1.25 mm per turn. The baked bean tin lid was fitted to the bolt head. (I told you to save the lid, didn't I?)

The lid was fitted with a paper disc marked into ten divisions. Thus, turning the bolt with attached disc one division moved the carriage with the razor blade by 0.125 mm. I could read the angle turned by the lid to one tenth of a division so I could measure the position of the razor blade to about 0.0125 mm accuracy. This is about ten times more accurate than needed to produce an essentially perfect mirror.

The Mount. The tester has to be solidly mounted. We are trying to measure the convergent point of light rays over a distance of two or three metres to an accuracy of about 0.1 mm. This means that the tester must be mounted rigidly on a concrete floor, the mirror must be held on its edge solidly and the air must be calm and uniform in temperature.

I know this sounds a bit fussy in view of the crudeness of my tester but it's true. Living rooms and bedrooms are out - I've tried them. The air current and flexibility of the wooden floors prevent the device working accurately enough.

I used my garage. It had a concrete floor, was enclosed and dark.

Once the mirror is set up and shadow patterns are found you are ready to start figuring the surface to a parabolic shape.

Shaping The Mirror Surface

The key to shaping a mirror's surface is to carefully polish away the 'hills' on the surface to reach a parabolic curve. We find where those hills are from the measurements with the tester.

All the books I have read give only qualitative advice. Usually this involves a lot of unnecessary work. What I did was to compute the shape of the mirror mathematically and work out the least amount of glass needed to reach a parabola.

I took the measurements of the change of radius of curvature as a function of radius across the mirror surface and converted these into a map of the hills and valleys on the mirror.

I did this using a computer program I developed for my BBC Micro which solved the calculus equations and plotted out the mirror surface profile.

The sketch on page 185 was generated by the BBC computer during the mathematical process of computing the mirror surface profile.

I will not go into the mathematics here because this would take a whole chapter to explain and be very boring.

I polished off the high points on my mirror using a small hand polisher.

Every book and article I have seen tells you never to use a small hand polisher because it will cause zonal defects and create hills and valleys in the wrong places on the mirror.

Above we see my hand polisher.

Here I am using the hand polisher! (Those spectacles were fashionable in the 1980s - honest!)

This small polisher was made by pouring molten pitch into the lid from a jar of coffee. Any brand of coffee may be used - it's not important.

Grooves were then cut into the pitch surface, as for the full size polishing disc.

For five to ten minutes I polished a zone where my BBC computer had indicated that there was excess glass, washed and dried the mirror, took it into my garage and performed another Foucault test.

I then typed the new values for the variation of radius of curvature over the mirror surface into the computer. The computer then displayed where on the mirror the glass needed to be polished away and I was ready for another few minutes of polishing.

Aluminizing Your Mirror

The final stage of mirror making is to aluminise the surface. This has to be done commercially.

Of course, real enthusiasts like the late and great Eric Mobsby built their own aluminising tank.

His apparatus is shown below.

He made this up from bits of a discarded X-ray machine donated by a local hospital.

The mirror surface was first scrupulously cleaned and mounted in the aluminium tank. At the opposite end was a small sample of pure aluminium.

The air was pumped out - Eric Mobsby built his own molecular pump to reach the necessary level of vacuum. The aluminium sample was then heated electrically until it evaporated and a thin film of aluminium coated the mirror surface.

Having got the mirror aluminised we have to mount it.

So - onto the next chapter after this short appreciation of a really great British amateur telescope and mirror maker.

Eric Mobsby.

Dear Eric - a man of unlimited generosity and interest.

As a child, Eric was playing on his family lawn with his sister when a sparkling, crackling ball of light descended between them. It was about the size of a golf ball. When it hit the grass it exploded in a shower of sparks and left a scorched area about six inches in diameter.

This was ball lightning - a very rare event and the only first hand case I know.

Eric was born in Lewisham in East London on April 11th 1904.

I first came across him when we both worked at the Atomic Energy Authority Establishment at Winfrith in Dorset in the early 1960s. By then he was close to retirement and I was a keen scientist in my early twenties.

He was a true engineer - by which I mean that he had half a thumb missing!

He was devoted to telescope making.

Indeed, he was a master mirror maker and his house was devoted to this pursuit.

He cast aluminium ingots in his own smelting equipment and turned them into telescope mountings. He built grinding, polishing and figuring machines - some of which you have seen earlier.

Eric's wife Freda whom he married in 1952 was a cheerful bubbly character who had an amazing tolerance for Eric taking over every part of their house with telescopic equipment.

When Eric was well into his eighties he had operations to

remove cataracts from both eyes. He went into Weymouth Eye Hospital and I visited him there. He was sitting up in bed covering pages with extremely laborious calculations. He worked to ten significant figures with a hand calculator.

He was working on a book dealing with mirror making and a sample page showing the laborious and detailed work that he put into his calculations is reproduced on the next page.

Despite my persuasions, he refused to use a computer to automate his calculations because none, at that time, worked to enough significant figures for his task.

One day we were in the middle of a deep discussion on the 'Mathematics of Conic Sections' when the Ward Sister walked in.

"What are you doing? Mr Mobsby is supposed to be resting, not working through all your numbers!"

I pointed out that it was Eric who was doing the calculations and discussing them with me. I was the innocent party!

On the next page is a table from Eric's book on telescope mirror making. There were about 200 more pages like this!

Eric's enthusiasm and talents were widely recognised in the world of amateur telescope making as well as in academic circles.

His particular interest was to develop a type of telescope mirror that was large, light, rigid and not prone to distortions due to changes in temperature.

These requirements contradict each other.

To be rigid conventional mirrors have to be thick. A rule of thumb is that the thickness of a solid glass mirror should be at least one-tenth of the diameter.

However, a disc of solid glass that thick would be heavy and would be prone to distortion caused by differential warming or cooling as the ambient temperature changed

.

y_p 2.4	y_p 2.8	y_p 3.2	y_p 3.6	y_p 4.0
x_p 0.08	x_p 0.108	x_p 0.142	x_p 0.18	x_p 0.2
Tan(2β)·13392857/	Tan(2β)·156502297	Tan(2β)·179193628	Tan(2β)·202020202	Tan(2β) 0.225
2β 7°628149668	2β 8°8947697	2β 10°1592172	2β 11°42118627	2β 12°6803836
y_s ·66613801	y_s ·77693853	y_s ·88763655	y_s ·998217555	y_s 1·10866702
x_s ·026169516	x_s ·035609423	x_s ·04649491	x_s ·058823108	x_s ·07259077
6 82°37185033	6 81°1052303	6 79°8407828	6 78°57881373	6 77°3196165
x_{NS} 8·497560766	x_{NS} 8·502146359	x_{NS} 8·507434167	x_{NS} 8·513422795	x_{NS} 8·52011066
Tan y ·078633838	Tan y ·091765799	Tan y ·10490993	Tan y ·118067985	Tan y ·13124172
y 4°496135339	y 5°243108574	y 5°988988578	y 6°733623731	y 7°476864/
ϕ 3°132014329	ϕ 3°651661126	ϕ 4°170228622	ϕ 4°687562539	ϕ 5°2035193
2ϕ 6°264028659	2ϕ 7°303322253	2ϕ 8°340457244	2ϕ 9°375125079	2ϕ 10°4070386
δ 1°364121012	δ 1°591447448	δ 1°818759956	δ 2°046061192	δ 2°273344 8
Tan δ ·023812902	Tan δ ·027783143	Tan δ ·031754015	Tan δ ·035725692	Tan δ ·03969818
γ 27·9999969	γ 27·99999508	γ 28·000016	γ 27·99999086	γ 27·999988 =
δ' 1°11412I012	δ' 1·341447448	δ' 1·568759956	δ' 1°79606II92	δ' 2°0233441
Sin δ' ·019443854	Sin δ' ·023410535	Sin δ' ·027376605	Sin δ' ·031342047	Sin δ' ·0353066
Cos δ' ·99981095	Cos δ' ·999725935	Cos δ' ·99962579	Cos δ' ·999508717	Cos δ' ·9993765
Tan δ' ·019447531	Tan δ' ·023416953	Tan δ' ·02738687	Tan δ' ·031357453	Tan δ' ·03532871
ε 0°227337231	ε 0°227326435	ε 0°227312508	ε 0°227301235	ε 0°2272836
Sin ε ·00396777727	Sin ε ·0039675843	Sin ε ·003967341 2	Sin ε ·003967/445	Sin ε ·0039668 3
$x_λ$ 34·2792585	$x_λ$ 33·21407333	$x_λ$ 32·4575207	$x_λ$ 31·89232537	$x_λ$ 31·454060
H'H 1·61115435	H'H 1·065785167	H'H ·756552628	H'H ·565195333	HH ·43816472
HE 6·28446319/	HE 5·220130152	HE 4·464275735	HE 3·90031957	HE 3·4627362
DE ·122194184	DE ·122206039	DE ·122216697	DE ·122243999	DE ·12225775
HD 6·283275113	HD 5·218699497	HD 4·46260188	HD 3·898403409	HD 3·4605772
Tan ψ ·0043647041	Tan ψ ·0043652226	Tan ψ ·0043656742	Tan ψ ·004366805	Tan ψ ·004367364
ψ 0°150077536	ψ 0°250107243	ψ 0°250133121	ψ 0°250197907	ψ 0°2502299
Sin ψ ·0043646625	Sin ψ ·0043657 81	Sin ψ ·0043656326	Sin ψ ·0043667633	Sin ψ ·004367632
OE 27·99625006	OE 27·9956405 6	OE 27·9957856	OE 27·994 1887	OE 27·993750
off-axis Focusing error	off-axis Focusing error	off-axis Focusing error	off-axis Focusing error	off-axis Focusing error
·00374994	·00435944	·0048144	·0058113	·00624968

Eric's solution to this problem was his invention of the 'Hairbrush Mirror'.

These would consist of a thin disc of glass and an aluminium circular plate. These would be glued together with thin aluminium rods [163], [164].

Air could circulate freely over all sides of the thin glass disc enabling it to reach a stable temperature quickly. However, the whole mirror was very rigid because the aluminium frame and the interconnecting rods formed a very stiff construction.

Eric with one of his 'hairbrush' telescope mirrors.

Earlier I described the Foucault Mirror Tester which I used for manufacturing my own mirrors.

There is another method of testing mirrors called the Ronchi Test [165].

This consists of parallel thin straight lines on a transparent sheet rather like a coarse diffraction grating.

When this is inserted between the mirror and the eye and is close to the eye the grating image shows curved lines. The shape of these curved lines can be interpreted to give an idea of the shape of the mirror.

Typical images seen through a Ronchi grating can be viewed at [166].

This test is largely qualitative because it shows up mirror defects (such as the notorious 'turned edge') but is difficult to interpret quantitatively.

Eric greatly improved on this technique by developing a screen of curved lines such that when the light coming from a mirror was viewed through this grating the lines appeared to be straight.

The eye/brain system can judge the straightness of lines far more accurately than it can judge the curved lines of a Ronchi grating.

This technique, which Eric described in 'Sky & Telescope' magazine in [167] was seized upon with enthusiasm by telescope mirror makers.

Indeed, Mobsby Null Gratings can still be purchased from specialist optical suppliers [168].

When he was in his late 80s his intellect was as keen as ever and he was making mirrors and designing special off-axis reflectors up until the day that he died.

On the 2nd of December 1987 he drove to Canterbury and back to discuss his telescope designs with an academic friend of his at Kent University. On arriving back in Shillingstone from this very long and tiring journey he fell asleep in an armchair and never awoke.

He was 83 years young.

7. Telescope Mountings

Having got yourself a good mirror, you next need to build a mount. This, like mirror making, is easy if you ignore those perfectionists who spend huge sums of money on a telescope mounting.

If you want to do precision, long time exposure photography then there is little option but to spend a lot of time and trouble getting a first class equatorial mount. However, for everyday viewing and in the absence of any desire to do photography through the telescope, a simple cheap mounting will suffice.

The choice is between equatorial and altazimuthal mounts. The latter, particularly the Dobsonian mounting, are cheaper and simpler.

In the late 1980s I made a Dobsonian mounting using only wood, two bits of plastic drain pipe and it cost about £25. What's more - it worked very well.

I started in the 1950s with an altazimuth. I then changed to equatorial mounts but then ended up with altazimuth types again in the 1980s.

My Second Alt-Azimuth Mounted Telescope

When I graduated from university and moved to Dorset 1960 I decided to build something more substantial and permanent than the simple telescope seen on page 4.

I made a 215 mm (8.5 inch) aperture mirror and an altazimuth mounting to hold it.

This was mounted in my garden on a brick pillar.

These two pictures show my homemade reflector on an altazimuth mounting. This was built in 1970.

Notice the counterbalancing pile of bricks in the second picture.

OK! So this mounting looked very crude but it was cheap (£1 for all the metal from a scrap yard) but it worked very well.

The control in altitude was extremely simple and crude. A string was tied to the horizontal frame, looped through a smooth hook on the tube and then brought down to wind around a Meccano spindle.

The tube is hinged and weighted so that it tries to rise up and thus keeps the string tight. The tube is lowered and raised by turning the spindle. The tube does not run away under its own weight because the spindle axis is small and friction holds it all stable. It was a very simple and smooth method of using gravity to control the altitude motion.

In the above picture the finderscope is made from a piece of builder's scaffolding pole and two convex lenses.

We can also see that a 1940s vintage Box Brownie camera has been clamped to the eyepiece. How crude is that!

To be honest, I never got a worthwhile picture of anything other than the Moon on a one second exposure; the longest this camera would provide.

This mounting worked beautifully.

In favour of the altazimuth mounting is the convenience of always having an eyepiece at a comfortable height above the ground and a convenient angle to look into. None of the disadvantages of an altazimuth ever worried me. Moving in two dimensions is easy with a bit of practice and I never was much interested in photographing celestial objects.

My 1960s setting circles consisted of card protractors which were set against the top of the brick tower and the side of the tube. I plotted out the relationship between hour angle/declination and altitude/azimuth using the Atomic Energy Authority computer - an IBM 7090 - at my work.

I worked out the local sidereal time from tables in the BAA Handbook, subtracted the RA to get the hour angle and Bob's your proverbial uncle! It only took a minute or two.

For my last and commercially built Dobsonian bought in 1996, I put on setting circles in altitude and azimuth as shown above.

At the start of each evening, I pointed the telescope at the Pole Star and stuck, with Blutac, little cards with arrows at zero azimuth and 50.4 degrees altitude. The telescope was then calibrated for the night!

My First Equatorial Mounting

Despite the obvious advantages of the simple first altazimuth, I decided to convert it to an equatorial mounting in 1972.

The picture shows the original metal square tube mounted in a closed yoke. The telescope tube could swing through the main frame to reach all parts of the sky with the eyepiece uppermost.

Despite this, the mount worked well.

The lower bearing was a ball race from a discarded washing machine found at my town dump.

The upper bearing was a piece of aluminium tube from a TV aerial resting on two washing machine wheels.

The main timber frame was left behind by the builder after my bungalow was completed in 1962 and the rest of the mount was bought for a few pounds.

You can see the start of a circular wall which was later to form the base of my first observatory - see the next chapter.

You can also see the remains of the vegetable garden which was sacrificed on my altar to astronomy.

There were times when I was required to straddle the telescope tube and lower polar bearing in order to get to the eyepiece.

If you achieve this feat, you will be too busy balancing yourself to use a clip board and record any observations. Of course, with a great increase in complexity, the entire upper part of the telescope tube can be rotated to bring the eyepiece to a comfortable position. This often does not work well because the delicate alignment of the optics will be affected as the top assembly rotates.

Overall I regretted the inconvenience of have an equatorially mounted telescope but, despite my misgivings, I rebuilt my optics into a new equatorial mount when I moved house in 1974.

My 'Plank' Equatorial Telescope

The photo above shows my second attempt at an equatorially mounted telescope after I had moved house and left my previous attempt for the new owners to puzzle over.

It can be seen that this is a 'minimalist' version inspired by Commander Henry Hatfield's 'plank' telescope.

Only two steel struts are used as the main 'tube' with the main mirror mounted on one end and the secondary mirror and eyepiece on the other. It worked really well apart from the inconvenience of still having to climb over the lower polar bearing.

It may seem odd having so little metal in a telescope when so many telescopes have closed tubes.

A closed tube can be a bad thing. Circulating air currents can set up in a tube as the ambient air temperature drops which cause image instability.

Wooden tubes are better than metal ones because wood is a good insulator but open tubes are best. As long as the optics are held rigid relative to each other, it does not matter how skimpy the body of the telescope is - but more stray light can enter the eyepiece from neighbours' lights

This picture shows the eyepiece end of the telescope which is a simple but strong contraption made of wood. Notice that the scaffolding pole finderscope was still with me after twenty years although I now had a cheap eyepiece in place of the simple magnifying glass with which it started!

My flirtation with equatorial mounts left me with no clear reason to own one. The advantage of being able to track objects with one - rather than two - movements was very small as I never got interested in telescopic photography and it was just as easy to find an object using azimuth and altitude scales on a Dobsonian as on an equatorial.

My Large Homemade Dobsonian

I decided, in the early 1980s to build a Dobsonian reflector using a homemade 305 mm (12 inch) aperture mirror.

The mounting was made entirely of wood apart from the main bearing in the base - which was the washing machine ball bearing from the first equatorial telescope made many years previously. The plastic altitude bearings were two drainpipe couplings.

I made a square tube for simplicity - there is no reason, other than historical convention, for having circular tubes.

This mounting worked extremely well for many years with no repairs or maintenance.

Note the scaffolding pole finderscope is still there!

This rear view shows the three bolts upon which the mirror rested. These could easily be accessed to adjust the mirror alignment.

The supporting bolts are on a circle with diameter equal to 70% of the mirror diameter. This minimizes the stress on the mirror

I devised a method of mounting the mirror on a hinged flap which gave easy access. I left the mirror in the telescope when not in use but covered it with a pad of dry cotton wool.

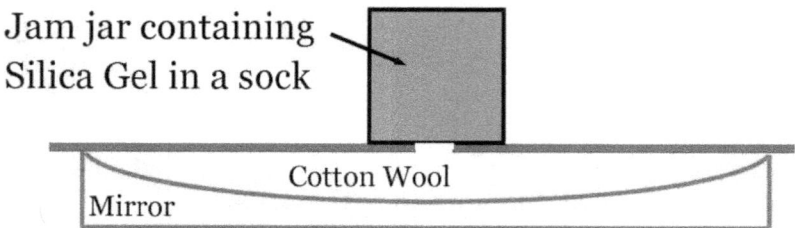

Jam jar containing
Silica Gel in a sock

Cotton Wool

Mirror

The air is dried by the silica gel in a sock (to keep the particles away from the mirror surface. The dry air circulates through a hole in the jar lid and keeps the cotton wool dry which is in contact with the aluminised surface of the mirror.

Incidentally, I noticed that cotton wool appears to make fine sleeks on freshly aluminised mirrors. This is because much 'cotton wool' these days contains a high proportion of viscose and this is the culprit.

Expert mirror maker Eric Mobsby used old, well washed flannelette but this is rarely available these days.

My most recent telescope was a 305 mm (12 inch) aperture Meade Dobsonian - my first commercial telescope.

It was a very convenient and portable Dobsonian and I was very pleased with it. I don't think that it worked any better than my cheap, simple home-made telescopes but I simply could not be bothered to build another one after my previous telescope had to be destroyed in 1989 - see page 226.

Simple Guided Photography

I said above that I had no interest in taking photographs through a telescope. However, I took many photographs of star fields using a simple camera mount. These were in support of my searches for novae as described in Chapter 13.

By far the easiest and most productive way to take pictures is the 'Scotch' or 'Barn Door' mounting, seen above, which I built in the 1970s.

I had seen one of the earliest examples of the 'Barn Door' mount demonstrated at about this same time at a BAA 'Winchester

Weekend' and I built mine in a couple of hours as soon as I got home. It cost under £1.

The idea is simple although some people have made it unnecessarily complicated over the years. Some people just can't help complicating a simple idea with little or no gain to its performance.

"If it ain't broken - don't fix it!"

Two boards are hinged down one edge and the hinge is pointed at the Celestial Pole. A camera is clamped to one board whilst the other is held rigid.

The boards are separated at the celestial rotation rate by turning a screw. The screw is arranged such that one turn every minute results in the camera turning at the same rate as the stars.

By matching the direction of a pointer on the screw head to the second hand of a stopwatch, the camera takes excellent guided shots of the stars.

A small red light operated by a torch battery illuminated the stopwatch.

In the picture above you can see the watch hanging inside a sheltered hole with a red light illumination which was a torch bulb painted with red poster paint - LEDs were yet to be invented.

The pointer was large and the disc was white so that it can be seen clearly in my urban light polluted garden.

When using a 200 mm lens on my camera the setting up and guidance had to be more accurate. Along the hinge I set up a tube such that the hole at the top end - farthest from the eye - subtended an angle of 1 degree. The pole star is about one degree from the celestial pole so I sighted on the pole star and placed it at the edge of the hole in the appropriate direction.

This aligned the hinge more accurately parallel to the Earth's axis.

I took hundreds of pictures with this simple arrangement in my nova hunting days and easily reached magnitude 9 with a 50 mm lens and fainter than 10 with a 200 mm lens. I also took the stripped down version shown below to the Canary Islands to photograph Halley's Comet in 1986 - see Chapter 18 for more details and to see some of the photos I took.

This particular camera guider was extra lightweight to carry in my suitcase and was a Braille version because I expected the night sky to be extremely dark.

The dial had nails sticking up at 'five second' intervals and I moved the pointer from nail to nail whilst listening to five second 'beeps' on a Walkman (a 1980s device like an iPod but using audio tape).

Recent 'improvements' reported in the magazines have included motorised screws running off mains power, screws bent to the shape of an arc of a circle, double tangent drives, etc. but these don't significantly improve the pictures.

Turning a screw once every minute following the second hand of a stopwatch gives perfect star images for exposures up to three minutes or more and could be built, like mine, from discarded junk in my garage.

Observing With Binoculars

Using binoculars is wonderful! The stars blaze out in wide-angle views. The drawback is having to stretch out on your back holding the binoculars above your head.

Some useful observing chairs have been devised to avoid this problem but I prefer to look down comfortably when observing.

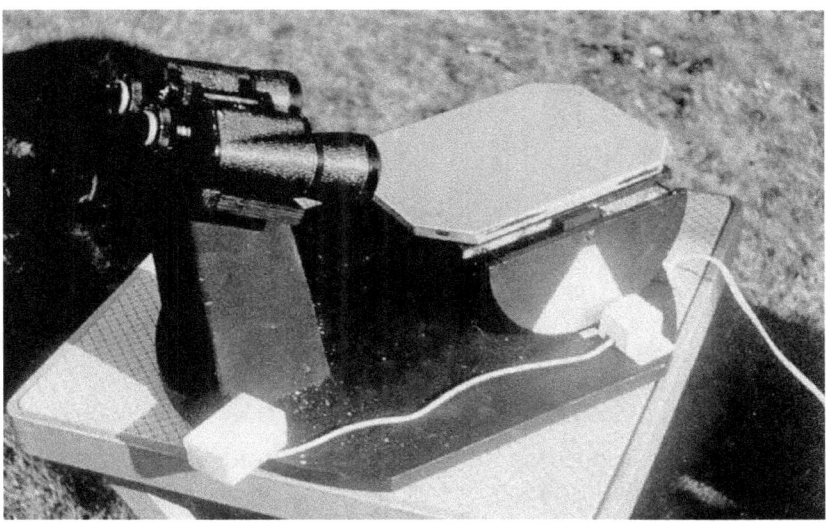

So, I built this binocular mount which stood on a brick pillar. The binoculars are at just the correct height to look down comfortably. The sky is reflected off a flat mirror which can be tilted and turned so that the whole sky is available to see.

The advantages of such a system are that one's hands are free to either write down observations or simply to keep warm in my pockets.

The observing position is comfortable.

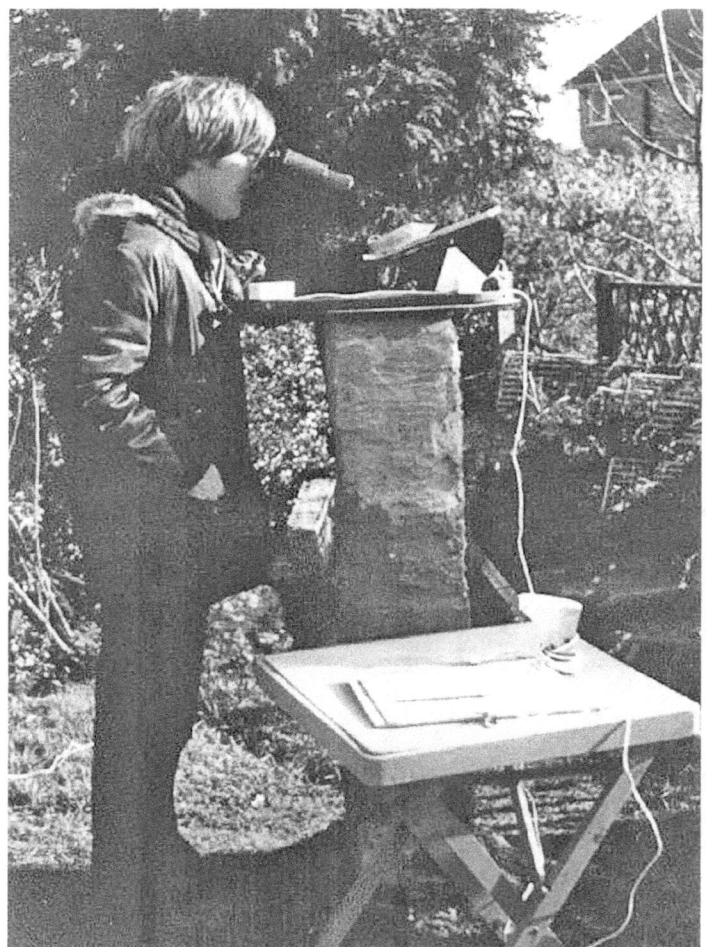

The complete mounting is shown above and you can see the extreme lack of complexity.

I was given the flat glass mirror by Eric Mobsby which he aluminized for me and the rest was wood costing less than one pound. I had crude setting circles to help me find objects but these were rarely needed.

The disadvantages are that the sky is reversed relative to conventional star charts and the mirror absorbs some light. However, the latter is counteracted by the gain because the binoculars are being held firm rather than being waved around.

In fact, I could see fainter stars via the mirror than when the binoculars were hand held.

Building a solid binocular mount does not have to produce an ugly tower of breeze block in the middle of the garden.

The picture below shows what a binocular mount can look like if you are anxious to keep your partner happy.

Even that example was minimal cost because the fancy bricks were left over after the construction of a house extension. The only cost was the mortar and the wooden platform on which the revolving binocular mount was placed.

A classy binocular mount.

8. Observatories - Good, Bad And Ugly

Having got myself a telescope I really wanted to get it into an observatory.

Of course, you may not have room.

However, in all my house moves the possibility of setting up an observatory in the garden was paramount.

When house hunting, I would go straight up every garden with a compass, check the height of the horizon, the proximity of street lights and the access to the site.

Only those houses that passed my checks were investigated further.

The first cry of anyone who thinks about observatories is that it's not possible to build anything decent that works.

Rubbish!

That view comes from seeing too many professional observatories with their superb domes and sliding shutters.

It also comes from seeing too many amateur observatories which are far too refined and expensive for the purpose of providing a convenient and secure observing site for the telescope and the observer.

My First Observatory

The above picture shows my first observatory. In your worst nightmare can you imagine a more ugly construction?

And yet it worked extremely well considering how little it cost.

It was built from eight sheets of chipboard covered in roofing felt.

It had a simple sliding shutter in the sloping flat roof and it ran directly on the top of a circular brick wall using eight nylon wheels of the type used in 1970s shopping trollies.

Like telescope making, some people make the objective of their efforts creating the observatory rather than using it.

Don't put more effort and money into the equipment than is absolutely necessary. There is no point in spending a fortune and years of labour on an observatory if you don't intend to use it for observing as soon as possible. The Grim Reaper may catch up with you whilst you are still refining it.

My Second Observatory

My second (and so far my last) observatory looked much more elegant and professional and yet it didn't work as well in many respects - one problem (later solved) being shown below!

Strong winds blew the dome off!

I am not an enthusiast for the 'run-off shed'. This consists of a shed which has one end wall which can be removed or swung out of the way. This is a bit of an engineering feat in itself since it means that the moveable end wall cannot provide any strength to the overall building.

The shed then runs on rails (where can you get rails these days?) and leaves the telescope and observer exposed to the wind, cold and light pollution whilst in use.

The shed performs none of the functions required of an observatory apart from keeping the telescope dry when not being used.

The principle of both my observatories was that the supporting walls were low. This was because the telescopes themselves were low to avoid having to climb halfway to the stars to reach the eyepieces.

You saw in Chapter 2 how I made the eyepiece of my 305 mm (12 inch) reflector never rise more than about 1.5 m (60 inches) above the ground. This meant that the observatory wall - the 'non-rotating bit' - had to be no more than about half a metre high.

After a long period of heavy rain I decided to put my building work on my observatory 'on hold' and relax with a little fishing.

All I caught was a cold!

The entry to the dome was through the observing shutter.

This caused a problem.

My observatory was built in the middle of the rose garden. After the observatory had been rotated back and forth many times in a night, I would lose my orientation somewhat. I would pack up the telescope, switch off the light and open the door - only to step out into the rose bushes.

This caused great amusement to visitors who watched me struggling to disentangle myself from rose thorns whilst trying to rotate the dome to a more convenient position. In the end, I painted a large white arrow on the observatory floor and I only climbed out when the door lined up with the arrow.

It is amazing how much difference an observatory makes. It keeps off the wind, protects the telescope, allows the telescope to be left in place avoiding the irritating process of setting up a portable telescope each observing session, provides storage for accessories and maps and the observatory keeps out stray light

It was 'snow' joke working under these conditions!

The more one uses an observatory, the more features one decides to include next time the urge to build strikes.

Having built two I suppose I have not had enough experience to describe the 'perfect' observatory. Like choosing houses, the selection is very much a personal thing and no two people seem to agree on what is the ideal abode.

One book well worth reading is *'Small Astronomical Observatories'* edited by Patrick Moore [169] and available from Amazon. Beware that the price when I recently checked was the outrageous rip off price of £72!

This is well worth reading but there are certainly cheaper books available.

Essential features I have discovered for my observatories are as follows.

1. Adequate space to walk around the ends of the telescope when it is pointing towards low altitudes without the observer and visitors knocking it.

2. Shelves on which to place star maps, books, cups of coffee, pictures, log book, laptop, etc. The more shelves the better and these should rotate with the dome so that the essential requirements are always at hand near the eyepiece.

3. Adequate headroom so that crouching is eliminated. This is especially true for hemi-spherical domes. The hexagonal rotating shed was good on this feature.

4. Any concrete flooring should be covered with wood or carpet to reduce the chilling effect on the feet. Because of the low eyepiece position of my telescopes, I sometimes had to sit on the floor to observe at low altitude.

5. The rotation should be as quiet as possible to avoid complaints from the neighbours. I made the wheels run on a sandwich of polystyrene ceiling tiles between hardboard sheets on the top of a brick wall. The polystyrene muffled the sound very well.

6. The shutter should be capable of being partially opened in the vertical plane. This keeps out more of the wind and stray light.

7. The dome should rotate easily. There is little worse than suffering a hernia a few minutes before having to make an important observation.

The rotation of a dome is usually the aspect that causes most problems. I don't think that there is a universal solution available to non-technical people like me.

Hedley Robinson had his dome supported on lots of golf balls running in a channel. That sounds excellent and he was very pleased with the arrangement. It is certainly worth a try.

The Hampshire AS observatory dome ran on roller skates - at least it did when I saw it being built in the 1970s. That also seems to have worked well.

Even cannon balls have been used!

I used supermarket shopping trolley wheels.

They were used on both my observatories and worked well. I tried fixing the wheels to the wall with the dome running on top but this did not work very well. I'm not sure why but it probably had something to do with the inadequately thick wood used to form the base of the dome.

It bent when the wheels were between the load bearing vertical strips. The effect was that the observatory dome bounced up and down like a ship in a force ten as it rotated. This also caused it to jump off the tracks trapping me inside.

The above picture shows the original plan for supporting the dome - which did not work. When I put the wheels on the dome beneath the vertical struts it worked well.

Here we see the assembly of the main vertical ribs which supported the sliding shutter. I used plywood but, by economising too far, all the dome ribs were too thin and flexible. In the end, the strength was mainly in the wood making up the panels rather than in the ribs. If I ever make another dome I will make the supporting ribs stronger and save on the panelling.

Here the ribs are mostly in place. The shutter had to fit snugly on the main parallel ribs and so these had to be arcs of circles. The other ribs were also arcs of circles and each was individually cut to fit between the horizontal shutter supports and the base.

I tried to get everything aligned properly but it was not to be. The panels were made of thin plywood. Each piece of wood was a unique and had to be cut to fit the hole available using newspaper templates.

Here we see the sliding shutter in position.

Free labour was provided by my three children

The dome was painted with primer and a top coat of aluminium paint. This was to make the whole thing look professional and keep the temperature down in the daytime. However, it soon peeled off and, in deference to the neighbours, I painted the dome green to help it to blend in with the rest of the garden.

The day of the observatory official opening by my daughter.

The only problems I really had, and these were minor, were the dome needing a lot of effort to turn and water leaking in where the plywood panels joined the ribs. The latter were sealed with waterproof tape, pitch, mastic, etc., but nothing really worked because the dome flexed so much as it rotated.

Although the observatory looked smart to begin with it weathered after a few years to become very tatty.

Here we see me working inside. Note the convenience of charts on the wall - always near the eyepiece - a pin board for the list of observations planned for each night, a mug of cocoa and headphones feeding Heavy Metal music of which I am a fan.

In 1989 I moved house and the observatory had to be abandoned and demolished. I also destroyed my homemade telescope - keeping the optics safe of course.

I sold the main and secondary mirrors to a friend who put them to good use in a Newtonian mounting.

An Unusual Mounting For A Refractor

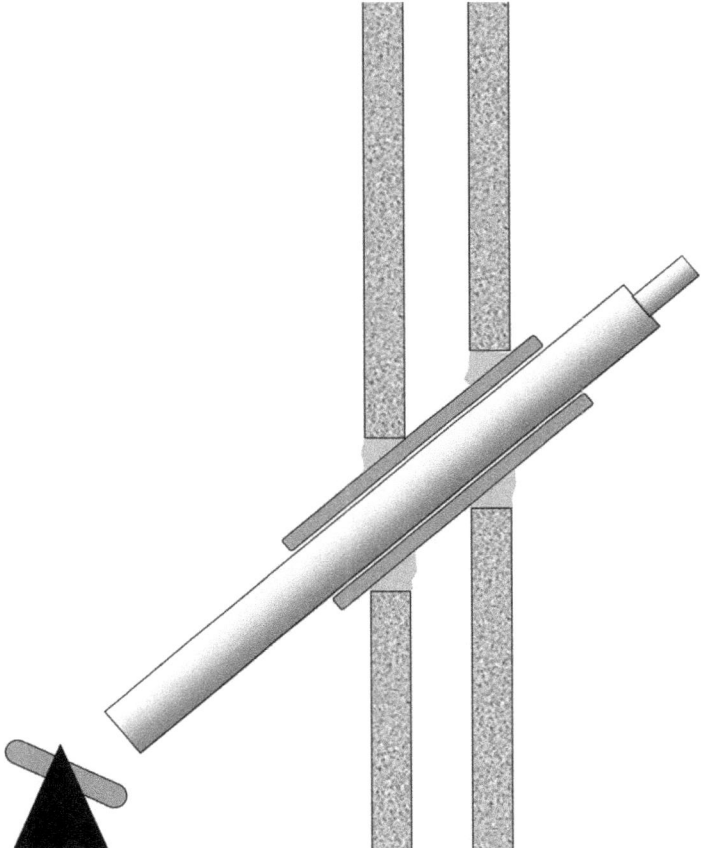

Hedley Robinson had a refractor which he wanted to mount in a solid manner with him not exposed to the weather when observing.

He came up with a dramatic solution as shown above.

His house had a living room facing south with a bay window. So, he smashed a hole through the cavity wall and fitted his refactor as shown pointing downwards.

The axis of the telescope pointed at the north celestial pole.

Outside the house was a mirror which could be remotely rotated in azimuth and altitude. This mirror reflected the night sky up into the telescope.

This system enabled Hedley to sit in a comfortable chair in his living room and observe the night sky by looking downwards into the eyepiece at a comfortable height and angle.

The disadvantage of this arrangement was that only one half of the available sky could be observed at any one time.

However, by choosing the time and date correctly more of the sky can be seen.

An observer with an unrestricted view of the sky above the horizon will have access at some time and date to 82% of the entire celestial sphere at Hedley Robinson's latitude in Teignmouth, Devon.

With Hedley's arrangement this was reduced to 70% so he was not losing a large amount of sky.

This fraction reduces somewhat when taking into account the shorter nights in summer.

I would consider this an ideal observing arrangement were it not for my wife's reluctance to have a large hole smashed through the living room wall!

9. Observing The Moon

Unaided Eye Observations Of The Moon

Have you ever looked really carefully at the Moon with the unaided eye?

When my children were young I would take them out when the Moon was high in the early evening sky and ask them to draw the Moon.

Sometimes the results were hilarious - my eldest son once claimed to have seen the saucepan lids of the Clangers on the surface - see [170] to learn about these knitted aliens.

I was surprised at how much genuine detail my children could see. They interpreted the large dark impact areas in different and fascinating ways.

Every culture interprets different lunar features in their own ways.

The website at [171] shows six examples of features formed in the minds of different cultures including a rabbit cooking rice cakes (East Asia), the two hands of Astang Mata (India) and a banyan tree with a woman called Hina making cloth (Hawaii).

There is also a woman called Rona, a Maori maiden who disrespected the Moon and now lives there as a penance. Also a screaming face of the Man in the Moon as I and fellow Westerners tend to see the Moon.

Some Muslims see the name of Allah written on the Moon.

Some also see Allah's name written on other planets and on the moons of the planets; specifically Enceladus [172].

Possibly requiring the most imagination and a lot of squinting is the nursery rhyme of Jack and Jill being played out on the Moon.

See [173] for a full description of this story being enacted every month. You will need a lot of imagination to see the figures outlined on the website in the real Moon!

The first verse runs

Jack and Jill went up the hill

To fetch a pail of water.

Jack fell down and broke his crown,

And Jill came tumbling after.

The imaginative story is played out by the patterns on the Moon between New and First Quarter supposedly showing the shape of a small boy carrying a pail of water.

Because the crescent Moon is tilted downwards towards the setting Sun it looks as though Jack is climbing up.

At Full Moon Jack and Jill are together on the 'top of the hill'.

Between Full Moon and Last Quarter the figure of Jill supposedly appears on the Moon she appears to be climbing the hill because the Moon is tilting down in the morning sky towards the rising Sun.

This story may have originated in ancient Scandinavia where the characters were named Hjuki and Bila.

Two millennia ago Mediterranean civilizations believed that the Moon was a mirror and the markings seen on its face were reflections of the land masses of the Earth.

At a time when the Roman and Greek empires had spread over only a small part of the Earth, the Moon's markings were much studied as a means of discovering, in the 'mirror' Moon, new lands and oceans beyond the explored horizon.

It is possible that the legend of Atlantis, an island in the Atlantic, arose because of a feature seen on the Moon.

Two excellent books for unaided eye and binocular observers of the Moon is *'Exploring the Moon Through Binoculars and Small Telescopes'* by Ernest Cherrington, Dover Books (1985) and *'Exploring The Solar System With Binoculars'* by Stephen James O'Meara, Cambridge University Press (2010)

The Visibility Of Earliest 'New Moon'.

Apart from watching the Moon's phases and making out imaginary figures on the Moon, another Moon-related unaided eye exercise is to find the earliest time after a New Moon that the thin crescent can be seen in the evening sky.

Not only is this an interesting way to watch some wonderful sunsets and gorgeous crescent Moons but it actually has some value.

The visibility of the crescent Moon has been a topic of much interest for centuries particularly so because the first sighting of the crescent Moon marks important times in several religious calendars.

For example, the first Eid was celebrated in 624 CE by the Prophet Muhammad after the victory of the battle of Jang-e-Badar. The annual festival begins with the first sighting of the crescent moon in the sky.

Muslims in most countries rely on news of an official sighting, rather than looking at the sky themselves, and some professional observatories provide a news service to alert Muslims that the crescent Moon has been 'officially' seen in the evening sky.

The months in the Jewish calendar follow the lunar period. The sighting of the New Moon was ordained to establish the beginning of the monthly cycle and therefore the first day of the month.

The first visibility of the crescent Moon is determined from the time in which it occurs in Jerusalem to ensure the uniformity of religious worship throughout the world.

Apart from the need to set religious calendars it is intrinsically interesting to check when the crescent Moon is first visible each month, there is a great joy to be experienced when a thin sliver of the Moon is sighted in a clear sunset sky.

Her Majesty's Nautical Almanac Office had long has an interest in collating crescent Moon first appearances and providing empirical equations to enable religious sects to be able to forecast, in advance, when each lunar month should be deemed to start [174], [175].

This is particularly important when a spell of cloudy days prevents observers looking for the crescent Moon.

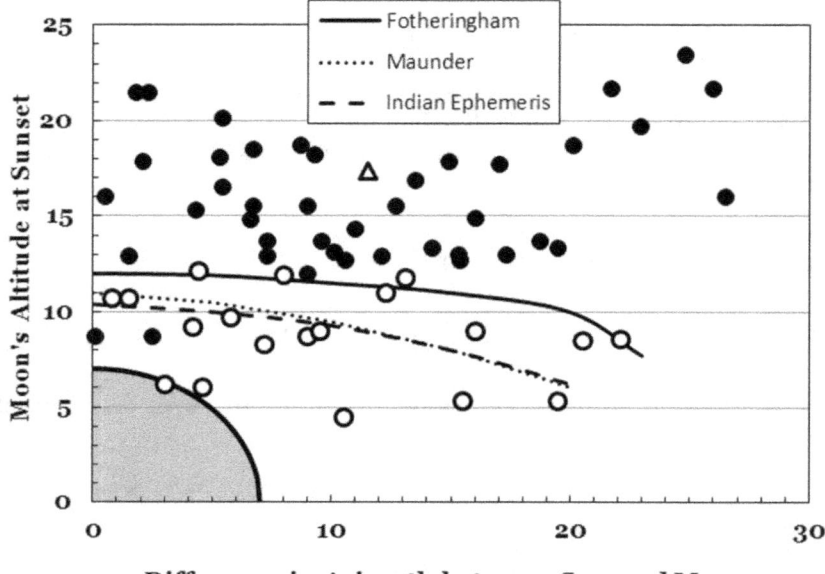

Difference in Azimuth between Sun and Moon

I have plotted the above chart using data collated by J K Fotheringham from various observers in the period 1859 to 1880 [176].

In this chart the visibility of the thin crescent Moon has been plotted as a function of the Moon's altitude at sunset and the difference in azimuth between the Moon and Sun also at sunset.

Open circles mean that the crescent Moon was not seen and filled circles that the Moon was seen with the unaided eye.

The single open triangle was an observation I made from Dorset on 11th February 2016 at 17.38 UT. The thin crescent and strong Earthshine was easily seen.

This is a somewhat empirical way to plot the results because there are so many parameters that will affect an observer's ability to see a thin crescent Moon including atmospheric clarity, visual acuity apart from geometrical circumstances.

Even so, plotting the results in this way does appear to separate out the observations.

Three empirical curves have been drawn corresponding to the researches of J K Fotheringham, E W Maunder of Greenwich Observatory (after whom the 'Maunder Minimum' in sunspot records is named) and the curve used in the Indian Ephemeris and Nautical Almanac.

Observing the first visibility of the crescent Moon is an interesting project for unaided eye observers and one to be enjoyed singly or as part of an astronomical society group activity.

Indeed, among some observers there is keen competition to see who can spot the crescent Moon earliest after the New Moon or latest before the New Moon.

Accounts from Scarborough and Durham in the North of England claimed that a crescent Moon was seen in the evening of 2nd May 1916 in a perfect sky by four people.

This corresponds to a Moon seen only 14½ hours after 'New Moon'; the latter being the time when the Moon is at the smallest apparent angular separation from the Sun in the sky [177].

However, these observations are doubted because they are so much at variance with other observations.

The great French astronomer André Danjon worked out by theory and experiment that a crescent Moon cannot be seen if the angle between the centres of the Sun and Moon is less than seven degrees [178]. This is because of the way that the shadowed walls of the craters on the limb of the Moon hide the lit inner walls.

The shaded area on the chart on page 232 shows the Danjon seven degree limit of crescent visibility.

The observation from May 1916 was at a separation of 7.5 degrees according to Stellarium [179] and so was unlikely to have been genuine.

The best time to watch out for very thin crescent Moons is in the Spring for observers of the evening crescent and in the Autumn for morning observers.

The same is true for southern hemisphere observers.

The First Moon Illusion

I have two Moon illusions to keep you entertained and no optical aid is required.

The first is the very old illusion that the Full Moon appears larger when on the horizon than when it is high in the sky.

In fact, the Moon is actually slightly smaller when on the horizon than overhead because the observer is the Earth's radius further away.

In addition, the actual size of the Moon varies as it goes around the Earth such that at its closest (perigee) the apparent diameter is 12% larger in the sky than at its furthest point (apogee).

If a Full Moon occurs at around the time of perigee the Moon does indeed really look large and this is referred to as a 'Super Moon'.

This description was used late in September 2015 when a total eclipse of the Moon occurred close to the perigee position in the Moon's orbit and a large blood-red fully eclipsed Moon was seen [180].

However, ignoring these orbital effects, it is widely accepted that a Full Moon close to the horizon appears larger than the same moon high in the sky.

The reason for this illusion is much debated and there is no consensus except that it is truly an illusion.

Patrick Moore once devoted a 'Sky at Night' programme to this illusion and we were treated to the spectacle of him and others standing on the roof of the BBC Broadcasting House observing the low Full Moon by facing away from the Moon and bending forward to observe it upside down between their legs.

Fortunately no recorded evidence seems to exist of this event. If it does then it should be destroyed as my memory of seeing Patrick bending over and looking between his legs is one that I can well do without!

Having said that, the experiment showed that the First Moon Illusion disappeared when no horizon was present.

It all seems to be down to the expectation that the Moon low down near the horizon should be further away and smaller - but it isn't. So the brain imagines that the Moon is larger.

This seems to me to be the best theory as objects do get smaller as they move towards the horizon. The brain is driven to believe that the Moon also does this by seeing perspective all around us, from clouds forming a flat plain above our heads to roads tapering to a distant blur.

Which 'Moon' appears largest in the picture above?

This Moon illusion disappears when the Moon is viewed through a cardboard tube of the type used inside toilet rolls. This supports the 'perspective' theory.

Having said all that, I have placed two identical 'Moons' in a diagram above. The left one is in a perspective field where our brains should be fooled into thinking that the left-hand 'Moon' is larger than the right-hand 'Moon'.

And yet, the left-hand Moon looks smaller - not larger - to me and to friends who have seen this diagram.

So where does that leave us?

As a project, try to observe the rising Full Moon over different horizons (such as the sea, fields, flat areas such as airports, etc)

How does the apparent size of the rising Full Moon appear to decrease as the Moon rises up the sky and if clouds come along to surround the Moon?

There are lots that can be done here with no optical aid.

The Second Moon Illusion

What I call the Second Moon Illusion is one that I had not encountered in the literature before I noticed it one bright, clear summer's afternoon.

In the west the Sun was about 20 degrees above the horizon and in the east was a gibbous Moon. The angle between them was about 120 degrees and the Moon was also the same altitude above the horizon as the Sun.

What startled me was the way the Moon was tilted.

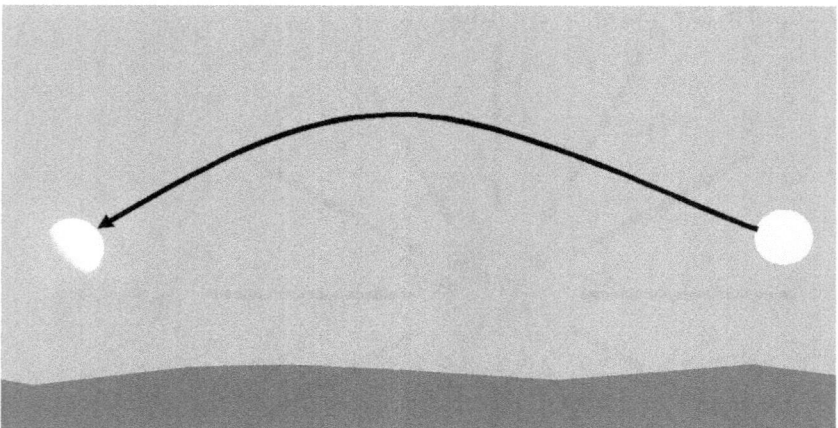

This diagram is not to scale but is used to illustrate the strange configuration of the Sun and Moon when both are the same altitude above the horizon and the Moon is gibbous phase.

It is obvious from the way the Moon is illuminated that the light from the Sun must be arriving at right angles to the line joining the lunar poles which is also at right angles to the terminator.

So, the sunlight is beaming *down* onto the Moon but the Sun is the same height above the horizon as the Moon.

The sunlight must be following the curved path shown above.

I swept my pointing finger from the Sun (CAREFUL!) to the Moon and the path travelled by the light certainly appeared to be bent upwards and then downwards.

Had the Laws of Physics changed and light no longer travelled in straight lines?

I then took a length of string and held it taut such that one end was near the Sun (CAREFUL AGAIN!) and the other end was on the Moon.

The string appeared to bend following the curve in the diagram opposite!

The reason that this illusion happens is because the light and the string are indeed following a straight line from Sun to Moon but the horizon nearby is fooling the brain into believing that the light path and the stretched string should be parallel to the horizon.

This is related to the First Moon Illusion in as much as the perspective of the flat horizon and landscape is forcing our brain to make a low Full Moon appear large and appear to be making a straight line curve up from the horizon.

The strength of this second illusion is greatest about ten days after New Moon. At other times the illusion is still there but is much less obvious.

Of course, this second illusion only works in temperature latitudes.

Near the equator the Sun and Moon travel almost overhead and the illusion does not appear.

Binocular Observations Of The Moon

It is surprising how much can be seen using binoculars.

The best size is probably 10x50 as these give a good magnification whilst being comfortable enough to be handheld without needing a mount.

'Exploring The Solar System With Binoculars' by Stephen James O'Meara, Cambridge University Press (2010) has a very useful chapter on observing the Moon with binoculars. This includes charts and photographs showing all the main dark 'mare' areas as well as 136 craters that are potentially visible using binoculars.

Some of the smaller craters in Stephen O'Meara's list might be a challenge for a binocular user but you can never tell. For example, Tycho is a relatively small crater and yet is quite easily seen in binoculars because of its bright and extensive rays that spread across the Moon's face.

On the other hand, some of the enormous craters are not obvious through binoculars because they are sometimes heavily cratered, have low walls relative to the crater diameter, are distorted by perspective when near the Moon's limb and are often misshapen by lava flows early in the Moon's history.

A problem with Stephen O'Meara's list of binocular target craters, albeit a minor one, is that the identification maps and photographs are on such a small scale in his book that it is not obvious which craters are which.

What I recommend is to buy a good Moon map such as Philip's Moon Map by Dr John Murray (2015) and look up all the craters you spot for yourself.

This will lead to a positive identification and lead you on to craters not on O'Meara's list.

There are also excellent apps for tablets which allow a detailed map of the Moon to be displayed.

When observing the Moon through binoculars try to imagine the scene many millions of years ago when the Moon was much closer to the Earth and being subjected to the 'Heavy Bombardment' of asteroidal bodies which created the dark 'maria' features that are so obvious to even the unaided eye.

My Telescopic Observations Of The Moon

The Moon was the first object I studied when I got my homemade telescope working on its 'orange box' mounting - see page 4.

I loved looking at the mountains, craters, 'seas' and valleys; it was like flying over the Moon in a space craft and the closest I would ever get to visiting the Moon's surface.

My earliest surviving observation is dated July 16th 1959.

I had attempted to draw an entire half of the Moon and obviously I had given up after a short time when I realised that it was a pretty stupid thing to attempt.

Patrick Moore's advice, as always, was full of wisdom.

"Concentrate on a tiny part of the Moon and draw it big"

I did and started to draw features on the Moon.

Observing the Moon with a telescope requires a map and the first atlas that I obtained was the 1960 H P Wilkins book based on his original map drawn at a scale of 300 inches (7.6 metres) to the diameter [181].

This was an enormously cluttered and complicated set of drawings as shown below [182].

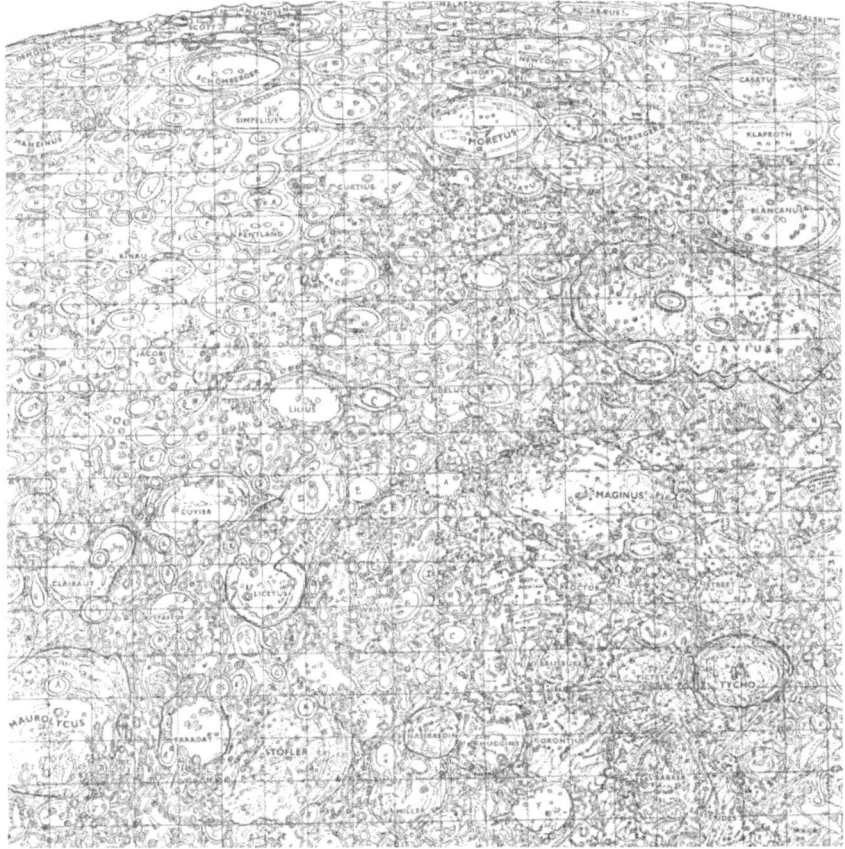

Can you see the huge crater Clavius in the above map sheet?

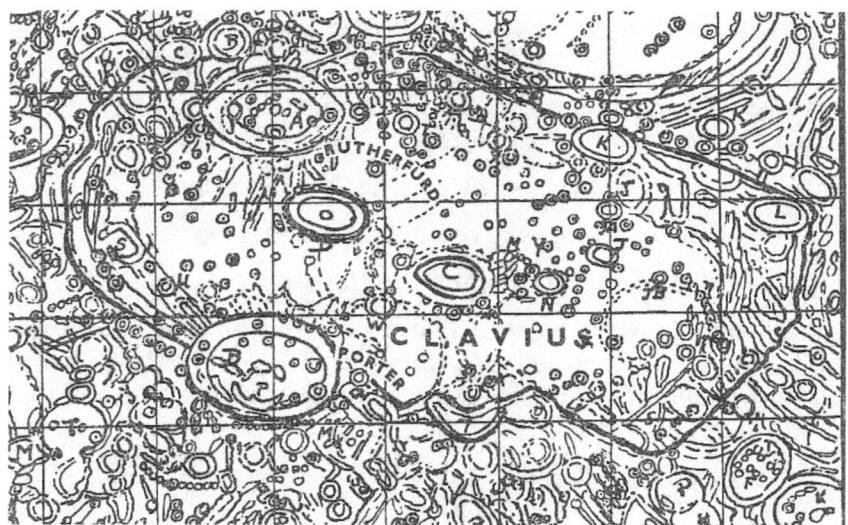

Now can you see it? Not easy is it?

Frankly, as a means of helping lunar observers to identify the features of interest it was far too detailed.

However, I was awarded the atlas as a prize at university in 1960 so it didn't cost me anything.

I cannot remember ever using it during my lunar observations.

The atlas was interesting as a book however. It was printed like a triptych - three pages could be read at the same time because it was spiral bound both on the left and right of the covers.

In this design the atlas could be opened with the gazetteer on the two pages on the left and with the appropriate map page on the right.

Patrick Moore had collaborated with Wilkins for many years mapping the Moon and they had been joint authors of the monumental book 'The Moon' [183] published in 1955.

It has been claimed by Patrick Moore that his mapping of the Moon with Hugh Percy Wilkins was an important input to the plans by NASA to land a man on the Moon.

However, there is no evidence of this although some Soviet astronomers did ask for material from Wilkins and Moore.

In fact, the Wilkins map has been heavily criticised on the grounds that much of the detail is spurious and the positions of the features are wrongly plotted.

I cannot comment on the former criticism but I can disprove the latter assertion.

I have carefully measured the positions of small named features on the 1960 Wilkins map and compared these with their official lunar coordinates.

I found that the discrepancies in position were about 20 kms on average which is tolerably good. This means that the craters likely to be seen by amateurs are pretty much accurately plotted in position.

What was a 'killer' for the massive Wilkins mapping project which had occupied him for decades was the fact that Gerald Kuiper in the USA was compiling at the same time a very detailed photographic atlas of the Moon.

This had many advantages over the work of Wilkins in that

-   the details recorded were real and not spurious,

-   the atlas gave a realistic view of the lunar surface as it would be seen by astronomers and space explorers,

-   the positions were unambiguously correct,

-   the level of detail shown was much as seen by amateur astronomers.

In 1968 I bought a very useful atlas of the Moon which was the newly published *'Amateur Astronomer's Photographic Lunar Atlas'* by Henry Hatfield [184].

This was all that a lunar atlas should be for amateur lunar observers like me.

The visible face of the Moon was divided into sixteen sections and the area of each section was reproduced in several excellent quality photographs taken with the Sun at different altitudes.

This allowed features to be shown under a variety of lighting conditions which was necessary because mountains and craters change in appearance very significantly during the course of a lunar 'day'.

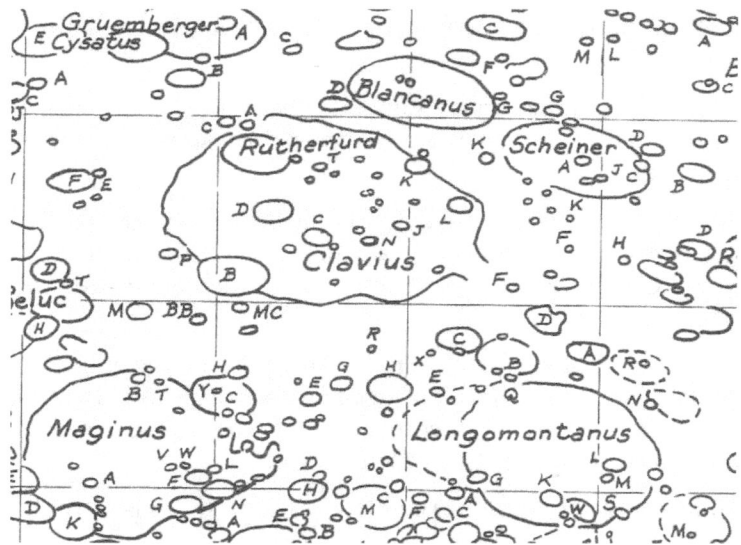

(Copyright Lutterworth Publisher. Reproduced with permission)

Above we see a section of Henry Hatfield's key map covering the area of Clavius [185].

Compare this with the Wilkins map of this area on page 239.

The Hatfield map shows clearly just the features within reach of amateur telescopes.

Above is the same area around Clavius as photographed by

Henry Hatfield.

Above is pretty much what I would see through my various telescopes and Henry Hatfield's atlas was much used by me.

A digitally remastered edition was published in 2012 and is available at the time of writing at the price of over £40 for both the paperback and the Kindle versions [186].

There have been other recent photographic Moon atlases [187].

My favourite atlas is the Phillip's Moon Atlas which contains over 300 named features and is very reasonably priced [188].

The area around Clavius on the Philip's Moon Map is shown below for comparison with other maps.

My only reservation is that the low contrast of the map and lack of distinctive shading on the mare regions makes the map a little difficult to use in low level red light conditions.

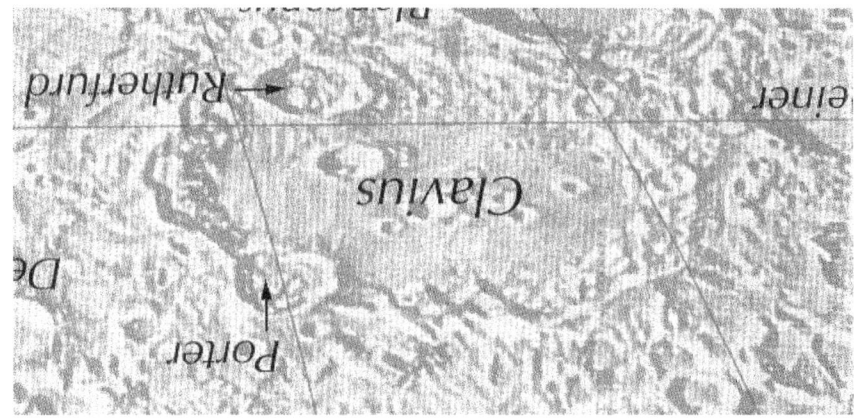

(Copyright Philip's and reproduced by kind permission)

Commander Henry Hatfield's Secret Life

At this point I am going to ramble off to write about Commander Henry Hatfield RN [189].

Henry was a delightful person; easy to get on with and also polite and helpful, especially to young astronomers like myself.

He was a frequent attendee at the BAA Winchester Weekends in the 1970s where we mostly met.

Henry was involved in the 'Great Schism' that rocked the BAA in 1987 resulting in the incumbent President, Storm Dunlop, being challenged (the first President to be challenged in the history of the BAA) and Henry taking over as President serving between 1987 - 1989 [190].

Commander Henry Hatfield RN directly entered my life in two ways.

Firstly, he built a telescope which he called 'The Plank'. This was because it was - literally - a plank being a flat rectangle of wood with the main mirror mounted at one end and the eyepiece and secondary mirror at the other end.

I copied this arrangement myself as seen in the pictures on pages 202 and 203.

The second way in which Commander Henry Hatfield RN directly entered my life was in the mid-1970s.

In those years I was a scientist working on naval weaponry for the Ministry of Defence.

I was involved briefly in the analysis of Naval Intelligence information being gathered as part of the ongoing Cold War.

I was called to a highly secret briefing at Northumberland House in London where a branch of the Royal Naval Intelligence Service (DI57) was then located. The building was later sold to The Welcome Trust and turned into student accommodation for the London School of Economics.

Early in the 20th century this building had been The Victoria Hotel where King Edward VII entertained his mistresses.

On entering the meeting room which was protected by armed sentries - so secret was the material I was there to discuss - I saw Commander Henry Hatfield RN!

Our eyes met, our jaws dropped and Henry pulled himself together to say

"So Geoff, my cover is blown!"

"And mine!"

I responded.

To the outside world Henry was a Royal Navy Hydrographic expert but, in secret, he - like me - was engaged in highly classified military operations.

Our mission at that time was gathering Intelligence on the huge Soviet aircraft carrier 'The Kiev' being built at the Black Sea construction docks in Sevastopol in the Black Sea.

Progress on the building of this heavily armed monster had been followed by spy satellites and the day was nearing for its launch.

A Royal Naval nuclear submarine was despatched to covertly spy on this ship on its maiden trial.

Amazingly, the nuclear submarine went right underneath the ship and took photographs from the bow to the propellers through the periscope. The mighty Soviet warship crew - 1,600 men - had no idea there was a huge nuclear submarine a few metres beneath its keel.

I was involved in the team analysing the photographs and trying to work out what sonars and weapons the ship was carrying.

Henry Hatfield's task was the operational and hydrographic support for the spying mission.

It was curious that in 2014 two programmes were broadcast on BBC television describing all the activities that Henry and I had been engaged in. The Soviet and British naval captains involved were talking openly about their work [191].

In fact the photographs stamped TOP SECRET of the Kiev's sonar domes that I had seen in a room guarded by armed marines were shown on national television!

Four KIEV class aircraft carriers were built in the 1970s.

Two of these huge and powerful ships were sold to China in the mid-1990s where they were converted to luxury floating hotels, one was scrapped and the other was sold to the Indian navy.

After forty years the work that Henry and I did was completely unclassified and being shown on national television.

After that diversion into the murky world of Military Intelligence, back to my observations of the Moon.

Are There Changes On The Moon?

Almost immediately on completing my first telescope I was collaborating with other BAA Lunar Section members looking for changes on the Moon and also charting the areas on the visible edge of the Moon.

There was, in the 1950s considerable controversy as to whether there were physical changes on the Moon.

Was the Moon really dead and unchanging?

I had recorded on July 16th 1959 that the small craters Messier and Pickering were apparently equal in size and that a North-South halo existed.

This extract from a 1937 scientific paper [192] suggests that real changes could be occurring.

"Professor William H. Pickering has recently studied two small adjoining craters, designated Messier and W. H. Pickering, and suggests that they go through a cycle of changes of size and shape at each lunation.

Following this work two observers, E. Martz, in Jamaica and Illinois, and Walter Haas, in Ohio, report that they have systematically observed and drawn these two craters during 1936, using various telescopes ranging in aperture from six to twelve inches.

They found changes in the general distinctness of the craters, a possible twilight effect in the crater Pickering, and the development of bright and dark spots independently of the sun's altitude."

Recent space probe photographs like the one below show that the pair of craters certainly looks odd.

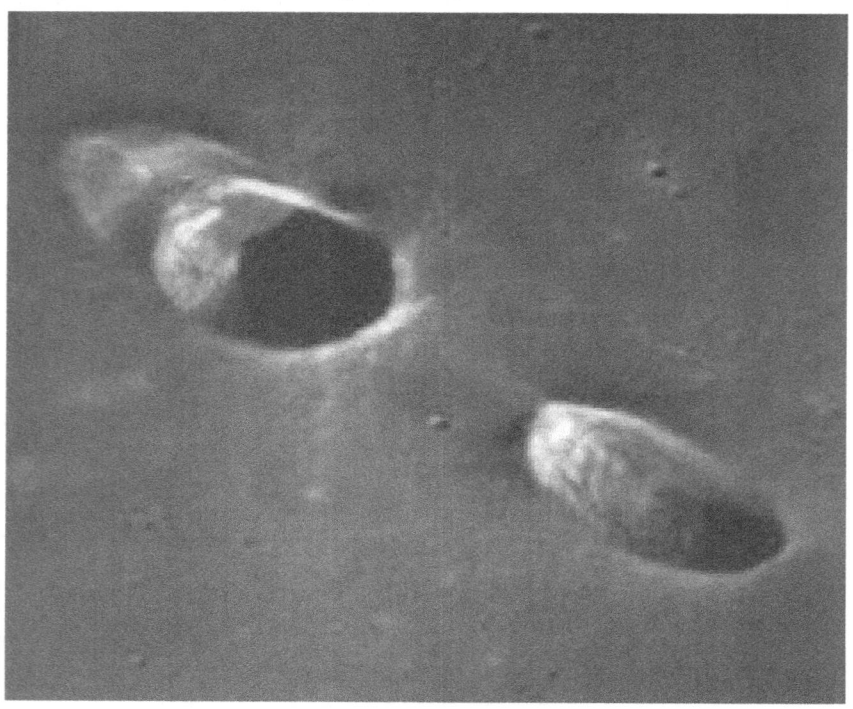

Craters Messier and Pickering - a tunnel on the Moon?

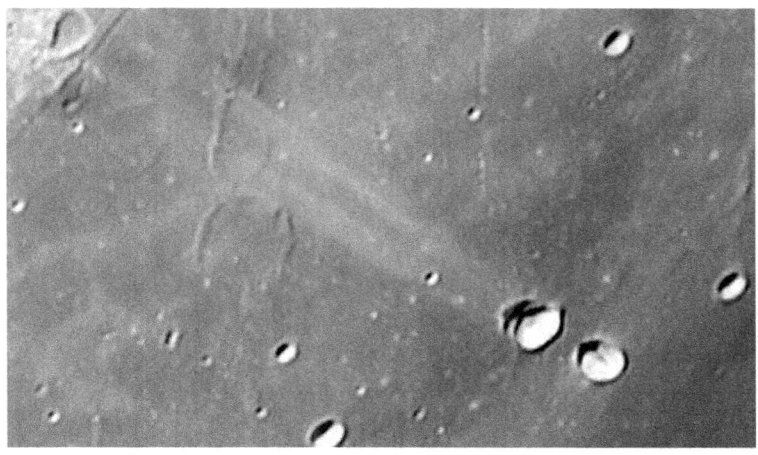

It is currently believed that a large object hit the Moon at grazing angle leaving a rare elliptical crater and a halo of debris on one side. The impactor skipped and hit the Moon again making the second crater.

The article at [193] is well worth a read if only to see that some observers have suggested that an impactor hit the Moon, travelled beneath the surface and burst out again!

Transient Lunar Phenomena (TLP)

The whole subject of changes on the Moon was a 'hot topic' when I started to turn my simple little telescope towards the Moon and I spent several decades watching the Moon for changes.

These alleged changes were grouped together under the name Transient Lunar Phenomena (TLP) and were a pet topic of Patrick Moore.

However, there were others who disbelieved these claimed observations of changes on the Moon (colourations, obscurations, mistiness - even volcanic plumes) and believed that the sightings were not real but due to straining to observe things right on the limit of visibility.

This is now called 'Pathological Science' [194] and I published an article on this topic in New Scientist magazine [195].

In Chapter 10 I deal in more detail with Pathological Science as I believe it was the reason why so many non-existent details were seen on the inner planets such as cloud features on Mercury, 'canals' on Venus as well as the 'ashen light' on Venus's unlit hemisphere, changes on the Moon (TLP) and the 'canals' of Mars.

Certainly there was much interest in detecting changes (whether real or imaginary) on the Moon's surface in the 1950s when I had set up my small telescope in my North London home.

A device was invented called a 'Moonblink Comparator'. This was fitted with two or more coloured filters and these could be quickly switched at the eyepiece.

The idea was that a small coloured area on the Moon would appear to 'blink' as the filters were switched whilst the rest of the Moon would not show the same change.

For example if red and blue filters were being switched, a red spot on the Moon would be practically invisible through the red filter but would appear as a dark patch through the blue filter.

Thus, as the filters were switched, the red spot would appear to blink on and off.

Exactly who invented this device was the subject of a vitriolic exchange of insults in the BAA Council which spilt out into a trade of insults in the pages of the BAA Journal.

None of this aggressive name calling bothered me as the Moonblink device was far too complicated for me. I fixed two coloured filters on a cardboard mount and simply slid the card back and forth between the eyepiece and my eye.

There is still a lively following for TLP studies and this is an area you could get involved in - if you want to spend years staring at features on the Moon waiting for nothing to happen [196]!

Did I ever see a TLP?

None so convincing that it was reportable. I saw plenty of odd things on the Moon but nothing I would claim was a definite change.

For example, I noted in my log book on 2nd June 1960 at 20.30 UT that the isolated mountain Pico [197] was *brilliant white - it almost appears to be glowing*.

This was not a TLP - I never thought it was - but the brilliance of the mountain was quite startling. I used to imagine what it must be like to stand on that mountain and see the Sun blazing down.

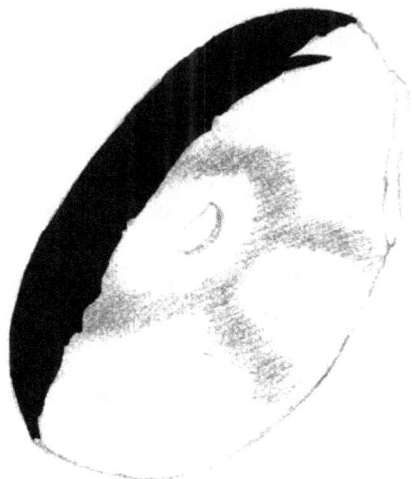

This drawing of Aristarchus was made at 22h 30m on 4th September 1968.

It shows the elusive radial 'spokes' in the crater which are considered to be a good test of a telescope optics. Obviously my home-made telescope was a good little instrument.

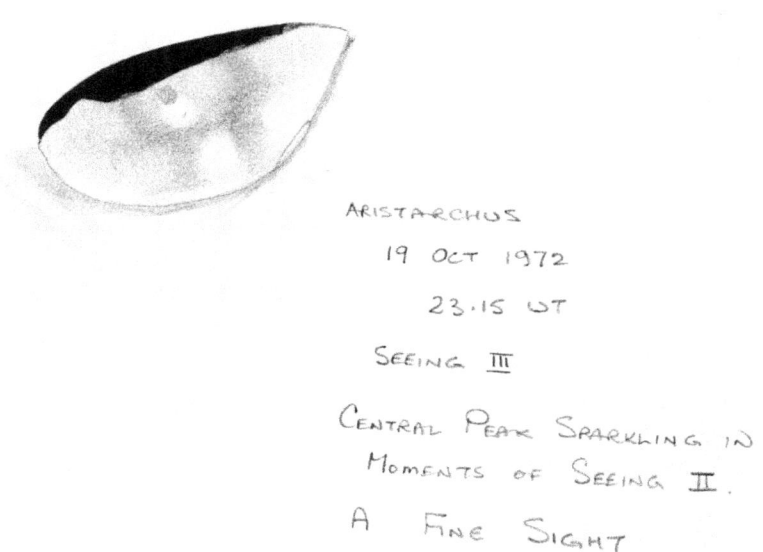

The above drawing of Aristarchus has been reproduced with my written comment that the central of the crater appeared to be 'sparkling'.

Aristarchus is the brightest spot on the Moon. During a total solar eclipse this can sometimes be seen illuminated by the light from the Earth

This crater has long been suspected of change - in particular, of TLP activity and the note above of 'sparkling' in the crater is the closest that I have ever come to seeing a TLP. Even then, I do not consider that my observation was anything more than a glinting of the sunlight inside the crater.

TLPs mainly clustered around the edges of the main Mare Plains and at the central peaks of craters. This correlation suggests some sort of seismic activity. Some craters were frequently reported to be active. Aristarchus was one such crater with reports of glows and even sparkling lights in its bowl.

When the Apollo astronauts flew over Aristarchus when it was lit only by Earthshine, they got very excited and described it as 'glowing'. In fact, it is naturally very brilliant and has been described in old Moon books as 'The Washbowl'.

One aspect of TLP searches was the thorny question of bias.

When an observer thought she had seen a TLP a telephone alert system would spring into action and other observers would rush out to view the suspect location.

Some care was taken to avoid bias by not giving the precise location and the nature of the suspect TLP.

However, bias inevitably crept in because experienced observers, if directed to look 'in the general area of Aristarchus' were bound to look at the crater and to expect brighter than usual appearances in the crater bowl.

After a few years observations appeared to show that the edges of the large lava plains - the 'Mare' - were amongst the more likely places to spot TLP activity.

Naturally, lunar observers, anxious to see a TLP would concentrate on these marginal zones of the Mare and this built up a bias against other areas of the Moon being searched.

Brief flashes have been recorded on the Moon and these are due to meteorite strikes - see [198] for a movie of a bright meteor strike on the Moon.

The Farce of O'Neill's Bridge

Of great interest to me in 1959 when I got my first telescope operational was the supposed existence of a huge bridge on the Moon. It was named O'Neill's Bridge after the American amateur who claimed to have discovered it in 1953.

A fascinating account of O'Neill's 'discovery' and subsequent investigations can be read at [199], [200] and at [201].

The 'bridge' was 'discovered' on July 29, 1953, using a 110 mm (4 1/4 inch) refractor. It appeared as a fan of sunlight streaming under an arch no less than 19 kms (12 miles) in span in the region of Promontorium Olivium and Promontorium Lavinium on the edge of the Mare Crisium.

A drawing by O'Neill can be seen in Martin Mobberley's book together with an account of the discovery and subsequent fallout from the period [202].

There is a modern drawing of the area which shows the illusion at [203].

O'Neill was somewhat cautious about his 'discovery' knowing that a natural arch of that size really was not feasible.

However, he spoke to Hugh Percy Wilkins in the UK who was then the Lunar Section Director of the BAA. Wilkins confirmed the bridge using a 375 mm (15 inch) aperture reflector.

An account and drawing for the 'bridge' by H P Wilkins can be seen at [204].

Wilkins claimed that he saw not only the arch with sunlight streaming beneath it, but also the bridge's shadow on the surrounding lunar surface.

This enormous span could not have survived even in the Moon's lower gravitational pull compared with the Earth.

Wilkins then did something that resulted in him resigning as the Lunar Section Director of the BAA and also resigning entirely in humiliation from the BAA.

What Wilkins did was to say in an interview with the BBC

"Now this is a real bridge. Its span is about 20 miles from one side to the other, and it's probably at least 5,000 feet or so from the surface beneath.

There's no mistake at all. It's been confirmed by other observers. It looks artificial. It looks almost like an engineering job."

Within a few years O'Neill's 'bridge' 'vanished' because it was never again seen and those that had backed up O'Neill and Wilkins changed their minds and kept quiet.

Nobody was willing to stand up and be associated with the infamous 'bridge' and especially not to be associated with Wilkins belief in Flying Saucers as alien spacecraft.

By the time that my first telescope was in use O'Neill's Bridge was not a subject of serious interest but I had a look at the area anyway.

Below is my drawing of the area where the two chains of mountains almost meet on the edge of the Mare Crisium.

I saw no bridge.

The complexity of the lunar surface and the way the sun illuminates and cast shadows can easily give rise to strange sights.

For example, on 19th September 1959 I spotted something odd on the Moon in a region close to the crater Hecataeus as seen above.

The way the shadows were falling that night gave me a strong impression of an arch - what do you think?

Of course, the rather strange configuration of the shadows was almost certainly due to two separate shadows falling either side of a raised elongated area of the Moon and the edges of the shadow just appeared to line up to make it seem as though the Sun was shining through an arch.

I did not report this drawing to anyone and this is the first time it has been extracted from my logbook for public viewing.

A whole world of mountains, craters, plains, domes - even volcanoes [205] - exists in a countless variety of forms and you never know what oddity will swim into the field of view of your telescope as you let it drift over the Moon's surface.

The Straight Wall

On 2nd June 1960 at 20.30 UT I drew the Straight Wall and the crater Birt on the Moon [206].

This was very exciting as this feature is visually outstanding.

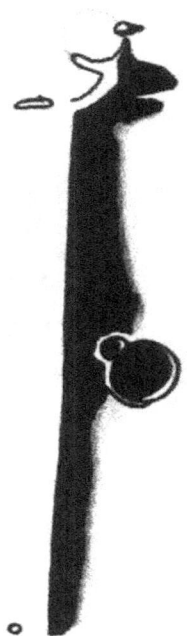

The picture below is an artist's impression of this feature taken from an astronomical encyclopaedia published in the 1920s.

In fact this is wholly misleading. The 'Straight Wall' is not a wall as such but has a relatively gentle slope of about 6 degrees which only looks severe when the Sun is at a low elevation.

The fault has a length of 110 km, a typical width of 2–3 km, and a height of 240–300 m.

By drawing the length of the shadow cast as the Sun changes position in the sky over this feature any amateur can show that the 'wall' is more of a gentle slope. This is a nice little project to undertake.

Mare Smythii

In the mid-1960s I got very interested in mapping the Mare Smythii in which there was a genuine interest at that time.

This small circular Mare was on the face not usually visible to Earth.

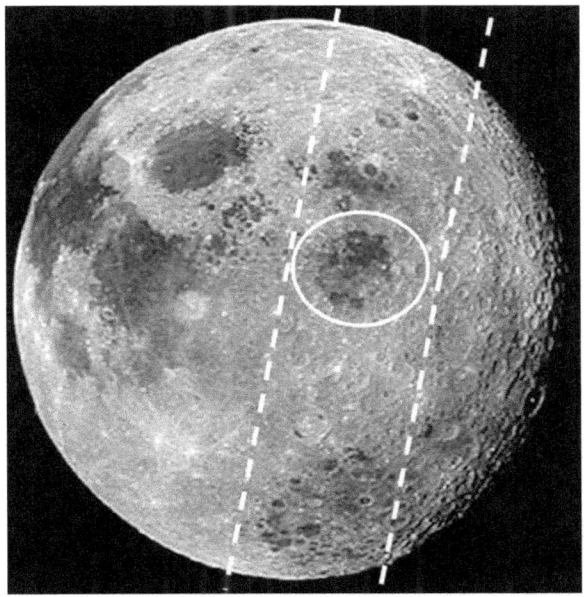

On the above oblique view of the Moon taken from Apollo 8 the dashed lines show approximately the edge features of the Moon which may be glimpsed as the Moon 'rocks' (librates) on its axis as seen from the Earth. Nothing to the right of the right-hand dashed line can ever be seen from the Earth.

The area between the dashed lines can be seen very foreshortened when the libration is favourable.

There is an excellent animation showing the libration of the Moon at [207].

The Mare Smythii, which I set out to study, is the dark basin within the oval.

I was part of a team led by Patrick Moore in the 1950s and 1960s which set up to deduce what features might exist on the 'hidden' side of the Moon from traces on the visible face and by observations during favourable libration.

Patrick had carefully charted features around the edge of the visible face of the Moon and had tried to deduce what features were on the hidden side. Remember, this was before the rear of the Moon had been photographed by space probes.

Patrick had, quite remarkably, successfully predicted that there were far fewer Maria on the hidden side and he had charted many craters by their rays systems which came over the limb onto the visible side.

The Mare Smythii only appeared briefly when there was a very favourable libration - that is when the Moon was turned in a favourable direction relative to Earth.

Very rarely, when the phase of the Moon, the libration and weather were just right, it was possible to watch the Sun rise or set over this elusive feature. Only at those times were any details visible.

The four drawings below show one such very rare occurrence.

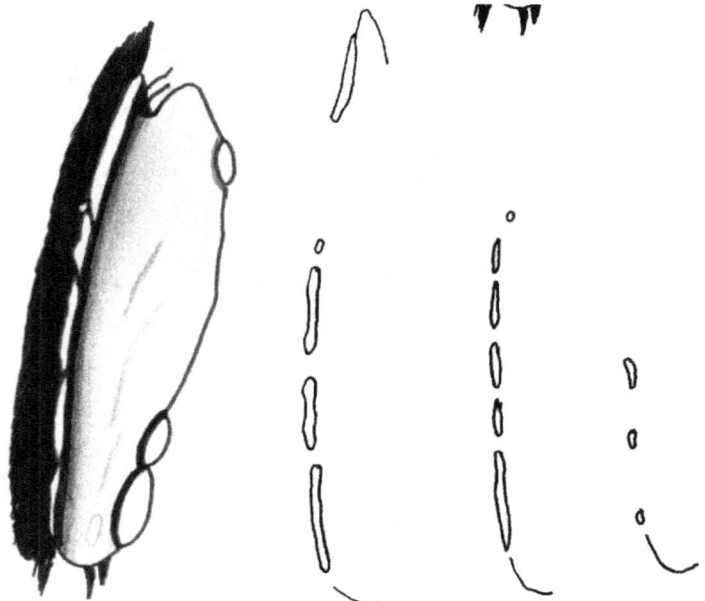

Perfect libration, perfect phase and, rarest of all, perfect seeing conditions all occurred that evening.

The left-hand sketch shows the appearance of the Mare as the Sun rose over it whilst the other sketches outline those mountain tops which were illuminated as the Sun's altitude changed.

On that single night I was able to see a very good view of the mountains around the edge of Mare Smythii and their profiles even though they were normally hidden from view around the rear face of the Moon.

I might have been the first person to have watched the Sun rise over the Mare Smythii and to have drawn this very rare event.

But then again I might not...

I did not care. I was just so very excited.

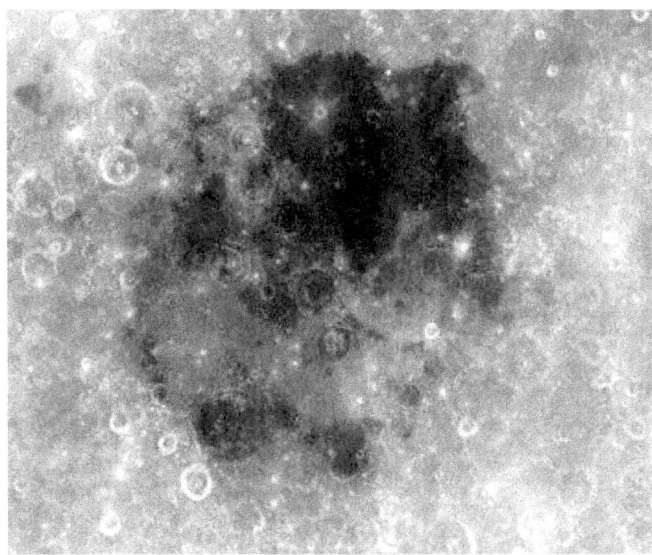

Above we see the Mare Smythii as pictured by a space probe.

A map I drew up in the early 1960s based on my many observations shows all the main features seen above and I was pleased to have contributed something to the study of our Moon even though space probes soon were to make all this work of mapping the Moon irrelevant.

Incidentally, this feature is named after William Henry Smyth [208] whose son Charles Piazzi Smyth was very much an 'Independent Thinker' who created a theory that Noah supervised the building of the Great Pyramid of Giza [209].

Smyth claimed that the measurements he obtained from the Great Pyramid of Giza indicated a unit of length, the pyramid inch, equivalent to 1.001 British inches that could have been the standard of measurement by the pyramid's architects. From this he extrapolated a number of other measurements, including the pyramid pint, the sacred cubit, and the pyramid scale of temperature.

To support this Smyth said that, in measuring the pyramid, he found the number of inches in the perimeter of the base equalled one thousand times the number of days in a year, and found a numerical relationship between the height of the pyramid in inches to the distance from Earth to the Sun, measured in statute miles.

Other Lunar Features

I got enormous pleasure from my little 150 mm (6 inch) aperture home-made telescope. It was crude but it worked and I particularly enjoyed drawing features on the Moon.

A drawing of the crater Gassendi below shows the amount of detail I could see.

Below is one of my sketches of the crater Plato with the shadows of the walls of the crater stretching sinister fingers across the solidified lava plain. Also shown is a space probe photograph.

Although the Sun was shining from a slightly different elevation angle over this crater it can be seen that my little homemade telescope on its greengrocer's box mount performed well.

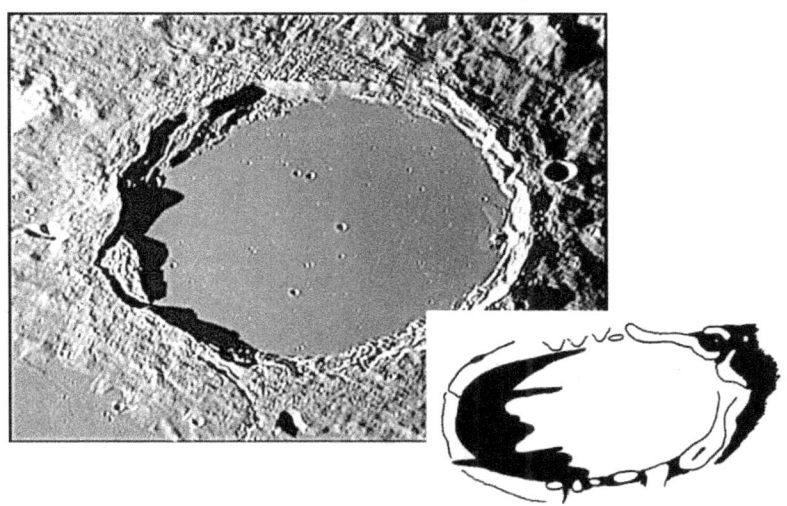

Some faint detail is seen in my original drawing on the floor consisting of four bright spots. There was - and may still be - much controversy about the permanence of features on the floor of Plato.

In the early 1800s herds of wild beasts and even the remains of cities had been reported here during the perpetuation of 'The Great Moon Hoax' in August 1835 [210]!

A scene showing the claimed inhabitants of the crater Plato is reproduced above [211].

Even in the middle 1900s, strange lights, bright varying patches and other changes on the floor were reported.

I made a special point of watching the floor of Plato but saw nothing strange; no cities, herds of Lunar Bison, mining activities, etc.

However, it was a creepy experience to scrutinise the crater believing that, at any moment, a bright glow might be seen and that I might be the only person to see it!

Incidentally, the belief that there are aliens living on the Moon is still very much in vogue; see just one of thousands of websites at [212].

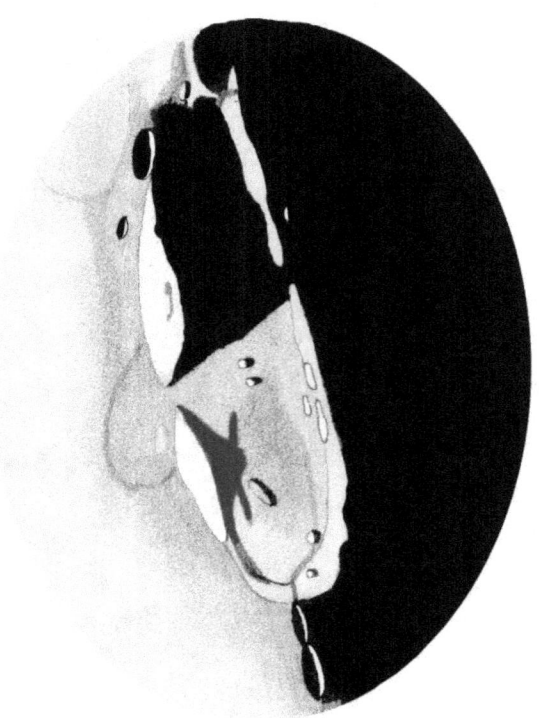

Above is one of my drawings of the crater Shickhard made on 3rd November 1968 at 17.00 UT.

Drawing craters such as this gave me enormous pleasure. I was not pushing forward the frontiers of lunar research but I was having fun with a simple homemade telescope.

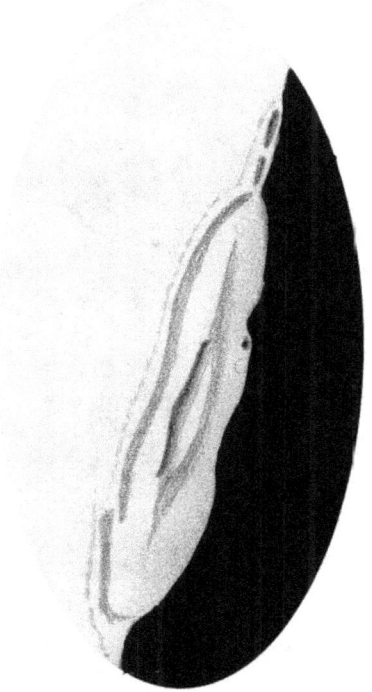

Above is the crater Caramuel (also named as Einstein in the USA) drawn on 20th May 1970 at 21.30 UT.

This large crater is often hidden around the limb of the Moon due to the rocking motion - libration - of the Moon as seen from Earth.

This crater is right on the limb of the Moon at a longitude of 89 degrees and is often impossible to see as it is out of sight.

However, this drawing was made when the libration was very favourable and is probably the best conditions under which the crater can be seen.

It has been claimed this crater was discovered in 1939 by Patrick Moore [213] but this is not true.

Certainly, the crater is very difficult to see being so close to the limb so I was excited to see it so well under rare conditions of favourable libration.

Above is a view looking down from the Obiter 4 spacecraft [214].

The above drawing was made on 6th October 1959 and shows the southern cusp of the crescent Moon. Note that South is conventionally drawn upwards in astronomical observations because this is how the image is seen in a telescope; at least from the earth's northern hemisphere.

This drawing shows the jumbled and confusing patterns of light and shadow created by a low altitude Sun in a complex mountainous area.

Note the isolated mountain peak on the right of the drawing where only the summit is still catching the Sun's rays.

This may not seem to be an exciting drawing to include but it is special in several ways.

It was the first part of the Moon that I ever saw through a telescope. Henry Wildey showed me this area in the late 1940s through the 150 mm (6 inch) Cooke refractor of the Hampstead Observatory [215].

The drawing is another illustration of what can be seen through a very simple homemade telescope.

Also, this area includes the relatively unexplored Leibnitz Mountains and craters where some deep parts are in permanent shadow and therefore extremely cold. Water ice has been detected in this region [216].

Whatever object I looked at in the heavens, whether with the unaided eye, binoculars or telescope I would think about the amazing things associated with what I am seeing.

For example, when observing the Andromeda Galaxy (M31) I think about the uncountable number of photons of light that have travelled for nearly two million years just to enter my eye.

When that light started its journey there were no humans to look up at the stars.

Project Moonhole

Project Moonhole was active in the United Kingdom in the 1960s but is now seemingly all but forgotten. This is a shame because it was a simple observational collaborative exercise which allowed amateurs with access to medium and large telescopes to work together.

There is an interim report available online [217] but Project Moonhole seemingly disappeared by the early 1970.

The interim reports starts by stating

"Because of difficulties over which neither the Lunar Section nor the British Astronomical Association had any control, the

project known as 'Moonhole' became into danger of folding up."

I do not recall the circumstances referred to but it sounds like a typically aggressive war between two factions; the sort of disgraceful vexatious fighting that plagued the BAA in the 1960s through into the 1980s.

I may be wrong - but I would be prepared to make a substantial bet that I have guessed correctly.

The aim of the project was to measure the bottom profile of small craters on the Moon.

This was done by observing a crater under conditions of different solar altitudes. The fraction of the crater floor covered by shadow was estimated.

The altitude of the Sun was computed from the identity of the crater and the time of observation by Howard Miles using the Elliot 803 computer at the Lanchester College of Technology.

A set of craters near the visible centre of the Moon were chosen with diameters between 10 and 20 kms.

The picture shows a typical 'Project Moonhole' crater.

It has a regular bowl shape without any internal structure and the crater rim casts a sharp regular shadow.

If I were estimating this crater for the project I would judge the fraction of the crater width in shadow to be about 55% and this result would be sent to Howard Miles.

At the time that the interim report was produced in 1967 over 3,000 crater shadow estimates had been obtained on nearly fifty craters.

Frankly, the results did not look good even though Howard Miles stated optimistically in his report that

"...many interesting facts are emerging"

For example, the depth of the crater Diophantus (18 km wide) was measured at anywhere between 2 km and 5 km deep with a large scatter of measurements between these extreme values.

Other craters produced better results, such as Bessel (16 km wide) where the depth was determined as 2 km with an uncertainty of 0.5 km but such relatively accurate results were rare.

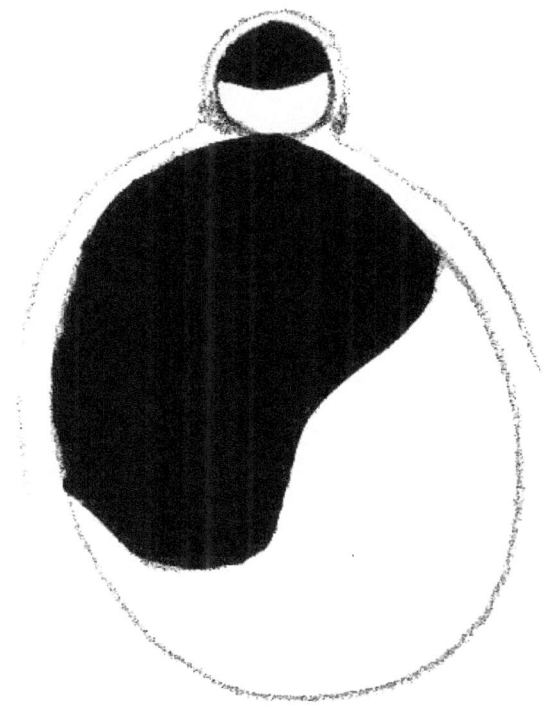

My drawing above shows the crater Birt which is close to the famous Straight Wall. This was one of the Moonhole craters.

It can be seen from my drawing that the shadow cast by the crater rim is not a regular shape and this would lead to unreliable estimates of the crater's depth.

Of course such measurements are now irrelevant because the Moon has been mapped in huge detail by various space probes but it was an interesting thing to be doing at the time.

Lunar Domes

The discovery and mapping of domes was very much a 'hot topic' in the early 1960s.

Unlike hills or mountains these domes have a very low profile and are only visible when the Sun is within a degree or so of setting or rising on the feature.

I saw a number of these elusive features.

Modern observations have confirmed that these domes, many of which have holes at their summits, are indeed shield volcanoes [218].

They are typically about 10 kms in diameter and about 200 - 300 metres high. This low profile shows why they are such difficult objects to see from Earth.

Most people consider volcanoes on the Moon to be extinct but a dedicated group of lunar observers believed in the 1960s that some domes were active volcanoes and a close watch was kept on the domes.

I watched several domes for a few months but watching an unchanging lunar feature soon becomes boring.

It is now known that the overwhelming majority of craters on the Moon were caused by the impact of large bodies.

It is difficult to see how a volcano could produce the shape of the larger craters which were usually a circular high wall, a flat crater surface inside the wall and - sometimes - a central hill.

It is now known that impact craters can explain all these features as shown in the website at [219].

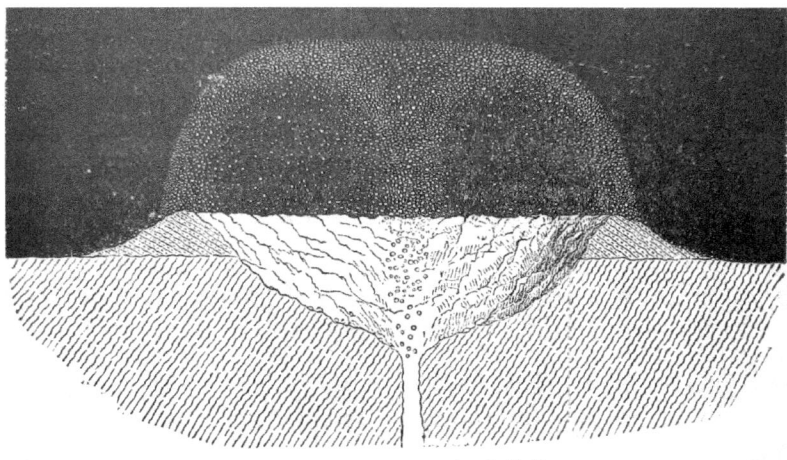

The diagram above from a 1920s encyclopaedia shows the concept of a lunar volcano throwing out gas and rubble so that it forms a rather precise ring.

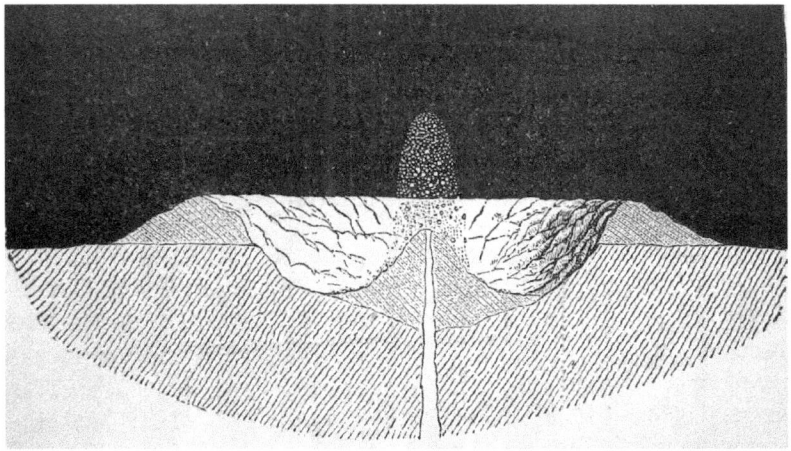

Then, as the outflow slowed the central hill was formed.

If you think about this for a moment you will see that such a mechanism is implausible even on a low gravity body with no atmosphere.

However, the evidence in favour of a volcanic origin was quite interesting.

For example, it was claimed that the central mountain that often occurred in craters often had a small pit on the summit like a volcanic vent. These tiny holes were really only seen by Patrick

Moore and H Percy Wilkins when observing with the huge telescopes of the Meudon Observatory.

If the central holes were made by meteorites what was the chance of the summit being hit?

Also, there is clear evidence of lava having flowed on the Moon in long past aeons. The rilles and broken lava tubes are there to see.

One of Patrick Moore's more incredible beliefs was that there might well be vegetation on the Moon; a view expressed in 1949 [220].

Rainfall And The Moon's Phase

Daily weather records have been collected in my home town of Weymouth in Dorset since 1927.

I decided on a whim to see if the rainfall showed any dependence with the Moon's phase.

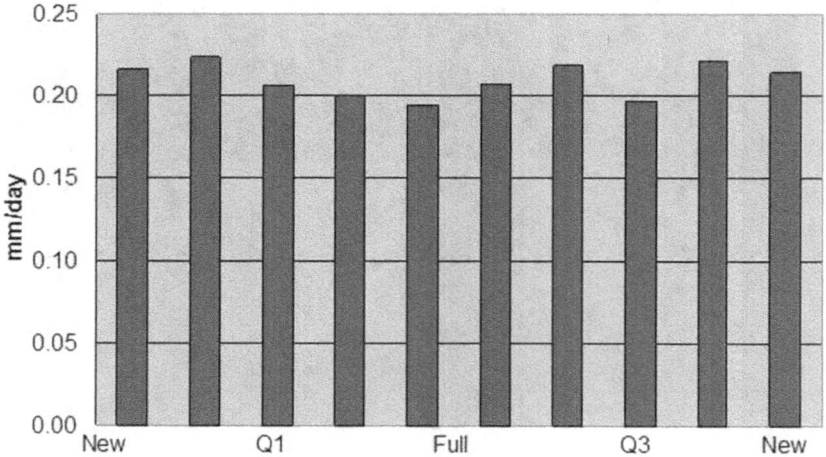

The above chart shows the average rainfall during the period 1927 to 2015 plotted against the Moon's phase.

Although there is only a fairly small variation, this chart is based on over 31,000 rainfall records which means that the observed changes are statistically significant.

Days in Weymouth just after New Moon and in the gibbous phase (between Full and New Moons) are wetter than other days.

10. Observing The Planets

Observing Mercury

The planet Mercury is a very difficult object to observe.

It never wanders more than 28 degrees from the Sun and that only happens when Mercury is furthest from the Sun in its orbit (aphelion) at the time of maximum elongation.

If Mercury is at the closest point in its orbit to the Sun (perihelion) when at maximum elongation then the angle between the Sun and Mercury is 18 degrees as seen from Earth.

Thus, in temperate latitudes Mercury is usually only visible to the unaided eye low down during twilight and is therefore difficult to see.

Although it was seen and known as a planet by the ancient civilizations there is a story that the great Copernicus never saw the planet.

So, if you have never seen it with the unaided eye you are in exalted company!

Mercury is best seen when the ecliptic - the path in the sky that the Sun, Moon and planets appear to approximately follow - climbs steeply up from the horizon.

This is in the Spring for the evening appearances of Mercury and

Autumn for morning apparitions.

I have seen Mercury with the unaided eye about a dozen times since the 1950s. I have not religiously set out to view it and I could probably have seen it many more times had I tried.

In general it is difficult to find initially with the unaided eye unless you are located in an area where the sky is exceptionally transparent.

However it is easy to find with binoculars and can often then be glimpsed with the unaided eye once the position in the sky relative to the horizon has been noted.

Mercury is rather boring when seen through a telescope. It is a small shimmering disc, slightly pinkish in hue, and showing phases.

Little if anything can be seen on the surface of Mercury.

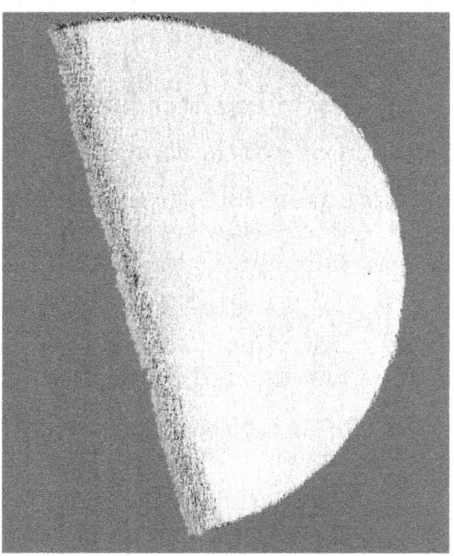

The above drawing is the only one I ever made of Mercury. This was in 1967 in a twilight sky and I saw no surface features.

Observers over many centuries recorded features on the surface of Mercury. The planet was believed to be covered with light and dark areas and even mountains and valleys.

In fact, most drawings of surface details on Mercury before the 1960s show about as much detail as one can see with the unaided eye looking at our Moon.

Throughout the 19th and early 20th centuries hundreds of drawings were made of the surface of Mercury.

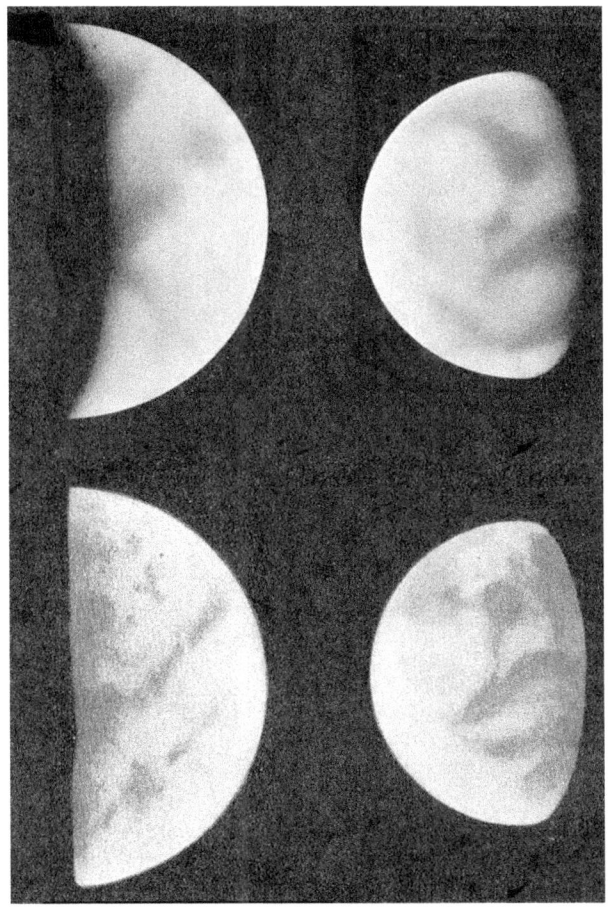

The above pairs of drawings of Mercury are typical of those produced from the 18th century until the mid-1960s.

The upper pair was drawn by a French observer in around the 1870s and the lower pair by a Scottish observer half a century later.

Look at the similarly of markings. Could there be any doubt that Mercury was locked into a rotation rate equal to its 'year' such that it turned the same face towards the Sun?

The map of Mercury below was drawn in around 1920 and was a distillation of hundreds of observations like those above.

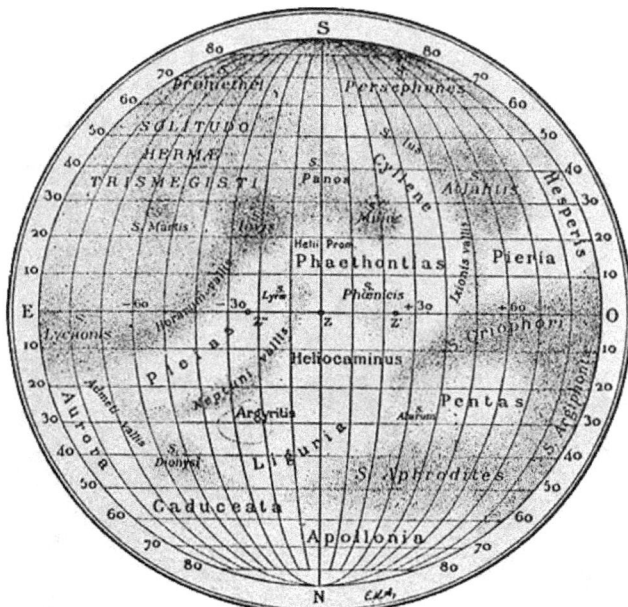

This map and others like it were produced on the assumption that Mercury rotated on its axis in 88 days; the same as the period in which it travelled around the Sun.

What wonderful names were given to the features - Liguria, Phaethontias, Solitudo Hermae Trismigisti, Hesperis, etc?

Could it be that these features never existed?

Surely not!

Decades before it was proved that Mercury was not locked in rotation like our Moon, mathematicians were casting doubt on Mercury's rotational period.

Some difficult mathematics were suggesting that the large eccentricity of Mercury's orbit (at aphelion, Mercury is about 1.5 times further from the Sun as it is at perihelion) made it more likely that the rotational period was shorter than 88 days.

Then in the 1960s suspicions that Mercury was rotating relative to the Sun mounted when it was discovered that the dark side of Mercury was very much warmer that it should have been. It was found that the night-time temperature was about −173 °C (−280 °F).

OK!

That's still very cold but, the back of the planet which was never exposed to sunlight should be much closer to absolute zero temperature at about -270 °C (-518 °F).

Between 1962 - 1965 Soviet and US radars were used to bounce signals off Mercury [221]. The observations of Mercury's radar Doppler spread showed that the rotation speed of the planet was more consistent with a rotation period relative to the stars of about 59 days and not the 88 days assumed when the above map was drawn.

Subsequent refinements in the measurement of Mercury's axial rotational rate - especially from the first spacecraft to visit Mercury which was NASA's Mariner 10 (1974–1975) - showed that the axial rotation period was actually $2/3^{rds}$ the orbital period.

So, how was it possible to produce such a detailed map of the surface of Mercury based on the assumption of a locked 88 day axial rotational period?

That mystery was explained - in part at least - by a contribution I submitted for Axel Firsoff's little book 'The Interior Planets' [222].

In this book Firsoff wrote

In the December 1967 circular of the Mercury and Venus Section of the British Astronomical Association G. J. Kirby draws attention to an interesting series of coincidences in the periods of Mercury, which may go a long way to explain the apparent observational evidence for synchronous rotation.

The synodic period of 116 days is very nearly four-thirds of the sidereal orbital period and twice the axial period of 58.5 days, as well as one-third of the terrestrial year.

Thus after a synodic period, within which Mercury repeats its phases, it will have turned twice round and so present the same part of the surface to the observer.

Moreover, every third western and eastern elongation is a favourable one, when most of the observations are likely to be made. These are three synodic periods and six axial periods apart, once more tending to confirm the spurious effect of synchronous rotation.

Reading that piece again forty-seven years after it was published I'm not sure that what I wrote entirety made sense.

However, a similar argument has much more recently been

proposed which says essentially the same thing [223].

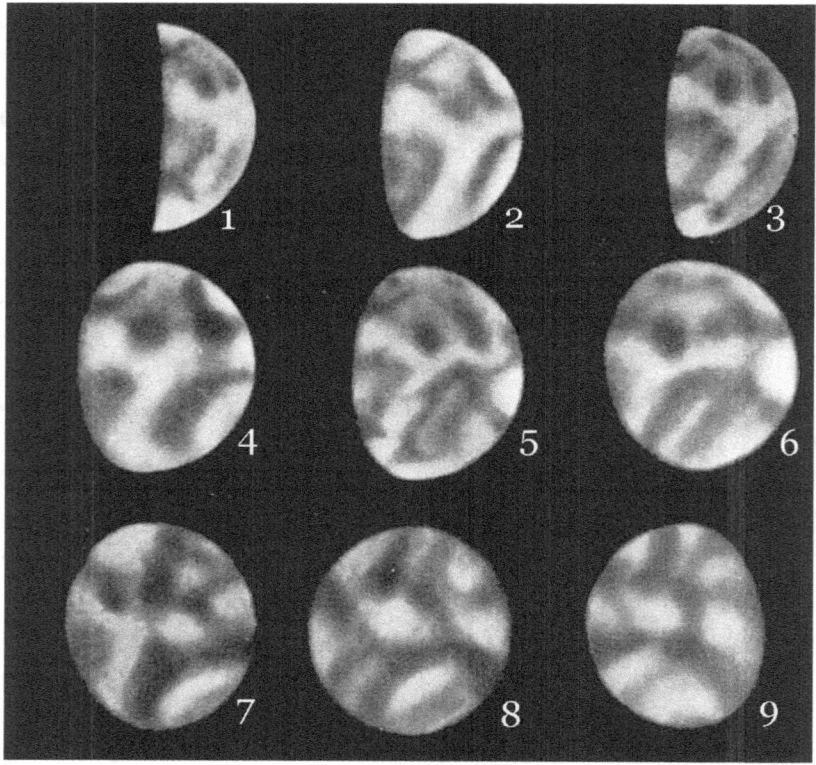

I have put together this collage of typical drawings made under conditions with large telescopes over a period of about eight years between 1942 and 1950 [224].

I have arranged the drawings by increasing phase irrespective of date of observation.

It can clearly be seen that the major features drawn - and specifically the prominent pair of dark patches in the upper (southern) hemisphere - do indeed appear to rotate in approximate synchronism with the sub-solar point, this being the point on the Mercurian surface where the Sun in overhead.

However, all these observations were made when Mercury was particularly favourably placed in its orbit. It therefore just happened that the above observations were made when the same face of the planet was most favourably seen from the Earth.

Such was the powerful belief in the theory of a tidally-locked Mercury showing only the same face to the Sun that observations made at less favourable times showing different features to those on the map reproduced on page 272 were usually rejected as spurious.

One reason given for the disagreements between drawings was the existence of clouds and mists in an extensive atmosphere.

As late as 1969 V A Firsoff was writing [225]

"...observations seem to show that the North Polar Region is more subject to cloudiness than the south, and so probably has a longer night"

Of course, we now know that the atmosphere of Mercury is extremely tenuous; scarcely one-trillionth of the Earth's atmosphere in terms of atmospheric pressure at the surface [226].

And so, the myth of a tidally-locked planet having a day length equal to one of its years was firmly entrenched until the 1960s when direct measurements disproved this assumption.

In 1967 two astronomers, Cruikshank and Chapmen, gathered together hundreds of visual drawings of Mercury and tried to get some sense out them by drawing on a large white globe and assuming an axial rotation period of 59 days.

They tried to reconcile the detailed drawings as typically shown on page 274 with the faster rotating planet.

Their attempt failed to produce a map of Mercury's features seen visually so that two hundred years of observation were effectively written off [227].

Since the arrival of Mariner 10 and the later MESSENGER mission to Mercury detailed maps have been produced and the pre-1960s visual observations have been largely consigned to the archives and ignored.

You will be wondering why I have spent so many pages discussing the planet Mercury in this astronomical autobiography when I have only observed it telescopically a very few times.

This is because the process of studying things that do not exist, such as the markings on Mercury, has insidiously invaded much of science.

In astronomy it has produced

- the Canals of Mars,

- the Spokes of Venus (never heard of them - then read on!),

- a lost moon for Venus, [228]

- O'Neill's Bridge on the Moon, see page 250

- the imaginary Planet Vulcan. That's not Mr Spock's home planet but the supposed planet orbiting the Sun inside the orbit of Mercury [229],

- the four lost moons of Uranus [230],

- the dubious visual drawings of the surface of Uranus by Leo Brenner [231]

- the discovery of several natural Moons going around the Earth by Dr Waltemath [232] and Frédéric Petit [233].

To this list of observational mistakes, misperceptions and downright frauds I am going to be highly contentious and add

- most of the features claimed to have been observed on the clouds of Venus,

- Transient Lunar Phenomena (TLP) which we met in Chapter 9,

- other changes on the Moon, such as in lunar craters such as Linné, Messier, Pickering, etc.,

- Unidentified Flying Objects (UFOs).

I know that the inclusion of UFOs will deeply upset some readers but there is more evidence for the existence of UFOs whizzing around our skies (see Chapter 21) than there is evidence for some of the other items in the above list.

The study of things that do not exist is called 'Pathological Science' and I will delve into the strange world of imaginary discoveries on page 283.

First I will describe my two decades studying the planet Venus and how I eventually came to believe that some, if not all, of the phenomena reported in amateur astronomy journals are probably just illusions.

Observing Venus

Having built the crude little telescope illustrated on page 4, I set out to make observations of almost everything except deep sky objects - they have never held much interest for me. Fuzzy blobs which do not change in appearance from night to night; there was not much excitement in systematically ticking off Messier and NGC objects on a long list.

On October 25th 1959 I made my first observation of Venus.

I took up observing Venus because it is easily located in the sky and changes from day to day.

My drawings and articles relating to Venus were published in the pages of the BAA Journal.

Venus is not an easy object to observe because of its relatively small apparent size.

It shows a large variation in apparent size as it moves around the Sun relative to the Earth.

On the far side of the Sun (Superior Conjunction) its apparent diameter is around 10 arcseconds.

When it is between the Sun and Earth (Inferior Conjunction) its diameter is a whopping 60 arcseconds - indeed, many people claim to have seen it as a thin crescent with the unaided eye.

However, when it is at its largest diameter it is an extremely thin crescent.

The chart below shows some characteristics of the planet's appearance as it travels one-half an orbit around the Sun relative to the Earth.

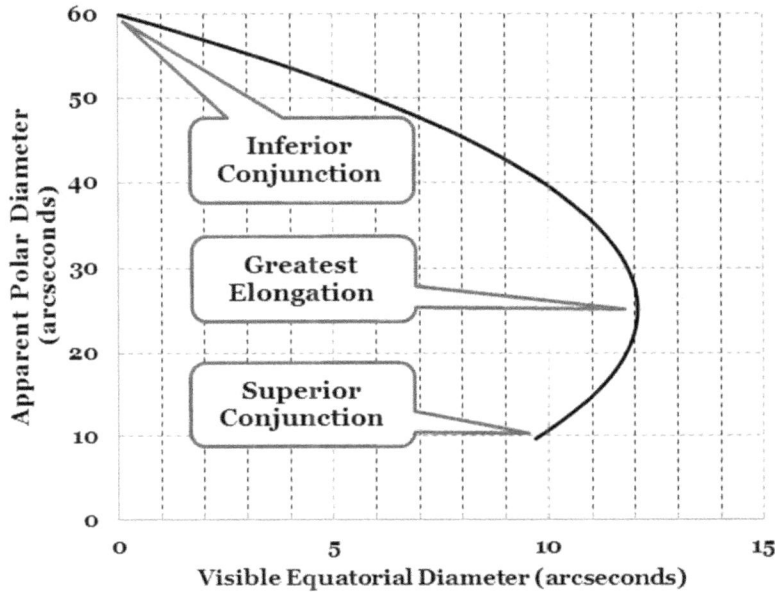

Plotted are the apparent polar and equatorial illuminated diameters of Venus as it moves around the Sun.

It can be seen from the above chart that the greatest illuminated equatorial width never exceeds about 12 arcseconds.

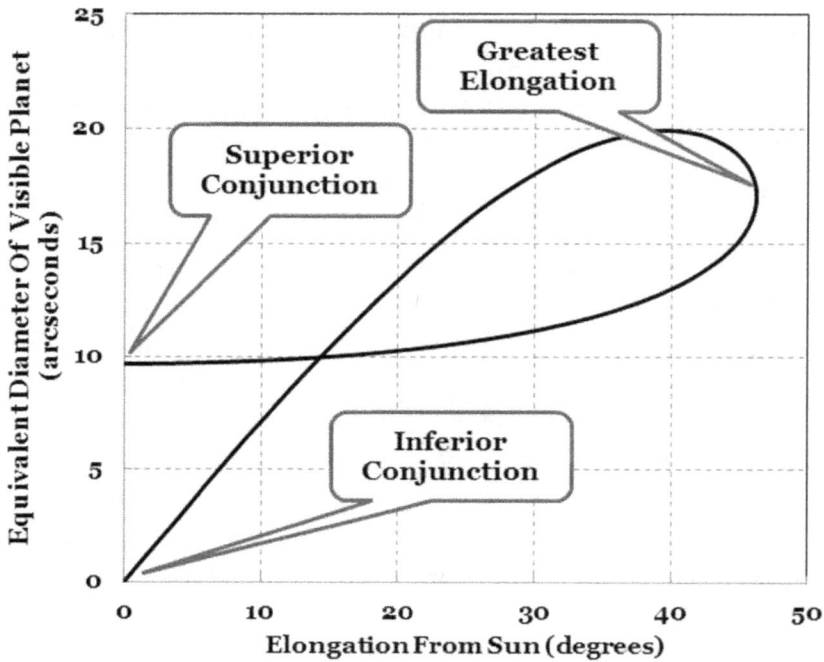

I have also calculated the visible area of Venus as seen from the Earth and related this to the apparent area of a disc at the same distance of Venus. These calculations are shown in the chart above.

At Inferior Conjunction the visible area is close to zero depending on how far Venus passes above or below the line of sight to the Sun.

Venus Maximum Values		Equatorial Width At Favourable Oppositions (arcseconds)		
Equatorial Width (arcseconds)	Equivalent Disc Diameter (arcseconds)	Mars	Jupiter	Saturn
12	20	24	44	18

The above table shows that, in terms of the angular width of the visible planet and also in terms of the visible area, Venus is smaller than Mars at the latter's favourable oppositions, substantially smaller than Jupiter and comparable to the globe of Saturn.

Thus, Venus is a difficult object to observe despite it having a large apparent polar diameter when it swings close to the Earth.

Venus is conventionally observed in a clear blue sky with the Sun well above the horizon.

Sometimes, around maximum brightness, it can be seen with the unaided eye.

I have seen it many times and have helped others to see it. I recall that in 1992 my partner and I were walking in the Dorset countryside.

I knew where Venus was to be found and spotted it easily in binoculars. The time was about 2 pm and the Sun was high in the sky. I moved to located Venus close to the top of a power line pylon and then found it with the unaided eye.

Sandra then found it and she was amazed to see this pure white star-like point of light in the blue sky.

It is reported that Napoleon glanced up into the sky whilst addressing his troops and spotted Venus. He was amazed and declared it to be a good omen for his planned invasion of England [234].

Observing Venus in daytime when near to maximum elongation from the Sun (about 47 degrees) there is relatively little danger of accidentally catching the Sun's light and heat in your telescope.

However, when Venus is close to the Sun there is a considerable danger of accidentally catching a glimpse of the Sun through the main optics.

I never caught a glimpse of the Sun whilst observing Venus - I still have my sight to prove it - but I was once looking for Venus with an open tube telescope. I smelt something burning and suddenly realised that the image of the Sun was coming out through the side of the telescope and was burning my hair as I peered down the eyepiece!

Ouch!

Hedley Robinson who was for many years the Director of the BAA Mercury and Venus Section permanently damaged the sight in one eye by observing the Sun through a small telescope using a faulty 'solar filter'. Sadly, although this blocked much of the visible light, it let through enough invisible infrared light to ruin his retina.

When the Sun is below the horizon Venus can be easily seen with the unaided eye. However, by then the glare of the planet is too great for observation. It is like observing a brilliant searchlight through a telescope and the eye is swamped with excessive light.

This reduces the ability to observe faint detail on the planet's clouds.

I have heard of some astronomers who observe Venus in a dark sky after shining a bright flashlight into the observing eye to reduce its sensitivity.

This seems to me to be a pointless and risky thing to do.

The point of observing Venus is to see faint details on the disc, whether these are cloud markings or terminator effects such as the Schröter effect (see page 288).

When Venus is observed in daylight, the whole disc is overlaid with a blue sky light which is typically about 10% as bright as the average disc brightness.

The overlying blue sky light will swamp out the fainter contrast features because the eye cannot detect differences in contrast between adjacent patches of the planet which are less than typically about 5%.

The BAA Mercury and Venus Section members experimented with observing through coloured filters in the 1960s and concluded that a Wratten 15 (yellow) filter enhanced contrast on the planetary disc.

A blue filter (Wratten 47) tended to make detail difficult to see and mostly unobservable - as expected if the blue sky light were diminishing the contrast of surface features.

A red filter (Wratten 29) blocked the blue sky light and gave the appearance of Venus in a dark sky but with less of the startling brilliance that one observes when viewing Venus in a black sky well after sunset.

Some observers can see further into the ultraviolet (UV) part of the visual spectrum than others.

The familiar 'Y' shaped cloud patterns that have been photographed in UV wavelengths - like that shown above - might be visible to those observers with some UV vision but these would be invisible to other observers [235].

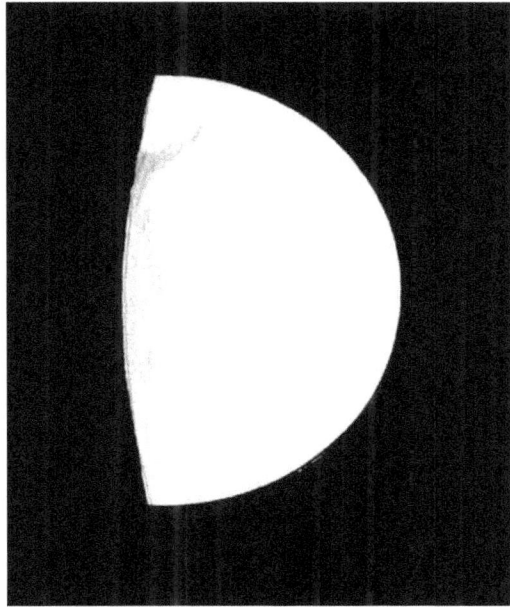

The drawing above shows the most distinct marking I observed.

This drawing was published in a memoir of the BAA Mercury and Venus Section.

The markings have been exaggerated for purposes of publication.

Personally I have always been a bit cynical about features seen on Venus. I have only suspected faint markings on a few occasions.

A huge menagerie of markings have been claimed - terminator extensions, cusp caps, cusp collars, various shaped dark markings, etc. I doubted that some - or all - of these existed.

True, some photographs in the 1960s gave a hint of markings or terminator distortions but they were always on the limit of the photographic technique and using film that had a sensitivity to ultraviolet light which penetrated deeper into the atmosphere of Venus than ordinary visible light.

I took the trouble to search through the archives of the BAA Mercury and Venus Section trying to find observations of markings made at very closely the same time by independent observers.

I found a number of such pairs of drawings separated by less than fifteen minutes.

They did not agree with each other at all.

The Section Director, Hedley Robinson, said that this not too surprising as

".....different observers had differing eye sensitivities and would therefore record different detail."

This appeared to me, but seemingly not to anyone else in the BAA Mercury and Venus Section, to be evidence that the claimed markings might be illusions.

At best, visual observations were unreliable and not reproducible. If different observers saw different markings at essentially the same time what value did the observations have?

In the late 1960s I suggested to Hedley Robinson that I organise a series of 'Venus Days'. Keen observers would be given a list of dates and times. They would then do their best to observe Venus at these times. Several such observing sessions would be organised because of the unreliable weather in the UK and because features were not often reported.

I would then compare the drawings and determine whether the features drawn were confirmed or not by two or more independent simultaneous observation.

Hedley turned the idea down although I never found out why.

I was willing to take on the whole of the organisation and analysis. Even today, four decades later, we cannot be sure that features claimed to have been seen visually on the cloud cover of Venus were real or not.

Pathological Science

Highly relevant to this problem of observing Venus is the whole area of 'Pathological Science' which is the study of things that do not exist - such as Martian canals, phantom satellites of planets, N-rays, cold fusion, etc.

I wrote an article on this topic which was published in New Scientist.

What follows is the substance of this article.

Reproduced from New Scientist 24th February 1990

(References and pictures have been added as they were not permitted in the original article.)

The term 'pathological science' was coined in 1953 by Irving Langmuir to describe the study of non-existent effects in nature. The term may be a little confusing but it can prove useful in describing certain kinds of work.

Pathological scientists believe every false 'observation' they make. They are not frauds. Heaven knows, there have been enough of them in the history of science! No, they simply convince themselves and others that they are observing something which does not actually exist.

The discovery of N-rays by the eminent French academic, René-Prosper Blondlot in 1903, was a classic case [236]. These mysterious rays were supposed to be radiated by a heated wire. They passed through aluminium sheet up to 70 millimetres thick but the thinnest iron foil stopped them dead.

How were N-rays detected? Only one method was reliable. When they were shone upon an object they increased the ability of the human eye to see it when the incident conventional illumination was just at the threshold for detection. The lighting in a room was lowered until the test object was on the verge of invisibility and N-rays then made the object easier to see.

Blondlot spent years studying these rays. He found that a house brick, left in sunshine for several days, stored N-rays and radiated them later. If the irradiated brick was held close to the head, the visibility of faintly illuminated objects was enhanced.

The rays were refracted by aluminium prisms and Blondlot measured the refraction index to a precision of three significant figures. Many papers were published on the curious characteristics of N-rays by Blondlot and others until one day, while Blondlot was demonstrating and measuring their refractive properties, a visitor to the laboratory pocketed the aluminium prism without Blondlot's measurements being affected!

N-rays were never heard of again. They were figments in the minds of those sad characters, pathological scientists.

Nowhere has pathological science been more active, caused more embarrassment and survived so virulently to the present day than in astronomy.

Consider the planet Mercury, for example. Mercury is notoriously difficult to observe. It is small and is always close to the Sun so it can be studied for only a few weeks each year.

Throughout the 19th and early 20th centuries many drawings of its surface were made. It was obvious that Mercury, being so close to the Sun, must be in locked rotation as it turned about the Sun; that it always turned the same face to the Sun, just as our own Moon always turns the same face to the Earth.

So textbooks into the 1960s described a world scorched on one face and unimaginably frigid on the other.

Features supposedly glimpsed through telescopes were carefully mapped and named. And what superb names! Apollonia, Liguria, Pieria, Cyllene, Aurora!

Then, in 1964, radar measurements showed that Mercury was not locked in rotation relative to the Sun. In fact, it rotated on its axis in 57 days while taking 88 days to travel once around the Sun. In desperation the best drawings made over two centuries were collected and mapped onto a globe using the true rotation period.

All that emerged were a few smudges; gone were Persephones, Heliocaminus and the rest - the accumulated figments of many imaginations.

Astronomers trying to observe at the threshold of their vision, driven on by a false assumption about the rotation rate, drew what they thought they ought to be seeing.

William Herschel, probably the greatest amateur visual observer of all time, stated over 200 years ago that an object becomes much easier to see once it has been seen for the first time. This, alas, is also true of objects that do not exist.

The Canals of Mars must also rank high in the roll of dishonour for pathological astronomy.

Seen only at the very threshold of visibility their observation and popularisation by Percival Lowell convinced many that Martians were conserving their precious water locked up in polar icecaps and feeding the melt water to make the deserts bloom.

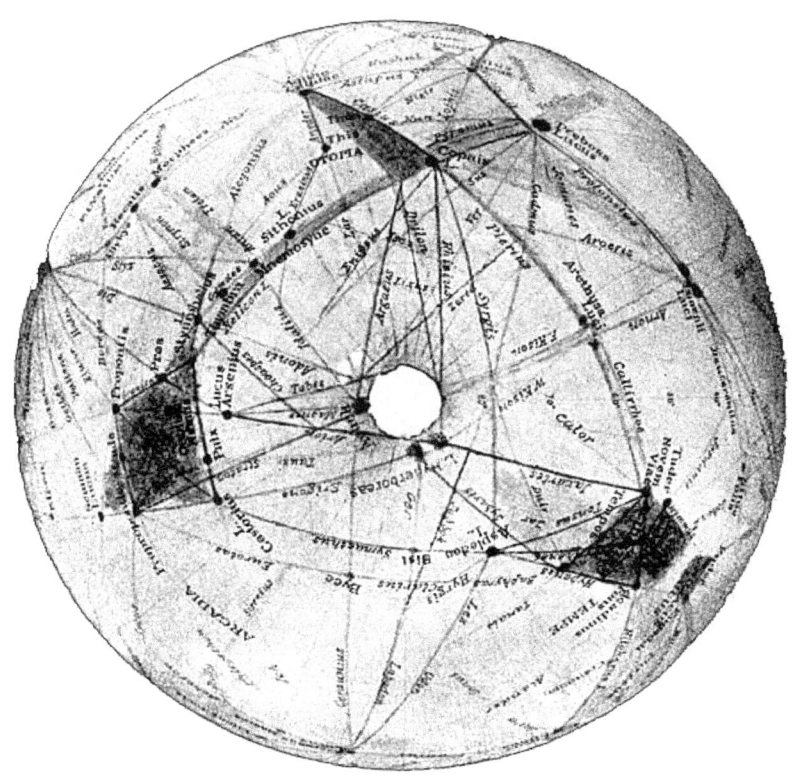

Percival Lowell's imaginary world of artificial canals

A modern view of Mars. Where did those canals go? [237]

Not until the space probes arrived at Mars and found no canals was it finally proved that Lowell and many others had been deceived by faint, irregular dappling on the Martian surface.

Imagination and the urge to believe in Martians did the rest.

It couldn't happen today, could it?

Alas it could and amateur astronomers have failed to learn the lessons of Mercury and Mars and are tumbling into the same old traps as did their predecessors.

For decades amateurs have been recording features on the cloudy face of Venus: cusp caps, limb shadings, bright spots and wisps of utmost subtlety. These observations were made right at the limit of visual detectability classical breeding conditions for pathological science.

And yet in 1896, in a series of experiments on Earth, a Swiss-born astronomer called Villiger made a number of observations through a telescope of featureless white balls at large distances.

He 'saw' all of the major features - cloud bands, cusp brightenings, terminator irregularities, etc., - that are still being recorded visually by amateur astronomers on the real Venus. Indeed, his drawings could not be identified amongst typical drawings of the real Venus [238].

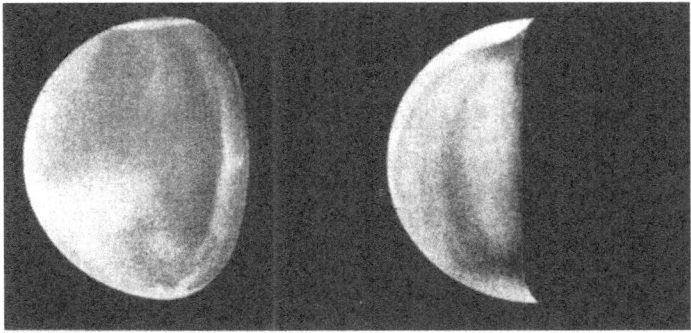

Venus observed by Villiger using a 140 mm Refractor (1895/6)

Observations of a featureless globe using a 130 mm Refractor

Observers of the planet Venus are still deceiving themselves.

The amateur astronomical fraternity is badly sold on pathological science.

Visual observations have been reported of cloud bands on Uranus and Neptune, markings on Jupiter's moons, changes in craters on the Moon [239], the mythical planet Vulcan [240] and a non-existent moon for Venus [241] - all the result of straining to detect at the limit of visibility and all having little or no credibility.

Bias and the urge to believe are strong in scientists; hence the use of the 'double blind' experimental technique. But the lone experimenter or observer has few controls to temper the imagination.

The lesson to be learnt, whatever your scientific field, is simple:

If it doesn't stick out like a sore thumb, forget it!

(End of article)

Amongst the more remarkable false - pathological - observations of Venus were the drawings by Percival Lowell from 1896 shown at the website [242]. These drawings show dark 'spokes' on Venus [243].

To make these observations he reduced the aperture of the Flagstaff, Arizona observatory's 610 mm refractor down to 75 mm.

This reduced the effective focal ratio and the size of the Ramsden Disc (see page 143) to such a gross extent (0.5 mm diameter) that what Lowell was seeing was the pattern of blood vessels and the shadows of 'floaters' on his retina!

In effect, the huge refractor was acting like a giant ophthalmoscope! [244]

My distrust of the observations of Venus cloud markings drew me away from struggling to observe markings which, in my opinion, were probably illusions.

Schröter's Effect

I observed and theorized on a better established and measurable feature of Venus known as Schröter's Effect.

Being closer to the Sun than the Earth, Venus shows phases which vary from a thin crescent to a full disc.

The curiosity is that the apparent phase is generally less than the theoretical phase. If we define the phase as the fraction of the apparent planetary diameter illuminated, then the observed fraction (or phase) is about 0.03 to 0.05 less than computed.

This is quite dramatic around the time of dichotomy which is the time when the planet should be exactly half phase, i.e. the phase is 0.5 and the angle Sun-Venus-Earth is 90 degrees.

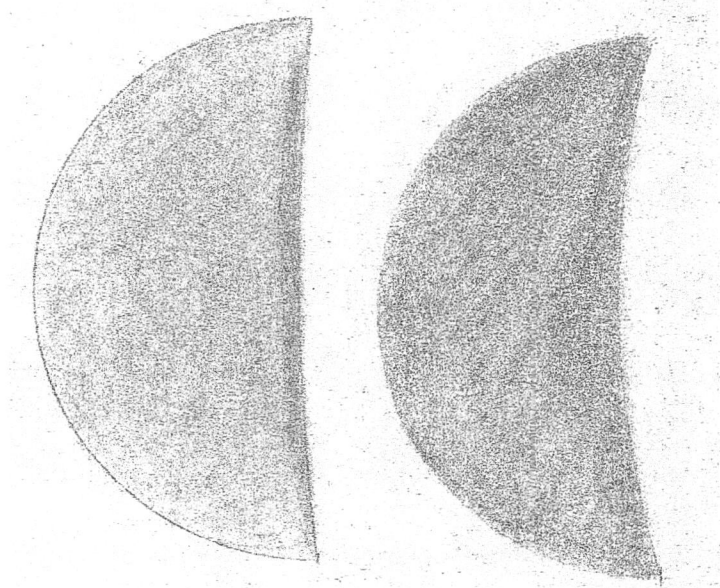

Above is a pair of drawings of Venus observed at 06.00 UT on 18 June 1969 through my home-made Newtonian with a magnification of 240x. The left image was observed through a red Wratten 29 filter and the right through a blue Wratten 47 filter.

The theoretical phase at this time was 0.50 so that the terminator - the edge of the sunlit hemisphere - should have been straight and the equatorial width of the disc should half the polar diameter.

Clearly some of the illuminated cloud cover on Venus is not being seen and this defect is greatest when the planet is viewed through a blue filter.

A great many papers and theories have been presented on this topic and I wrote quite a few of them in the BAA Journal in the period 1968 - 1971.

The theories of Schröter's effect can broadly be broken down into the following mechanisms:

a.   The illumination of the terminator is less than expected due to absorption and scattering of sunlight in the clouds of Venus. High clouds shadow the lower clouds nearer to the terminator [245].

b.   The effects are due to bias and physiological effects which cause the phase to be incorrectly estimated. Even observing a table tennis ball was shown to give the wrong apparent phase [246].

c.   The overlying sky light obscures the fainter parts of the terminator so that the apparent phase is less than expected.

Because Venus was always observed in daylight it followed that the planet was being viewed through the veil of overlying blue sky light. This would wash out the fainter parts of the terminator and make the phase appear smaller especially if a blue filter was used.

The fact that the phase was smaller when viewed through a blue filter than through a red filter - as seen in the pair of drawings above - confirmed this mechanism as far as I was concerned since the red filter would cut out the overlying blue light and give a phase estimate independent of the effect of blue sky light.

The observation that the apparent phase in red light was still less than the theoretical value showed that daylight masking was not the whole explanation because the blue sky light was blocked by the red filter.

In fact, all three mechanisms must be present to differing extents so the problem was to sort out their relative importance.

I got intrigued because it was a problem that should be solvable by a mixture of observation and theory.

However, there was a problem of a non-scientific nature however.

I was working for the Ministry of Defence in the 1960s and 1970s.

A lot of excellent measurements and theorizing about Schröter's effect was being conducted by a scientist called Bronshtehn in the Communist Soviet Union [247].

Hedley Robinson was keen for me to communicate with him but it was too dodgy for me to start up a private correspondence with a representative of a Communist country!

Anyway, I did some work and believed that the effect was a mixture of all three mechanisms in the above list.

Some micrometer measurements of the phase of Venus were reported in the BAA Journal in the 1960s and these also showed a phase deficit. These measurements were not subject to physiological problems but were subject to the other two possible sources of phase diminution.

In theory, by comparing the table tennis ball observations, micrometer and visual observations using coloured filters it should have been possible to sort out the relative values of the three mechanisms but, to my knowledge, this has never been done.

What I did was to measure the effect of daylight masking by a method that I thought was ingenious - but then I've never been known for my modesty!

The idea was actually simple.

I wanted to find the faintest illumination that could just be detected when overlaid by daylight. This would allow a figure to be computed for the phase defect due to blue sky light masking.

What I did was to rack out the eyepiece of my telescope to turn Venus into a large, extra-focal disc of light and I noted the eyepiece movement needed to just make this disc invisible against the sky light.

This could then be directly related to the brightness at the terminator which will just be invisible - since the brightness of the skylight is unaffected by putting the eyepiece out of focus.

The critical brightness of the terminator could be determined by simple geometry. No need for complicated photometers, etc.

These measurements showed that the masking of the faint clouds near to the terminator by the overlying sky daylight would reduce the observed phase by only 0.005 at most.

This meant that daylight masking was reducing the apparent phase of Venus but by so small an amount that this effect could be ignored for Venus.

As this had been my favourite mechanism of the list on pages 289 and 290 I had to say that I was wrong and did so in an article in a BAA Mercury and Venus Section Circular in the early 1970s entitled

"Schröter's Effect: Farewell Daylight Masking Theory!"

It is now widely accepted that Schröter's Effect is predominantly due to the dimming and scattering of sunlight in Venus's atmosphere.

This means that continued observation and monitoring of

Schröter's Effect may tell us something about the structure and behaviour of the planet's clouds.

Apparent Phase Defect for the Moon

In my pursuit of the mechanism of Schröter's Effect I also demonstrated that daylight masking could reduce the apparent phase of the Moon in a blue sky.

This was done by letting the Moon drift across the field of view of a small telescope and timing how long the visible part took to cross a wire in the eyepiece.

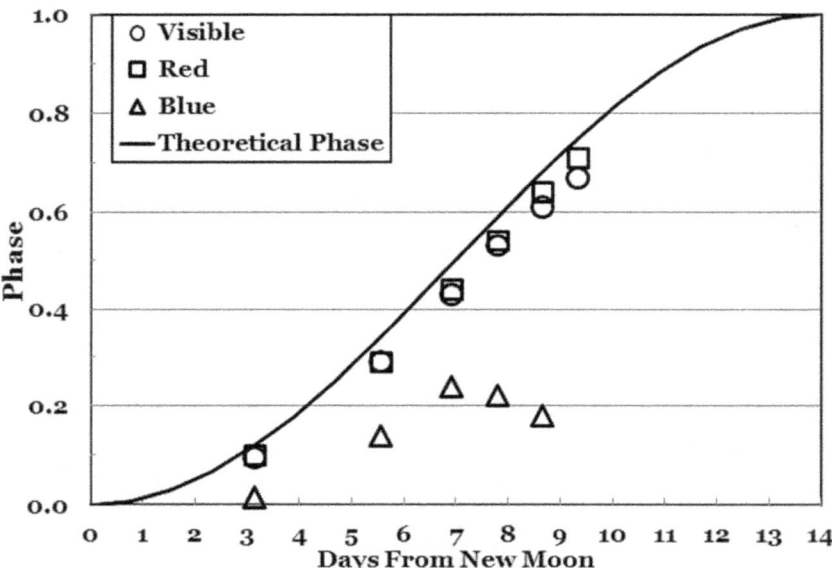

I made these measurements of the apparent phase of the Moon in blue clear skies as it progressed through the phases from near New Moon to near Full Moon. I made these measurements visually as well as with red and blue filters, all in daylight conditions, much the same as for Venus.

It can be seen that there is a clear 'daylight masking' effect and I took this as evidence that this effect must also be working for the planet Venus.

My quantitative measurements were free from the bias of physiological bias and, as the Moon has no atmosphere, the effects of clouds were not present.

The phase defect measurements on the Moon must have been due to daylight masking only.

The effect was similar to the results for Venus with a phase defect in red and visible light of about 0.05 around the half Moon phase (dichotomy).

However, using a blue filter rendered the Moon difficult to see and there was a remarkable large phase defect when the Moon was about half phase.

Indeed, I found the gibbous Moon impossible to see through a blue filter when more than about nine days from New Moon.

To me the chart above showed that there was much work to be done linking the phase effect for Venus and the Moon.

I tried to get the BAA Mercury and Venus Section involved in this sort of study but to no avail.

The group who thought that Schröter's Effect was due to effects in the clouds of Venus pursued their own agenda and the other group that thought the effect was an illusion created by the brain went off on their path of research.

Without cooperation, lack of aggression and a will to pursue science rather than personal vendettas the true cause of Schröter's Effect would never be found by the BAA Mercury and Venus Section.

It just plodded on gathering phase estimates without any sort of nationally directed programme. Everything I suggested was greeted with lukewarm interest and eventually I left the Section in the mid-1970s.

The Ashen Light

The Ashen Light was one of the phenomena involving Venus that I tried to observe.

Around the time of inferior conjunction when Venus shows a thin crescent there have been reports that there is a faint glow that may occur on the normally unlit hemisphere of Venus [248].

On 23rd September 1959 I recorded in my log book that for an hour around midday - hence observed in a bright blue sky -

"...the terminator appeared to continue around the whole disc. However, this was not visible when the seeing improved for brief periods indicating an optical illusion."

The phase of Venus was just 14% so this was a very thin crescent.

From my log book for April 7th 1969 I again recorded that I may have seen the Ashen Light.

The 19th century Bavarian astronomer Franz von Gruithuisen believed that the Ashen Light glow was due to massive fires lit by the inhabitants of Venus to celebrate the coronation of a new emperor [249].

However, it now thought to be either an auroral effect or a contrast effect in the observer's eye.

One paper provides evidence that the Ashen Light is probably a real glow, auroral in origin, in the dark atmosphere of Venus and that it might be within the range of visual visibility [250].

On the other hand, observations of the unlit hemisphere of Venus whilst the lit part was covered by the Moon's dark limb found no evidence of the Ashen Light [251]. However, this could simply mean that it is intermittent in character.

The Ashen Light is supposed to be more easily seen when a Wratten 45 (purple) filter is used and auroral lights would presumably be purple in tinge. I was using such a filter at the time of the 1969 observation.

The Ashen Light is, when claimed to be seen, right on the limit of visibility which makes it a classic candidate for Pathological Science - see page 283.

Today opinion is still divided.

That there are auroral glows on the unilluminated hemisphere of Venus is not disputed.

What is still unsure is whether these auroral glows can be seen visually.

Some commentators believe that there is a reliable body of observation of the Ashen Light; mainly from the archives of the BAA Mercury and Venus Section.

However, as I have shown with regard to observations of cloud features, there was never a coordinated programme in operation in the BAA Mercury and Venus Section such that independently confirmed observations were on record.

Personally, I will go out on a limb and say that the Ashen Light exists but at such a low level of intensity that it cannot be seen visually and that claims of it being seen visually are Pathological.

Indeed, I will go further and suggest that many claims to have seen visually markings on the cloud cover of Venus are and always have been, pathological.

My Interest In Observing Venus Fades

By and large, I became disenchanted with the Mercury and Venus Section of the BAA. This was no fault of the Director Hedley Robinson. He did the best he could to keep the programme going.

I faded out of Venus observing in the early 1970s.

THE BRITISH ASTRONOMICAL ASSOCIATION

```
                    Helmington
                 21, Inverteign Drive
                    Teignmouth.   TQ14 9AF
                    1978 Dec 20

   Dear Geoff,

        It may   come as a surprise to  hear that for
   a variety of  reasons I intend retiring from the
   directorship of the Mercury & Venus Section
   next summer, and there is a move on  foot to
   amalgamate with the Mars Section at   the same
   time.  The Mars Section work is much less than
   that for Mercury  & Venus.

        Knowing your interest and capabilities, I
   am wondering if I may put your  name forward to
   Council as the new Director of the joint section.

        It will involve some time in  correspondence
   and preparation of reports and  circulars,  but
   Bill Leatherbarrow would be  willing to help in
   printing and issuing circulars   as  heretofore.

        Please give the matter your careful thought
   and advise me accordingly.

                    Best  wishes etc.

                    Sincerely,

                    J.Hedley Robinson.
```

Imagine my surprise when, in 1978, I received the letter, reproduced above, offering me the Directorship!

I was flattered but turned down the offer.

Hedley's opinion of the large membership of the BAA Mercury and Venus Section must have been very low to have offered the post to me, over everyone else, despite me having been inactive in the Section for several years previously.

I made no further observations of Venus until April 1996 when I bought my large Meade Dobsonian and picked up Venus in daylight as a test of my ability to use altazimuth setting circles on it.

Even so, I made no drawings or phase estimates. My interest had waned such that I could see no point in continuing with the phase estimates and I really did not believe that there was much - if any - reality in the recorded cloud markings.

Observations of Mars

The winter of 1962/3 was terrible. Snow was so deep that my home town of Weymouth on the English south coast was cut off from the rest of the country for over a week before snow ploughs could break through.

My simple telescope seen on page 4 was propped up in the small back yard of the apartment my newly wed wife and I were renting and I made some observations of Mars when it was at opposition (closest approach to Earth) in February 1963.

The seeing was very poor due to the hard frosts disturbing the atmosphere and the heat rising from the buildings around me; my apartment was in the centre of the town.

It was also only 200 metres from the sea which created strong air currents due to the temperature difference between the sea and land being about 20° C.

The temperature dropped so low that, in a thermometer hung up in the backyard, the mercury went off the scale and retreated into the bulb.

All the water pipes froze solid including the mains water pipe in the road.

The water in the kettle on the kitchen hob froze overnight.

But, I was out there observing Mars.

Madness!

I have never found Mars to be particularly interesting as it is, except around a very favourable opposition, a tiny pink wobbly boring disc.

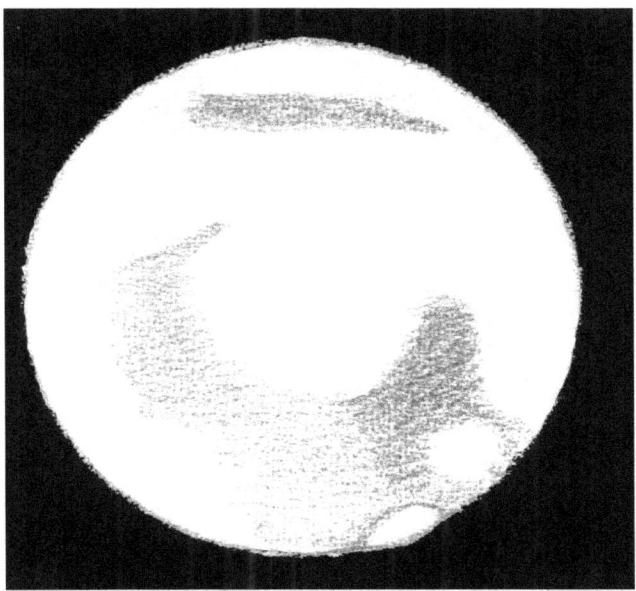

This is probably the best drawing I achieved of Mars. It shows the polar cap on the lower limb of the disc.

Observations Of Jupiter

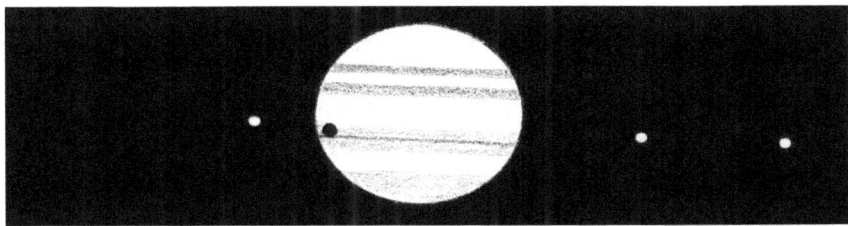

The above drawing of Jupiter with three of its 'Galilean Moons' on show was made on 6th April 1968 at 23h 10m (UT). South is up.

The moons of Jupiter from left to right were Io, Europa and Ganymede.

The shadow of Io is falling on the cloud belts of Jupiter.

Callisto was behind Jupiter at this time.

During the course of that evening I watched the dance of the four moons as they either moved across the face of Jupiter, cast their shadows on the clouds of the planet or emerged from being hidden in the shadow of Jupiter which was extending to the right of the planet in the above drawing.

For three hours I was transfixed by this display.

Anyone with a half decent telescope can watch similar displays almost any time that Jupiter is in the sky for three hours or more.

The phenomena of Jupiter's Moons are listed in some astronomy magazines and can be predicted using planetarium software such as Stellarium [252].

I was thinking during these observations that the timing of the phenomena of Jupiter's moon was once proposed as a method of solving the 'Problem of Longitude' [253].

This required Greenwich Mean Time to be known in distant locations around the Earth.

Local Time was easily determined by observing the position of the Sun or stars in the sky.

The difference between these two times gave a navigator his longitude.

The plan was to time the moons of Jupiter as they moved across the face of Jupiter, disappeared at Jupiter's limb or emerged from Jupiter's shadow.

The predicted times of these events would be published in an almanac and observations of these would give Greenwich Mean Time.

However, it is one thing to observe Jupiter's moons doing their entertaining progressions around the planet from solid land. Imagine trying to make these observations from a rolling ship!

In the end, Harrison's chronometer won the race to determine longitude reliably at sea [254].

Another scientific advance was made using observations of the transits, eclipses and occultations of Jupiter's four largest moons.

The Danish astronomer Ole Rømer noted that the orbital periods of the moons seemed faster when the planet was in the morning sky compared with when in the evening sky [255].

He correctly deduced in 1676 that this was because the time taken by light to travel between Jupiter and the Earth varied by as much as 17 minutes and he deduced the speed of light from these observations.

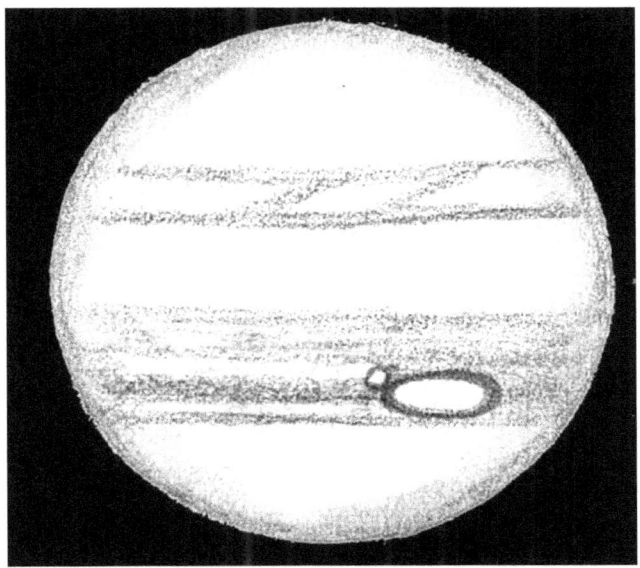

I spent a few years drawing the cloud belts on Jupiter and measuring the position of the Great Red Spot.

There was an interest in determining the positions of features on Jupiter's cloud belts which I engaged in during the 1960s.

The technique used was the method of transits where the time was carefully noted when the feature - such as the Great Red Spot - was central on the visible disc.

This timing was converted to position on Jupiter's face using tables published in the BAA Handbook [256].

There was much interest in these pre-space probe days in such measurements because the Great Red Spot and other features such as 'white ovals' drifted around Jupiter and the structure of Jupiter's clouds was still being elucidated.

Overall, I was not a keen observer of Jupiter but I frequently turned my telescopes to the planet just to watch the moons moving around the planet and to see if the Great Red Spot was still visible as there were times when it faded, changed colour or was prominent and colourful.

Observations of Saturn

Saturn is a difficult planet to observe - the apparent diameter of the globe is only about 18 arcseconds, but it is worth the effort if only to see the ring system.

So, where is the ring system in my drawing below?

This drawing was made on 9th November 1966 when the ring system was almost edge on to the Earth. South is up.

What we see is the shadow of the rings on the clouds of Saturn and the rings themselves were invisible. This happens because the thickness of the rings is a few hundred metres. If Saturn were the size of a football the thickness of the ring system would be 1/250th the width of a human hair.

In the above drawing Rhea is the moon on the far left with Dione next.

Mimas was in front of Saturn (in transit) but I did not see it and Enceladus was in Saturn's shadow on the left of the globe. This emerged from the shadow at about 20 h 30 m UT.

On the right of the globe is Tethys. Titan, the brightest of Saturn's moons, was beyond the right-hand side of the drawing.

At this time I checked the quality of the atmospheric seeing by observing the Pole Star (Polaris). This is a double star where the main and the fainter stars are about 18 arcseconds apart. I estimated the blurring due to atmospheric turbulence to be about 3 arcseconds.

This is a useful test of atmospheric seeing as Polaris is always available to see on a clear night north of the tropical regions of the Earth. The two components are also easy to see in moderate telescopes.

This resolution was typical for my location and the blurring was about one-sixth of the diameter of Saturn's disc. I would not therefore have expected to have seen much detail on the planet even if there had been anything to see.

Saturn's cloud belts and swirling patterns are subtle and I normally only saw the two main cloud belts as shown above.

The 'edge on' appearance of the rings gave rise to a curious incident in the career of Sir Patrick Moore which I well remember.

When the rings are invisible or appear as a very thin line there is an opportunity to see faint moons of Saturn which otherwise would be too dim to see in the glare of the ring system.

In October 1966 Patrick observed and recorded several times a faint moon on the edge of Saturn's rings [257].

Later the moon, subsequently named Janus, was discovered on 15th December 1966 by Audouin Dollfus and the moon Epimetheus was discovered on 18th December 1966 by Richard Walker Jr.

Patrick claimed to have made several pre-discovery observations of Janus and this set the astronomical world buzzing.

However, as Martin Mobberley has shown in his careful analysis of Patrick's drawings, there are confusions of recorded times and positions of the object reported by Patrick. Furthermore, Patrick mis-identified some of the known moons.

However, modern research has shown that the two tiny moons - Janus and Epimetheus - share the same chaotic orbit and the moons swap orbits on an irregular basis.

Because of the chaotic nature of their orbits it is not possible to calculate the positions of the two tiny moons at the time that Patrick Moore was making his ground breaking observations.

Just what Patrick Moore saw is still uncertain but he saw either Janus or Epimetheus back in 1966. This is a testament to the expertise and fine vision that he possessed.

Patrick Moore had a very rare talent. He was a celebrity who was also a 'hands on' expert.

11. Great Balls Of Fire!

Meteor observing is very much the province of the amateur enthusiast.

It can involve spending hours on freezing cold nights staring into the sky waiting for meteors to flash into view.

Of course, there are amateur astronomers who set up elaborate high-speed webcams to record meteors and sit in a warm house letting the equipment do its work.

Well, that's one way to do astronomy but it was never my way.

Meteors are the remnants of comets and asteroids plunging to their extinction in the thin air above you. It is one of the many staggering things in astronomy that a piece of dust, the size of a pea, travelling for millions - possibly billions - of years heats up so quickly and brilliantly in our atmosphere that it is seen one hundred kilometres away as a streak of light across the starry background.

When I have been out with non-astronomically inclined friends at night and they see a 'shooting star', it always gets them excited. It may be because a meteor is about the only thing in the sky which actually changes whilst you watch it.

The Role Of The Amateur

The role of the amateur in serious meteor studies is quite clear.

It is basically to count and track meteors that belong to periodic showers as a function of time so that statistics can be worked out. From this the orbits of the particles can be computed.

However, just for fun, it is also possible to work out the actual tracks of the meteors through the atmosphere and, since this involves groups of well separated observers working in unison, it is a very sociable pastime.

All you need to enjoy this pursuit is a length of string, a pencil, a simple star map, a ruler and a watch.

You then relax in a sun lounger watching the sky.

This is a very good opportunity to become familiar with the constellations - and your companions.

When a meteor appears you note the time and then hold up a piece of string against the stars to match the start and end points of the meteor trail and the path. Carefully note these positions and the track on the star map.

You will end up with a chart like that below.

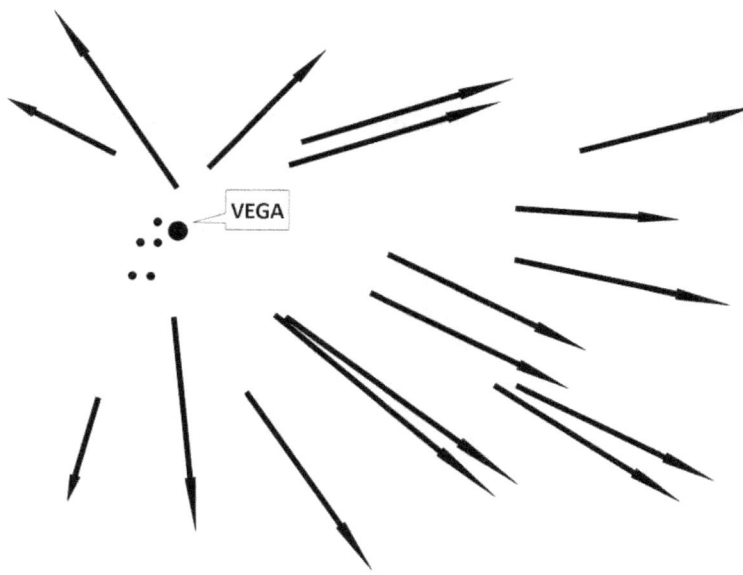

A Typical Group Meteor Watch

A typical group observation session that I participated in was of the April Lyrid shower in 1974.

The Wessex Astronomical Society split into two groups. One, led by me was based in my back garden in Weymouth. The other group was set up at Hengistbury Head about 50 kms away to the east.

We spent the night huddled in sun loungers watching the skies. When a meteor appeared which came from the direction of Lyra we carefully marked its track on a plastic laminated star chart from Norton's Star Atlas. The start and end points of the meteor's visible tracks were particularly noted.

The time was recorded as accurately as possible and the visual magnitude.

By knowing the start and end points of the two tracks of the same meteor, it is possible to work out by (relatively) simple triangulation the height of the meteor at the start and end and the positions over the Earth of these two points.

The diagram below shows the computed track of one of the meteors.

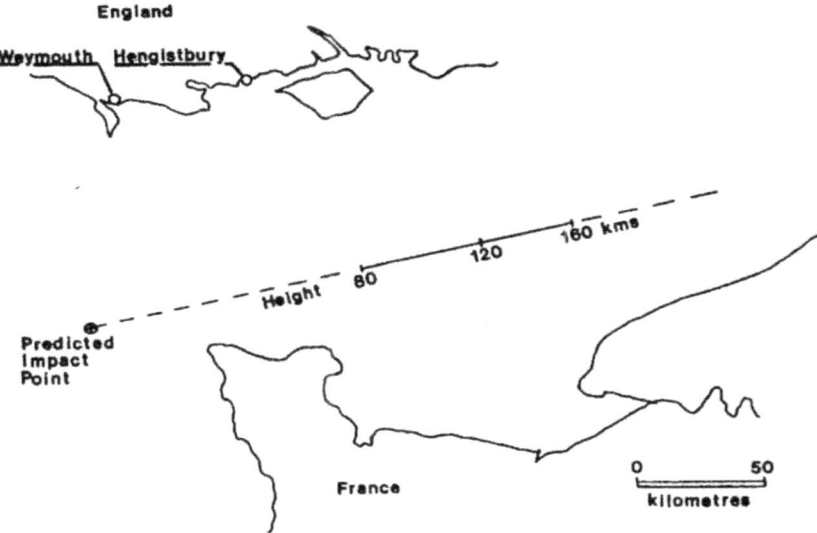

During this meteor watch I spotted a streak low in the northern sky - about 10 degrees above the horizon.

I later found that the same meteor had been seen by a lady observing in Derbyshire well down in her southern sky.

We exchanged data and I was able to show that the meteor had been overhead Bedfordshire at a height of about 80 km when it burned up.

The fact that the same speck of interplanetary dust had been seen by two observers about 300 kms apart really impressed upon me the distances over which these tiny glowing objects can be seen.

On one occasion I observed meteors from Weymouth until 3 am not knowing that the Hengistbury Head group had packed up and gone home before midnight!

On another night a small group including me observed from Hardy's Monument, high above the village of Portesham in Dorset. It was January 4th - the predicted night of the peak of the Quadrantid meteor shower.

Two friends and I huddled up in the freezing and biting wind on sun loungers. The stark grey stonework of the Hardy Monument towered above us faintly lit with a ghostly yellow light from the streetlights of Weymouth about 12 kms away.

Various cars arrived at this remote spot. Strange noises came from them as they rocked violently and then they were driven away; the occupants presumably returning to their wives and husbands.

One car remained long after the others.

The noise from that car was much louder and was a strange non-human squealing noise.

Eventually we realised that the squeals were coming from the car starter motor.

It must have taken enormous courage for the driver to walk over to us three astronomers, stretched out in our sun loungers in the dark, remote Dorset countryside with frost glistening on our boots. Each of us was holding clipboards and pens.

"Excuse me. Could you give me a hand to start my car?"

he asked.

We all went to the car.

Inside was a young lady wrapped in a coat and very obviously wearing nothing underneath as all three of us eagerly established by peering in through the steamy car window.

She looked sheepish - as well she should. The silly girl was missing the meteors!

We tinkered under the bonnet and found a loose wire on the distributer cap. We tightened this up and the car burst into life.

With much gratitude the driver departed, the girl trying - without much success - to retain what little modesty was left under her coat.

We three got back to our lonely freezing vigil recording meteors.

More BAA Trouble

During the course of other observations with a telescope I would often see very faint meteors crossing the field of view. These were about magnitude 8 or fainter and seemed to be moving much slower than unaided eye meteors.

At a BAA meeting I once asked whether the recording of these telescopic meteors was of interest.

The person to whom I addressed this question wagged his finger at me and grinned.

"I am sure you are being provocative knowing the current conflict in the BAA Meteor Section!"

I didn't know anything about any conflict but, from the knowing sniggers from some of the audience, I realised that I had stumbled across yet another internecine power struggle within the BAA of the type that erupted so frequently.

I never did get an answer to my innocent question.

Observing meteors is a very social opportunity to involve members of a society - anyone can do it. Indeed, at a friend's barbecue held at the time of the Perseids (August) I had everyone - all non-astronomers - watching the meteors shooting across the sky and getting enthusiastic enjoyment from the experience.

Bolloids! No really, it's true!

What cannot be planned is the observation of very bright sporadic fireballs called bolloids.

These can be brighter than the Moon and even brighter than the Sun as was the case in the super-bolloid that flew over Chelyabinsk in Russia in 2014.

The nearest I ever came to seeing such an event was from my garden in 1974. I was standing in the dusk when suddenly a friend grabbed my arm and said

"WOW! Look-at-that! Look-at-that! Look-at-that!"

I turned and saw a fireball heading from east to west, sparks coming from its head and a long sinuous smoky trail behind it. It was brighter than Venus.

I reported this sighting to the BAA and it was found to have travelled over Surrey, Somerset, the Bristol Channel and on towards North Cornwall. It may have dropped as a solid body into the Irish Sea.

I contacted other observers through the local press.

One fisherman off Chesil Bank said that it passed about 100 feet above his head and dropped in the water nearby!

I contacted the local coastguard stations. None replied - presumably they were too embarrassed to admit that they had failed to see what could well have been a distress flare!

Lost Star Constellations

I mentioned earlier the Quadrantid meteor shower. This is named after the defunct constellation of Quadrans Muralis - the Mural Quadrant.

You may well be surprised to learn that there are over two dozen constellations which are no longer marked on star atlases. Some lasted only a few years whilst others lasted many centuries.

A list of some of these obscure and curious relics from history are listed below [258].

Hirundo	The Leech	**Aranea**	The Spider
Lumbricus	The Earthworm	**Limax**	The Slug
Gallus	The Cockerel	**Felis**	The Cat
Anquilla	The Eel	**Patella**	The Limpit
Tigris	The River Tigris		

Machina Electrica	The Electrical Machine
Lochium Funis	The Log and Line
Officina Typographica	The Printing Shop
Telescopium Herschelii	Herschel's Telescope
Antinous	The boy lover of the Roman Emperor Hadrian

And   many   more!

12. Celestial Hide-And-Seek

Celestial hide-and-seek is played out when one astronomical body passes in front of another.

Types of occultation events vary from the relatively common to the extremely rare.

The types of occultation events that occur in broadly increasing rarity or difficulty of amateur observation are:

Occultations of stars by the Moon

Occultations of stars by stars

Occultations of the Sun by the Moon

Occultations of stars by asteroids

Occultations of stars by the moons of planets

Occultations of planets by planets

Occultations of stars by their planets.

These events will be described in that order in this chapter.

Occultations Of Stars By The Moon

These events occur every month as the Moon moves in its path around the Earth.

The rarity of the event increases with increasing star brightness.

There are uncountable numbers of stars passing behind the Moon every night but their visibility is limited by the glare of the Moon.

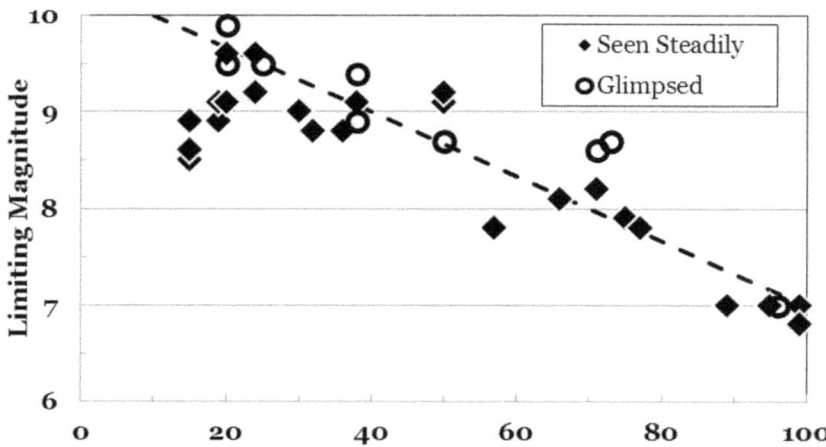

During my years as an active lunar occultation observer I recorded the faintest star that could be reliably seen close to the dark limb of the Moon with my 215 mm (8.5 inch) aperture reflector.

The above chart shows how the magnitude of the faintest star that can be reliably seen varies with phase of the Moon.

The line is approximately the dividing boundary between easily seen stars and those that are only glimpsed. The latter would not lead to reliable timings because the star would not be visible continuously.

Obviously, the nearer the Moon is to full (100%), the more difficult it is to see stars reliably.

The above chart shows why far more observations of lunar occultations are made around the half-Moon phase.

When the Moon is close to the Sun in the sky the stars cannot be seen due to the twilight glow and when the Moon is near the 'full' phase only brighter stars can be seen in the glare of the Moon's light.

It is for this reason that the observations of occultations during a lunar eclipse are important.

I have been fortunate to observe on many occasions the totally eclipsed Moon passing through the Hyades star cluster. Stars were popping behind the Moon and reappearing again in rapid succession. There were so many star occultation events to time that I could hardly cope.

It was at times like this that astronomy at the telescope became so exciting; not just because of the spectacle but also because I knew professional astronomers would be delighted by my measurements.

Predicting Lunar Occultations

When I was observing lunar occultations in the 1960s and 1970s the BAA Handbook published a two-page list of bright stars which were predicted to be occulted by the Moon as seen from Greenwich and Edinburgh. A simple linear extrapolation formula was given to enable the times to be worked out for other nearby locations.

The predictions covered only the brighter stars which are easy to see with almost any telescope.

I observed and reported my observations for a year or so and, when I was accepted as a proficient observer, I received much more detailed predictions for my observatory site covering thousands of stars.

The computer predictions arrived in November each year from Washington DC and weighed about a kilogram!

Today, observers are no longer wholly dependent on obtaining predictions from professional or keen amateur groups. The International Occultation Timing Association (IOTA) [259] will supply predictions via the internet and there are also software packages readily available to do this chore.

The Stellarium astronomical software [260] can be used to search for lunar occultations visible from your location.

I have crosschecked the predicted times of occultations published in the IOTA website against timings predicted by Stellarium for many stars and locations on the Earth. The differences rarely exceed 20 seconds.

This is perfectly adequate for getting you out at the telescope in a timely fashion to observe an occultation event.

So how did I make observations of stars being occulted by the Moon and have the techniques changed?

Stopwatches

In the early days only mechanical stopwatches were available.

The most popular type had a second hand which swept one rotation in a minute. These could be read to an accuracy of about 0.3 seconds.

I discovered my random error was much smaller than the precision with which I could read the stopwatch. So I invested in a more accurate watch.

This was a Heurer and cost me £35 in 1972 - equivalent to about £200 today. The second hand swept one rotation in ten seconds and I could read it to an accuracy of 0.01 seconds.

By the early 1980s digital stopwatches appeared. These had the great advantage that they did not suffer from significant variations with time other than a steady systematic drift which was very small - usually a few seconds in a week.

Thus, it was immediately possible to time occultations to a precision of about 0.01 seconds without all the hassle that mechanical watches had produced. In addition, a digital watch then cost only about £20 and the price has fallen since to less than about £10.

Although in the 1970s I could read my new mechanical watch to an accuracy of 0.01 seconds there was a delay between seeing a star disappear and pressing the watch button.

This is called the reaction time.

Determining Reaction Time

Each observer was - and still is today - required to estimate her reaction time. If no estimate was provided then a figure of 0.3 was assumed by the professionals.

I investigated my reaction time.

There have been many techniques developed for measuring reaction time. Some involve complicated electronic circuitry but I have one method that is accurate and simple.

This uses either a digital stopwatch which can be bought very cheaply or the stopwatch mode on a mobile 'phone.

Set up the display so that it is running in stopwatch mode - i.e. displaying hundredths of seconds. The units and decimals of seconds are then covered leaving only the tens of seconds and above visible as on the display above where the obscured digits are shown shaded.

Look away from the display for a while. Look back and stop the 'lap' function button as soon as the tens of seconds digit is seen to change.

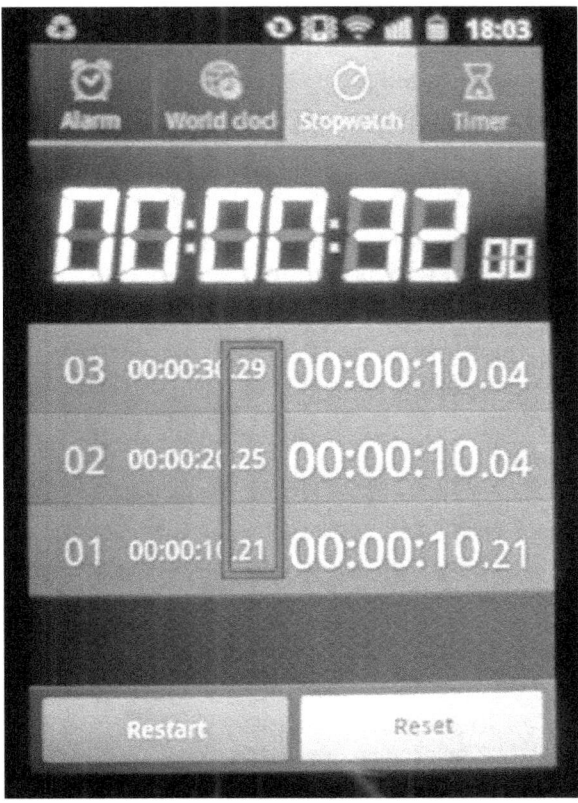

Look at the whole display. The decimal part of a second displayed is the reaction time as shown in the lower picture on the previous page.

After doing this several times my reaction times were displayed as the decimal parts of the 'lap' times as shown above.

Reaction time is affected by cold and discomfort. It is also affected by stress and, surprising as it may seem to any 'deep sky' observers reading this, timing occultations is stressful.

Occultation timing is *stressful*?

Yes, indeed!

Usually, the darkened limb of the Moon is visible at an occultation before full Moon. The dimly lit edge can be seen moving very slowly (the rate is 0.5 arcseconds per second) towards a star and the star appears to hang right on the limb for ages before it snaps out.

The heart races, sweat beads the brow and the hand shakes.

Sometimes the stopwatch button is not pressed because the stress level is too high. I took to listening to soothing music through earphones to help me relax.

Referencing Watches To Universal Time

To fix the timing to Universal Time we need to check the watch against a time reference. The easiest one to use is the telephone speaking clock using a landline. This has a claimed accuracy of 0.005 seconds [261] which is more than adequate for occultation purposes.

If a landline telephone is not available then there are various radio stations around the world that transmit time signals.

The National Physical Laboratory can advise on these [262].

Some words of warning!

Do NOT use clocks and watches that are linked to an hourly radio signal to keep them accurate. These are normally updated every hour on the hour. Because it is easier to put a clock on 'hold' until it is correct rather than run it faster, these clocks usually gain between updates. They are not as accurate as claimed at any one time - indeed, they are worse than digital watches in the short term.

Beware also of using the 'Greenwich Pips' on the radio and

television.

There can be time delays as long as eight seconds between the signals heard from an FM radio, a digital radio and radio stations broadcasting over the internet both on terrestrial and satellite channels. These delays are much larger than the accuracy required for timing occultation events [263].

Telescope Position

Observers need to know their telescope position in latitude, longitude and height on the Earth's surface. This used to be obtained from Ordnance Survey maps.

These days a simple GPS watch will give latitude, longitude and height to an accuracy of better than 10 metres and this is adequate for occultation work.

Remember that an error of ten metres in telescope position produces a similar error in determining the position of features on the limb of the Moon in space - which is a pretty astounding accuracy.

Lunar Grazing Occultations

Although observing occultations was interesting and useful - observers actually saw something happening in the sky as they watched - grazing occultations were even more fun as they were a team effort.

So far I have written about stars being occulted at the dark limb of the Moon as seen from my observatory. Stars can also be seen grazing the north or south polar regions of the Moon. This is particularly exciting as the star appears to flash on and off as it passes behind the mountains and valleys on the edge of the Moon.

This phenomenon can only be seen from a narrow strip of land which is the projection of the lunar limb irregularities on the Earth's surface.

This strip of visibility is a few kilometres wide. Observers within this strip at the right time should see a star flashing as its light is repeatedly interrupted by the mountains and valleys of the Moon edge.

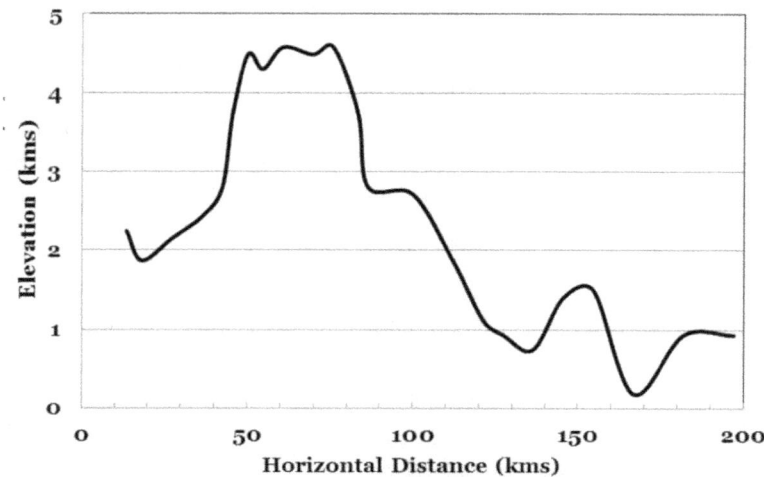

The diagram above shows an actual limb profile predicted for an occultation in 1972. The vertical and horizontal scales are very much exaggerated. However, allowing mentally for this exaggeration, you can see that the Moon's limb is quite 'bumpy'.

Between 1927 and 1956 Chester Burleigh Watts worked away photographing the limb of the Moon for all values of libration [264].

He built up a collection of 700 high precision photographs of the Moon's limb providing a detailed knowledge of the ups and downs of the profile of the Moon as seen from the Earth.

When published four years after his retirement in 1959, his measurements filled 951 pages in an astronomical journal [265].

The chart above is a sample from the Watt's charts.

If a lot of observers are strung out across the predicted track in which a grazing event will be observed they will each see a different pattern for the star blinking out and back on - sometimes the star will briefly disappear over a dozen times.

How exciting is that!

Using the timings of each event, it is possible to compute the predicted position of the star for each observer and event relative to where the Moon is expected to be.

The diagram below for a grazing occultation in the 1970s shows a comparison between where the star blinked on or off and where the Moon's limb was expected to be in the sky relative to the star.

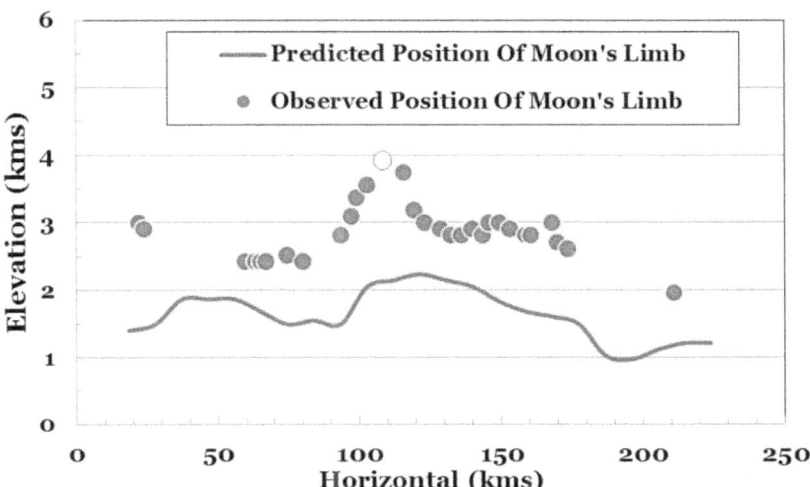

The open circle shows where the star blinked out for about one second. This accurately identified the summit of the lunar mountain on the limb.

It can be seen that the predicted relative position of the Moon and star was in error. In fact, in this case, the star's catalogued declination was about 0.8 arcseconds in error.

In some observations of lunar grazing occultations observers only fifty metres apart have seen quite different patterns of star brightness changes showing that detail down to that size can be resolved on the Moon's limb.

One of the fascinating things about grazing occultations is that an observer with a small telescope can indirectly detect and measure even large boulders on the Moon's surface.

Sometimes a star disappears in two steps at the limb of the Moon. This proves that the star is a double star. Many binary stars have been discovered in this way.

If a star disappears in two steps about 0.4 seconds apart then the separation of the two components of the binary will be about 0.2 arcseconds separation.

A pair of bright stars this close would be very difficult to detect visually through most telescopes and yet the binary nature of this star would be obvious during its two-step disappearance at the lunar limb.

During a grazing occultation of a double star the two stars have been seen individually flashing in and out of sight.

Overall, the most fascinating part of grazing occultation observations is the social nature of the events. Groups of enthusiasts have to travel to sites which may be inhospitable and remote and work together to produce a group observation.

One example of this type of expedition was organised in 1974 by the Wessex Astronomical Society.

Some Grazing Occultation Adventures

A grazing occultation was predicted for quite a bright star at a reasonable time of the night in 1974. The track went across England and the southern Hampshire coastline at Hayling Island.

A group of four friends and I set off with our telescopes and scoured the beach for suitable sites.

I set up my refractor next to a corner of a ladies toilet which was marked on the large scale Ordnance Survey map and friends observed and made timings from other nearby readily identified landmarks.

The night was fine but cold - it was the middle of winter and the beach was deserted - no surprise!

Everything went very well and all three groups got several timings using small telescopes and battery operated audio cassette recorders to backup our stopwatches.

When added to the observations made by other groups across the country the results plotted on page 317 were obtained.

Above and below we see my friends and me posing with the

Moon in the background after our very successful observation of this grazing occultation.

We took these photographs to celebrate our success and retired to the nearest pub to check our results and have a drink. Whilst there, we noticed a group of warmly clad people at a near-by table poring over a detailed map of Hayling Island.

We discovered that they were from another astronomical society and that they had made observations from sites between ours - but we had not seen them!

The Wessex Astronomical Society made quite a few expeditions to observe grazing occultations. They were always enjoyable both socially and scientifically. None were uneventful!

On one night a grazing occultation of a faint and difficult star was predicted for 4 am on the morning of September 23rd, 1973. The track passed over my observatory in Weymouth so I couldn't miss this one!

I had a debilitating attack of 'Man Flu' at the time but I gamely crawled out of bed and dressed. (I don't always dress. I once observed Comet Bennett from my garden at 5 am dressed in pyjamas, slippers and a ski hat.) The thin crescent Moon was high in the East and the winter stars were a grand sight.

As soon as I swung my telescope round to the Moon my heart sank. The row of runner bean plants that my wife had nurtured all summer was obscuring the view! With only twenty minutes to go I started to pull the supporting canes out of the ground - not easy because she had secured them well in the Spring.

The plants were thoroughly intertwined and I soon became enmeshed in a huge mass of vegetation. As the minutes ticked by I became more desperate, the canes snapped with loud cracks and I stamped the plants into the ground. At every noise I expected to see my neighbours' bedroom lights flash on and to hear a police car siren approaching from afar.

At last the beans were flattened and I then noticed to my disgust that a cloud had covered the Moon with only five minutes to go until the predicted event. Despite a slight breeze which caused my nose to run freely, the Moon did not appear again that morning.

Another example of the perils and excitement of trying to observe grazing occultations came one night when a friend and I decided to observe a grazing occultation from the tiny village of Little Bredy in West Dorset. I met my friend off the train at Dorchester and we drove to an ancient and quaint village bus shelter.

We set ourselves up with a telescope, a short-wave radio giving out timing bleeps from the East German DIZ long wave radio station and an assortment of torches and vacuum flasks brimming with hot coffee.

The sky was clear and the Moon bright.

During our wait we heard footsteps approaching.

We were not visible in the shelter so all the approaching person would have seen would have been an apparently empty bus shelter from which came loud 'bleeping' noises. The footsteps paused and retreated somewhat faster than they had approached.

As the moment of the occultation approached a high jet aircraft streaked across the sky leaving a condensation trial behind it. We watched in horror as this ghostly white line drifted towards our view of the Moon. Sure enough, at the exact moment of the occultation the contrail obscured the star!

Occultations Of The Sun By The Moon

OK!

Maybe I should have described these events as Solar Eclipses but they are aspects of lunar occultations.

Arguably the most spectacular of astronomical events, keen 'eclipse chasers' travel the world to place themselves inside the narrow strip on the Earth's surface from which they will observe this event - or not if it is heavily clouded.

I have not observed a total eclipse of the Sun but, on 11th August 1999 I stood just one kilometre north of the edge of the path of the total eclipse which was clouded out over the southern parts of England from which it could have been observed - see [266].

I have already described this event on page 92.

Occultations Of Planets By The Moon

These events are fairly commonplace in as much as most years there will be an occultation of a planet by our Moon.

A failure to see a planet passing behind the Moon in any year is more likely to be due to bad weather than the lack of opportunity.

For example, there were eight opportunities to watch the Moon hide a planet in 2015 [267]. Five involved Uranus, two Venus and one for Mars.

The occultation of Saturn by the Moon on 24th February 2014 can be simulated using the Stellarium software - other planetarium programs are available.

This event can also be seen on YouTube as a movie [268].

Over the seven decades of my astronomical adventures I have seen many such events with Saturn, Venus and Mars being hidden by the Moon.

The most memorable was a grazing occultation of Mars by the Moon.

This rare event occurred on May 15, 1972. This involved Mars skimming along the southern pole of the Moon which would be a very interesting event to watch.

The track of visibility passed through Dorset so the Wessex Astronomical Society mounted an expedition to observe and time this event.

We split up into groups along the deserted road running through the army tank testing range at Bovington Army Camp in Dorset.

Well, not quite deserted because the army were holding tank

manoeuvres that night. One group of amateur astronomers was accosted by the police; another was accosted by a tank.

"Have you seen a broken down tank anywhere?"

one group was asked.

My wife had found it.

It was parked without lights on a part of the road that she was travelling along at about sixty miles per hour to rendezvous with me in a quarry. She just managed to swerve around the massive machine leaving rubber for yards along the asphalt.

Meanwhile, in the quarry, I was rigging my portable refractor to my car's roof-rack when I noticed a darkened car a few yards away.

Thinking that the occupants might not appreciate me waving such a large telescope around, and that the male occupant might be seven feet tall with muscles like steel bars, I drove off quickly with the telescope dangling precariously, to a new position where my wife and I watched the sky clearing rapidly.

Anxious to see this spectacle through binoculars I leaned on a parked car and peered at the clouds scudding across the face of the Moon.

Suddenly the car window where I was leaning was wound down and a very angry man shouted

"Oi! What do think you are playing at?"

I had mistakenly thought this was a car belonging to a friend in my group but it was a stranger's car and beside him was a very scantily clad young lady trying to hide her extensive cleavage as I tried to hide my binoculars.

Another hasty retreat was made which saw me finally setting up my telescope a few minutes before Mars was due to skim along the Moon's limb.

Then a cloud covered the Moon.

Five minutes after the time of the occultation the cloud cleared and superb views were had of Mars hanging on the lunar limb.

Although we missed the occultation it was well worth the effort even though I nearly had my face punched in twice within the space of twenty minutes!

Occultations Of Stars By Asteroids

In 1997 I started observing occultations of stars by asteroids. What I should have written was that I *tried* to observe them!

These were relatively rarely predicted back then because the computers of the time were not powerful enough to make the extensive calculations.

However, with the rapid increases in computing power it is now potentially possible to see an asteroid causing a star to blink out several times a year provided you are willing to travel a reasonable distance to put yourself with the shadow track of the asteroid.

The map below shows the track across England and Wales within which an occultation of a relatively bright star by the asteroid Sapientia (275) was predicted to be visible on 30th September 2015.

The diagonal line shows the predicted centre of the track within which the star should have been covered by the asteroid. However, there was sufficient uncertainty in the predictions to make it possible to see an occultation from with the wider shaded zone.

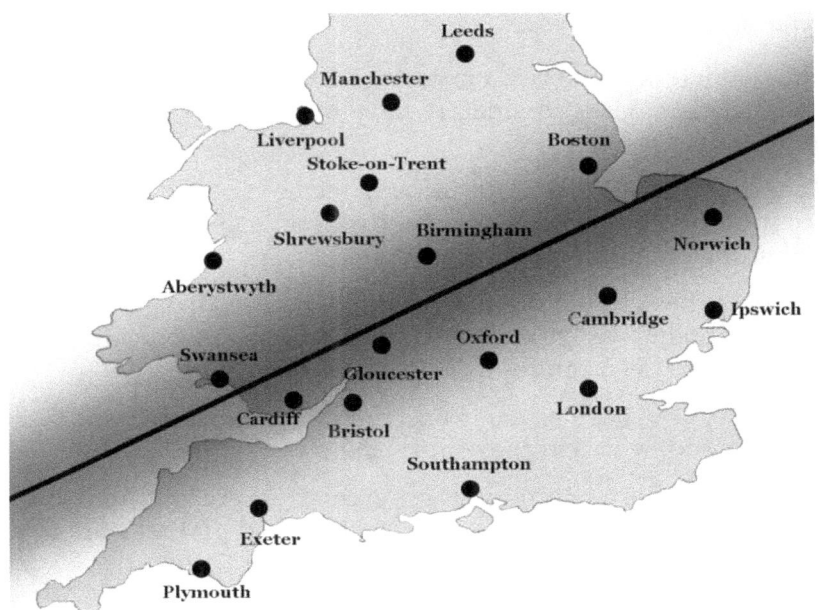

To put this prediction into context the width of the zone of uncertainty is about 100 kms.

This means that the predicted path of the asteroid across the sky is known to a positional accuracy of about 0.1 arcseconds; a remarkable testament to the precision of modern computational power in celestial mechanics.

This occultation was successfully videoed by David Arditti and can be watched on YouTube [269] where the star is seen to disappear for about 11 seconds.

Tim Haymes also videoed this event and recorded the star dimming for 14 seconds [270].

By the time that I observed my first star being occulted by an asteroid only two such events had previously been successfully observed from the UK.

There are many reasons for the rarity and difficulty of achieving a success - apart from the uncontrollable effects of the weather.

An occultation of a star by an asteroid occurs when the shadow of the asteroid cast by the light of the star passes over the observer. Because the star is so much further away from the observer than the asteroid the width of the shadow track over the surface of the Earth is comparable with the diameter of the asteroid.

The diameters of the vast majority of asteroids were unknown in past decades. These diameters have been progressively refined in recent years.

A fairly good idea of an asteroid's diameter can now be gained by accurate photometric measurements of an asteroid's brightness and surface composition.

Even so, the shape of an asteroid may be very irregular for the smaller bodies as shown by the space probe images of the asteroid Eros as it rotates - see page 64.

Thus, we often have an asteroid's shadow of unknown width and shape sweeping across the countryside.

Add to this the uncertainties in predicting the position of the asteroid in space and the precise position of the star being occulted we then see that the uncertainty, especially in the 1990s, was great.

These uncertainties meant that the chance of seeing a star blinking out as its light was blocked by a passing asteroid was low even if the observer was in the optimum location predicted for the event.

These days the situation is better.

Extensive studies of asteroid orbits, prompted by the risk of asteroidal collisions with the Earth, have greatly improved the accuracy with which the positions of asteroids are now known.

The number of asteroids being accurately tracked through space is now approaching 100,000 compared with the few back in the 1960s when I was starting to make serious observations of the heavens.

In addition, the positions of many stars are now known to greater accuracy than two decades ago because of the results of stellar mapping projects such as HIPPARCOS [271].

Computing power has also increased greatly - see Moore's Law [272].

This means that far more events can be predicted for ever more asteroids and ever fainter stars.

For example, in March 2016 a star was predicted to be occulted by an asteroid numbered 84200 in the catalogue of asteroids. (The track of this occultation passed right over my back garden but - it was cloudy! Spode struck again!)

In that same month there were three occultations of stars by asteroids over England for stars bright enough to be seen easily in binoculars.

The uncertainties in the prediction accuracies significantly reduce the chance of seeing an occultation by an asteroid.

This is where team work is important.

By setting up observers spanning the predicted track of an asteroidal occultation it is certain that one or more will see and time an event.

"They also serve who stand and wait"

said John Milton (1608-1674)

By deliberately putting themselves at locations with a lower probability of seeing the occultation than others in the team everyone shares equal glory when a successful observation is made.

A failure to observe an occultation is important as it sets an upper limit on an asteroid's size.

Observers outside the zone of higher probability have detected possible tiny moons circling the asteroid - as we will now see.

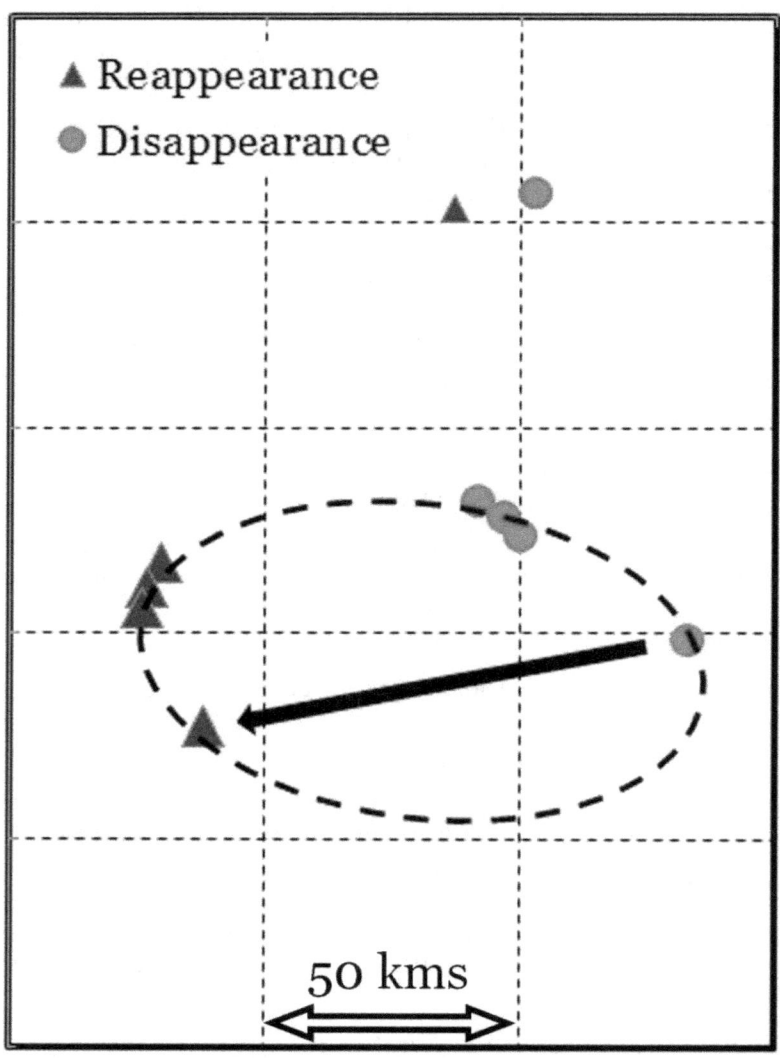

The diagram above shows the results from my second successful observation. It was the occultation of a star by asteroid Ani (791) on April 7th 2000.

The symbols show the position of a star at disappearance (filled circle) and reappearance (filled triangle) converted to spatial position at the distance of the asteroid.

My observations are marked by the arrow showing the apparent direction of motion of the star relative to the asteroid.

An outline of an elliptical shape has been superimposed to give an indication of the possible shape and size of the asteroid. The squares are 50 kilometres on each side at the distance of the asteroid giving a size of the asteroid as about 110 kilometres on the long axis and 65 kilometres on the shorter axis.

This corresponds to an apparent size of about 0.001 arcseconds.

Hence it cannot be seen as a physical object through any telescope.

It is remarkable that the size of an asteroid this small can be determined by amateur astronomers to an accuracy of about 10 kilometres.

This is another example of the power of amateur astronomy.

The above results are rather odd because one observer saw the star disappear well away from the expected position of the asteroid. This may have been a disappearance caused by a tiny moon going round the asteroid.

The above picture of a different asteroid (Iris) taken by a space probe shows such a tiny moon (Dactyl).

Occultations by asteroids do not necessarily require large

telescopes although the larger the aperture the more events can be seen.

For example, on March 20, 2014 the bright star Regulus was occulted by the faint asteroid Erigone (163).

Thus, a bright star easily seen by the unaided eye star would have disappeared from unaided eye view for over ten seconds for observers in the right place and time under a clear sky.

Unfortunately, it rained and no reports were received that this extremely rare and interesting event had been seen by anyone.

Spode Struck Again!

In the decade and a half since I made the above observations many things have changed a great deal and all for the better. A major innovation has been the development of highly sensitive equipment for recording star occultations by asteroids on a webcam.

The importance of the observations has also greatly increased due to the need to characterise and chart the presence of asteroids which might one day threaten the Earth.

Finally, have a look at the observation at [273] where you will see how fifty observers of the occultation of the star LQ Aquarii by a binary asteroid successfully determined the shapes and sizes of the two bodies when the best image from Earth through the giant Keck Telescope shows the binary asteroid as two indistinct blobs.

If that doesn't make you want to observe asteroidal occultations then nothing will!

Go to the IOTA website [274] now and sign up!

Occultations Of Stars By Stars

We would not normally call this an occultation but it is an event that can be easily observed even with the unaided eye.

We call such events the eclipses of binary stars when one star passes in front of another as the two stars move under their mutual gravitational pull.

I explain how to observe eclipsing binary stars in Chapter 14 and I will not spend any further time here on this topic.

Occultations Of Stars By The Moons Of Planets

On July 3rd 1989 the star 28 Sgr was occulted by Titan, the

largest moon of Saturn. This was an extremely rare event visible throughout England using binoculars. I observed out of the window of my rented house on Portland, Dorset.

I saw the disappearance and reappearance but it was all very difficult as Saturn was low in the smog polluted sky near the horizon.

Observers placed such that the star appeared to pass centrally behind Titan saw a flash of light briefly encircling Titan as the light from the star was refracted by Titan's atmosphere acting like a gigantic lens.

There is an interesting report on this very rare event at [275].

Occultations Of Planets By Planets

A very rare example of celestial hide-and-seek is when one planet passes in front of another.

This has only been observed once using a telescope. This was seen by John Bevis on the evening of 28th May 1737. In the twilight and using a huge six metre focal length non-achromatic refractor he observed Venus pass in front of Mercury and hide it.

Such mutual occultation events are very rare. The following list shows some events and is taken from the computations of Albers [276] supplemented by other events found by Larry Bogan [277].

Each of these predictions has been checked by me using the Stellarium planetarium software and the times of mid-event have been added by me.

2 BC 17th June 17 hr. Venus occulted Jupiter. This would have been a spectacular sight from the Middle East.

578 12th May 19 hr. Mercury occulted Mars.

1170 12th September 21 hr. Mars occulted Jupiter. Recorded by Gervase of Canterbury and observers in China [278].

1210 17th September 11 hr 32 min. Venus occulted Jupiter. This occurred very close to the Sun so this event would have been invisible to the unaided eye.

1278 25th August 15 hr 16 min. Mars occulted Neptune.

1387 22nd September 01 hr 03 min. Mars occulted Jupiter.

1427 30th August 08 hr 32 min. Venus occulted Neptune.

1477 9th October 16 hr 05 min. Mars occulted Saturn.

1522 28th January 08 hr 01 min. Mars occulted Saturn.

1570 28th May. Venus occulted Jupiter? Although reported by Albers this event is not confirmed using Stellarium as Venus and Jupiter are nowhere near each other on this date. Instead Stellarium shows a likely occultation on 5th February at 07 hr 40 min.

1590 13th October 05 hr 50 min. Venus occulted Mars. This event may have been observed by the German astronomer Michael Maestlin at Heidelberg [279]. This is a probable pre-telescopic observation.

1613 31st July. Galileo made a pre-discovery observation of Neptune amongst the Galilean moons but failed to realise it was a new planet and recorded it as a star. Stellarium shows Neptune amongst the Galilean moons but no occultation is predicted by Jupiter or its moons.

1623 15th August 18 hr 00 min. Jupiter occulted Uranus.

1702 19th September 15 hr 00 min. Jupiter occulted Neptune.

1708 14th July 14 hr 00 min. Mercury occulted Uranus.

1708 4th October 14 hr 34 min. Mercury occulted Jupiter.

1737 28th May 22 hr 40 min. Venus occulted Mercury. The only telescopic observation. Made by John Bevis at Greenwich.

1771 29th August 20 hr 34 min. Venus occulted Saturn.

1793 21st July 06 hr 34 min. Mercury occulted Uranus. Only visible from parts of Australia and Japan.

1808 9th December 20 hr 35 min. Grazing occultation of Saturn by Mercury. Only visible from the South Pole.

1818 3rd January 21 hr 47 min. Venus occulted Jupiter. Only visible from the Far East. No observations recorded.

2020 21st December. Jupiter and Saturn will come within 6 arcminutes of each other.

2037 15th September. Mercury will skim past Saturn but not occult it.

2065 22nd November 12 hr 37 min. Venus occults Jupiter. Too close to Sun to be seen.

2067 15th July 12 hr 52 min. Mercury occults Neptune. Only visible in daylight from the North Pole.

2079 11th August 02 hr 23 min. Mercury occults Mars.

2088 27 October 13 hr 35 min. Mercury occults Jupiter. Too near to the Sun.

2094 7ᵗʰ April 11 hr 40 min. Mercury occults Jupiter. Too near the Sun.

2100 23ʳᵈ August 10 hr 42 min. Mars will pass close to Neptune but not in front of it.

2123 14ᵗʰ September 16 hr 18 min. Venus occults Jupiter. Visible over Pacific Ocean. Stellarium suggests that Venus may occult one or more of Jupiter's Galilean moons.

2126 29th July 16 hr 58 min. Mercury occults Mars.

2133 3ʳᵈ December 13 hr 58 min. Venus occults Mercury. Very close to the Sun.

2173 16ᵗʰ July 02 hr 26 min. Mars will pass very close to Mercury but not in front of it.

2223 2ⁿᵈ December 12 hr 19 min. Mars occults Jupiter.

2478 29ᵗʰ August 23 hr 43 min. Mars occults Jupiter.

Although (obviously!) I have never observed a planet passing in front of another planet I have included this list here because it is fun to run a planetarium program (such as Stellarium) and watch these events being simulated.

I urge you to have a go!

Set the date into the program and start at 0 hours UT. Then lock onto one of the planets involved in the event and let the software run in fast time.

Occultation Of Stars By Their Planets

A further important type of occultation that cannot be observed by amateur astronomers is when a planet in orbit around a star passes across the face of the parent star.

It is from observing the very small dip in light level from stars that the presence of a transiting planet has been deduced using data from the KEPLER space probe as well as from ground-based telescopes.

If viewed from a great distance beyond the solar system, Jupiter would diminish the light of the Sun by about 1% which is about 0.01 visual magnitudes.

However, I doubt that this field of occultation observation will become mainstream for amateur observers in the foreseeable future.

Having said that a project was set up in 2009 by the University of London's Observatory to use its 350 mm (14 inch) aperture telescope to search for planets passing in front of stars and an amateur group is planning a similar venture at the Charterhouse Centre Observatory in Somerset [280].

13. Hunting For Novae

Why Hunt For Novae?

In my search for new topics to explore I decided, early in 1977, to start nova hunting.

Novae (plural of nova which is the Latin for 'new') are stars that can brighten rapidly - sometimes within a few hours - taking them from obscurity to fame.

Their discovery was highly competitive amongst amateurs in the 1970s and still is a popular aspect of amateur astronomy.

This was a time before automated amateur telescopes scanned the skies whilst their owners sat indoors watching television and drinking hot chocolate.

In the 1970s nova hunters spent hundreds - some even thousands - of hours scanning the skies with binoculars searching for a 'new' star amongst the thousands that drifted through their field of view.

People sometimes wondered why amateurs bothered to search for novae when there were so many professional observers routinely photographing the skies all the time.

The high success rate over the past four decades of amateur

searches answers that question.

A dedicated amateur can spot an intruder star immediately if a visual search is being made. A professional may take days to check observations.

In fact, in the 1970s and 1980s amateurs discovered the great majority of nova which brightened to above about visual magnitude 8 or so.

The UK campaign was - and still is - organised by 'The Astronomer' (TA) magazine.

The UK Nova Patrol Scheme

Back in the 1970s interested observers were allocated sections of sky, about 50 square degrees in area, and they searched their 'patrol area' visually on every available night.

This is quite a good scheme as long as the 'owner' of an area is conscientious. The areas were spread over the Milky Way because most novae appear around the Galactic Plane.

This scheme has paid off several times for UK amateurs.

For example, John Hosty - a postman - discovered a nova in the constellation of Sagitta on 7th January 1977 with half a pair of broken binoculars proving that expensive optical aid is not needed [281]. He had memorised all the stars in his area and immediately spotted the 'new' star.

John became nationally famous on television and in the media showing his broken binoculars which he used so successfully.

Sadly John died in 2001 aged only 51.

Retired schoolteacher George Alcock memorised about thirty thousand star positions covering a lot of the sky and discovered several novae as well as comets using binoculars, a deckchair and his fantastic memory.

Later in life he observed from inside his house looking out of a window [282].

A picture of George Alcock with the Nova Patrol Coordinator Guy Hurst can be seen at reference [283].

Dave Branchett was another early member of the team and discovered a nova in the constellation of Scutum. Unfortunately, his sighting was never confirmed professionally and there is a mystery that is unsolved to this day [284].

Dave emigrated to the USA and I heard rumours that he was very popular because of his amazing impersonations of Benny Hill who has remained far more popular there than in the UK.

I joined the TA Nova patrol in early 1977 inspired by the successes of John Hosty, George Alcock and Dave Branchett.

I was allocated an area to search which was quite well north of the galactic plane well away from where most novae appeared.

I spent many hours searching my designated zone.

However, I soon became restless.

I worked out the likely number of novae bright enough to be visible in binoculars that would appear in my (probable) remaining lifetime. I converted this figure to the probability of a nova occurring in my patrol zone in my lifetime and the probability of me being the first to see it. This came to about 1 chance in 20,000 of me ever discovering a nova.

From the mid-1970s the Japanese were rapidly becoming successful in discovering comets and novae. They had a large number of amateur observers concentrating on the western sky as soon as the sky was dark enough to search.

This gave them an advantage of about nine hours over observers in Western Europe and fourteen hours advantage over east coast observers in the USA.

This is why UK observers also searched the western horizon first because they then had a chance of seeing a nova that had burst into sight after the part of the sky containing the nova had set in Japan and before US observers had a chance to scan the skies after dark.

Searching in the eastern sky meant that we were looking at stars which the Japanese had already been searching.

My Photographic Nova Search

Having worked out that the chance of me finding a nova in my zone before I died was very, very small - especially as my zone was well out of the Galactic Plane, I decided that I was not going to devote hundreds - even thousands - of hours staring at a zone of sky within which no nova would probably ever erupt in my lifetime.

Nova hunting had to be done more effectively or not at all. I would search all the Milky Way every clear night using photography.

I had a Praktica SLR camera and I built a 'Barn Door' mount as described in Chapter 7.

The simple device seen on pages 208 and 209 built from junk wood, a screw, a stopwatch and a small red light was perfectly adequate to take sharp images of stars in three minutes exposures and reach down to about visual magnitude 9.5 on 1000 ASA roll film. This was fainter than my contemporaries using binoculars could achieve.

Below we see one of my typical Nova Patrol pictures.

My plan was to photograph the entire visible Milky Way every clear night, develop the pictures and search every one for a nova by dawn.

Each three minute exposure photograph taken with a standard 50 mm lens showed about two thousand stars on average and it needed about twenty pairs of pictures to cover the large area I wanted.

Thus, every clear night I would set out to photograph all the Milky Way and its environs in the hope of finding a nova lurking in the tens of thousands of star images.

It was great challenge but I estimated that my chance of finding a nova was increased greatly compared with a binocular search of a small zone.

I would have to search through about 20,000 to 40,000 star images (and many film defects) during the course of each night to find the one that was the nova I had set my heart on.

So, how was I to search all those pictures so quickly?

As soon as the sky was dark I would take overlapping pairs of pictures covering as much of the Milky Way as possible. I then developed the film in my downstairs toilet.

No!

Not in the toilet bowl but in the wash basin.

I had to blacken out the window and door with tape and black bin liners cut into strips on account of the street light outside my house.

I was using 1000 ISO monochrome film which was very sensitive to stray light and had to be handled in totally dark conditions.

Once developed and fixed I dried the film by holding it in front of an electric fire.

"Yeeooow!"

I can hear keen photographers screaming...but I was after speed, not finesse. Even as I was drying the film some keen nova hunter could be finding *my* nova.

Once the film was dried I cut it and mounted each picture in a slide holder. I had a small slide projector which I fitted in the spare bedroom.

I projected the slide vertically down onto a flat board. On the board I placed a page from the Smithsonian Astrophysical Observatory (SAO) Star Atlas. This covers the entire sky in about 170 A3 sheets.

I bought the SAO Atlas whilst in Washington in 1973. I was taken by a reluctant taxi driver to the Government Printing Office deep in the ghetto of that otherwise charming city. I stepped out from the building into the street holding a huge box.

The streets were full of the poorest slum dwellers eyeing this idiot with a huge heavy cardboard box.

I was sweating.

The temperature was 95°F and I was wearing a smart black suit.

However, all those eyes following me with the box were the

cause of my sweat. I was so relieved when I reached the Central Park area and relative safety of a taxi.

I projected the negative star pictures onto these atlas sheets directly and at the same scale.

I was able, with a bit of jiggling, to get my star images superimposed onto the atlas stars. I then moved the atlas sheet a little. Any star on my photograph which was not matched on the atlas visually jumped out of the page as it was a single image and not a pair. I was surprised at how one single star amongst thousands of pairs of stars was found quite easily.

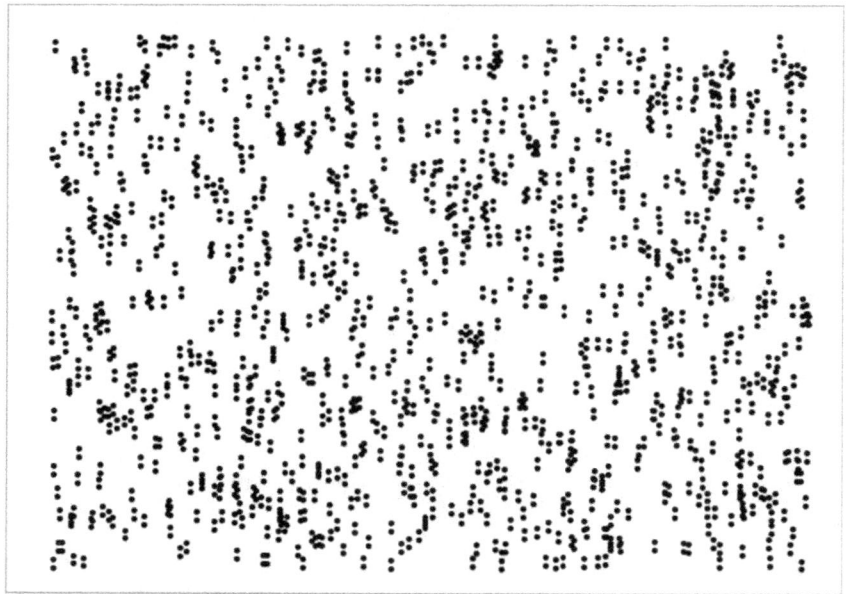

You can try this for yourself.

Above I have simulated a star field with 1000 dots.

OK! So it does not look like a real star field because the dots are all the same size but this exercise is merely to illustrate what I did every clear night.

Alongside 999 of the 'star' images is another dot simulating the stars on the SAO Atlas sheet.

So, on the picture above there are 999 pairs of dots and one singleton.

What I had to do with every photograph was to search the atlas page to find the lone single dot.

Can you find the single dot amongst all the pairs in the above picture?

The answer is at the end of this chapter on page 343.

Using this projection technique I was able to search each picture, some with well over 2,000 star images, in a few seconds. I marked with a soft pencil on the atlas sheets any stars on my pictures that were not on the atlas.

Each picture had many 'stars' not on the atlas sheets.

These were mostly small image defects on the film. The second image of the same area eliminated all of these film defects.

My First Photographic 'Nova'

On my very first night I found a star of about visual magnitude 6.5 in Perseus not far from α Persei (Mirfak).

My heart pounded!

A 6.5 magnitude nova on my first night of searching!

The atlas limiting magnitude was about 9.5 - much the same as my pictures and the new star was like a beacon on both of the duplicate photographs. Feverishly I measured the position. I telephoned Guy Hurst who was the national Nova Search Co-ordinator.

It was 3 am.

The telephone rang for a long time. It was picked up and a sleepy voice said something like

"Ghhuy Hsttt spleekin - heep glunge tchooo!"

I wondered if I had the right number – this didn't sound like Guy Hurst.

"Guy! Guy! Wonderful News! It's Geoff Kirby from Weymouth! I've found a 6.5 mag nova in Perseus! Please announce my fame immediately."

There was a sort of gurgling noise down the phone. I assumed it was a fault on the line.

"I'm suffering from the 'flu and I'm really bad but, hang on, I'll get my star charts out and check."

said a very ill and sleepy Guy.

After a quick transfer of co-ordinates he went silent for a minute or two.

"Atlas Borealis has a 6.5 star at the position you quote. No nova there. What atlas are you using?"

I started to get a bad feeling about my nova.

"Smithsonian Astrophysical"

This was an omission in my atlas.

And so began a long saga that kept me busy for years thereafter.

I had discovered that the Smithsonian Astrophysical Observatory Atlas had thousands of stars missing when compared with other atlases. Almost every night I found lots of 'novae' which were simply omissions from the atlases. I started to compile a catalogue of these omissions.

They ran into thousands and my database of atlas corrections became quite famous. When the American Association of Variable Star Observers Atlas of Variable Stars was being compiled my database of atlas errors was used in its production.

There were various reasons why atlases had stars missing.

The Smithsonian Astrophysical Observatory Atlas was compiled from the Smithsonian Astrophysical Observatory Catalogue.

This was a huge four-volume listing of the precise positions and proper motions of stars above about 9.5 visual magnitude.

The reasons that stars were omitted included the following

1. If a star had no measured proper motion it was left out of the catalogue and was not therefore in the atlas.

2. The atlas sheet masters were drawn by hand and there were human errors.

3. Some missing stars were variable stars. For example, on December 29th 1977 I found a 6.5 magnitude star on one of my nova patrol photographs which turned out to be the variable star W And. This had a visual range of 6 to 14. Because the middle of this range is below the limit of the S.A.O. Atlas it was not recorded.

4. The star catalogue upon which the atlas sheets were drawn was compiled from several other catalogues.

Not only were errors compounded in the copying process but swathes of the sky were missed out due to the catalogues having been compiled from different epochs such as Epoch 1900 and Epoch 1950 which were still both popular reference coordinates for catalogues and star charts in the 1970s.

For example, whilst I was compiling my list of star omissions in the SAO Atlas I found a disproportionately large number of missing stars around declination +50° as shown in my chart published in the BAA Journal [285] and reproduced below.

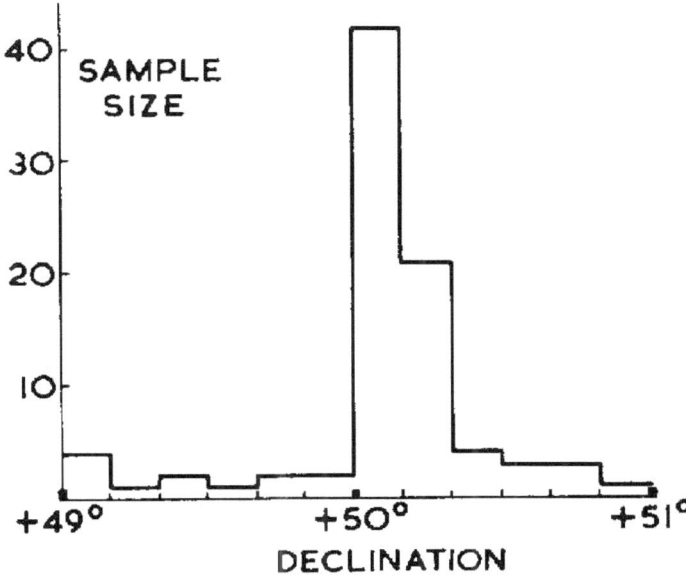

The excess of missing stars close to declination +50 degrees

The Outcome Of My Nova Hunting Project

I never found a nova and I eventually gave up searching by the start of 1978 with 543 pairs of photographs having been taken.

There were several problems that caused me to give up this highly labour and time intensive pursuit.

The greatest was the need to devote an entire night to photographing, developing, searching and checking films. Often I would have problems blacking out the bedroom where I checked the photographs because of the dawn glowing in the East. I was not at my best when going to work after having had no sleep the previous night. Astronomy is an anti-social pastime at the best (worst?) of times but giving up an entire night's sleep was really not acceptable.

The other was that my third child was born in 1978 and I no longer had the energy or inclination to give up so much time to astronomy.

I 'Discover' Nova Cygni 1975

I went out into the garden on the evening of 30th August 1975 to see if the sky was clear. I looked up at Cygnus and there was a bright star near Deneb.

I rushed indoors to my wife's astonishment gibbering

"Where the hell's my Norton's Atlas! There's a nova up there!"

I swept through the house, dressed in anorak, boots, scarf and red woollen hat, scattering books everywhere as I searched for my star atlas.

Having found my Norton's Star Atlas, I plotted an approximate position and made a rough estimate of brightness. I tried to calm down.

I telephoned Guy Hurst who was the Nova Discovery contact in the UK but his 'phone was engaged so I called his backup, Jim Muirden in Exeter.

"Hi Jim! I suppose you know about the nova in Cygnus?"

I tried to ask casually but with my voice tremoring.

"Oh yes. It was discovered by a Japanese amateur about twelve hours ago." replied Jim.

I suppose it was crazy to think that I would have been the first person to spot this nova out of all the observers in the world.

However, I experienced, just for a few minutes, the excitement of making a major discovery in the sky.

This nova reached approximately visual magnitude 1.5 and was a bright interloper in the familiar pattern of Cygnus.

A chart at [286] shows the estimates of the brightness of Nova Cygni (1975) - also known as V1500 Cyg. It can be seen that the rise to brilliance was extremely quick - probably within minutes rather than hours - and it faded relatively quickly.

I was lucky to see it.

A few nights of cloud and it would have faded beyond unaided eye visibility.

Nova Hunting Success Rate - Zero!

So, after all those years of searching I never found a nova.

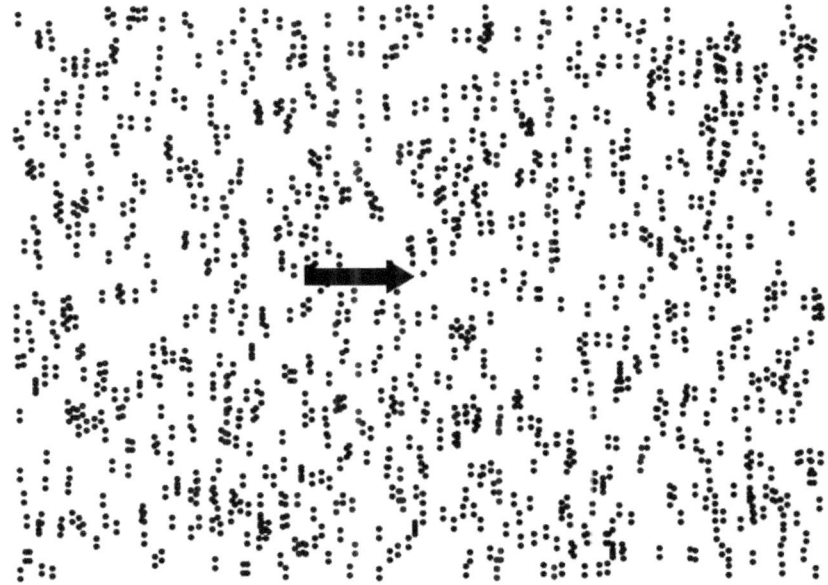

Answer to the challenge of finding the 'Lonesome Dot'.

14. Observing Variable Stars

Binoculars Or Telescope?

I started observing variable stars in the early 1970s. The appeal was the belief, possibly misplaced, that I was doing something useful for professional astronomers.

However, you must always pursue astronomy for the pleasure it gives; not out of a possibly misguided belief that professional astronomers are desperate for your observations.

Many variable stars can be observed with binoculars and even a few with the unaided eye. With a telescope the number of variable stars accessible to study becomes vast.

However, I observed far more variable stars with my binoculars than I observed through my telescope. I used the binoculars a lot and I could make many dozens of estimates of variable star brightnesses in one evening.

The advantage of observing binocular variables is that the stars are easy to find and observations take a short time resulting in many brightness estimates being obtained on every clear night.

I recorded many nights when I estimated the brightness of over eighty stars in one evening and I was still back indoors at a reasonable time.

Typically I would record over several thousand brightness estimates for over one hundred different stars each year.

In fact, the observation of binocular variable stars is one of the best projects for the amateur to follow. It is easy, convenient and may make a contribution to the research programmes of professional astronomers.

I urge you to take up the observing of variable stars using binoculars.

How I Observed Variable Stars

The idea is very simple.

The variable star is surrounded by other stars whose brightnesses are constant (hopefully!) and well known. The brightness of the variable star is then estimated using these comparison stars and recorded. This only takes a few seconds if you are well practised.

What all observed variable stars have in common is that a star chart is available with comparison stars marked that enable the observer to estimate the brightness of the variable star.

Variable star charts, along with guidance on observing variable stars, are readily available from the BAA Variable Star Section [287] and from the American Association of Variable Star Observers (AAVSO) [288].

I concentrated on variable stars on the BAA Variable Star programme lists. For these I was provided with charts for over one hundred variables.

There are also books showing how to observe variable stars visually such as [289] and [290].

Although I had some fairly large telescopes 97% of all my variable star observations were made using binoculars.

Some of the stars I observed showed significant changes but others appeared never to vary - such as P Cygni which I observed for well over ten years and which never varied by any detectable amount during that time.

In fact, P Cygni showed nova-like outbursts in the 1600s but, since 1715, it has hardly varied by an amount that can be detected by eye.

So, I hereby nominate P Cygni as the most boring 'variable' star I ever observed!

If a star does not vary this may seem like a boring result but it is actually very important; as important as if I had found a variation.

Professional astronomers need to know if a star has not varied just as much as knowing that it has varied so do not get downhearted if you make hundreds of observations of a star that stays resolutely the same brightness.

Whatever stars were observed, it was great fun and my results were entered into the BAA database and used by professional astronomers in their research.

There are three broad types of variable stars;

1. Those which vary intrinsically because their internal structure is unstable

2. Those whose variation is caused by two or more stars in orbit about a common centre of gravity such that the light of one or more components is eclipsed, and

3. Pairs of stars which do not show eclipses but the stars interact through processes such as by matter transfer.

Confusingly, a group of two or more stars engaged in eclipses is referred to as a single star - such as Algol the *'Demon Star'* of the mediaeval Arab observers - even though there are three stars in orbit in this system.

The intrinsically variable stars are subdivided into many types. Some have a fairly regular and repeatable variation of brightness with time - such as the Delta Cepheid stars.

Other stars are erratic and their future brightness is largely unpredictable.

This is not the place to go into a description of these different types of variable star because there are excellent descriptions available of variable star behaviour and observing techniques.

The AAVSO has, as I write this, over 30 million variable star measurements in its database [291] and the BAA has 2.5 million [292].

Intrinsically Variable Stars

These are stars that vary in luminosity due to physical and

thermal processes in the star's structure.

Some of the variable stars that I observed were more interesting than others.

For example, the charts below show my estimates of eight variable stars selected from my results for well over one hundred variable stars that I observed and recorded.

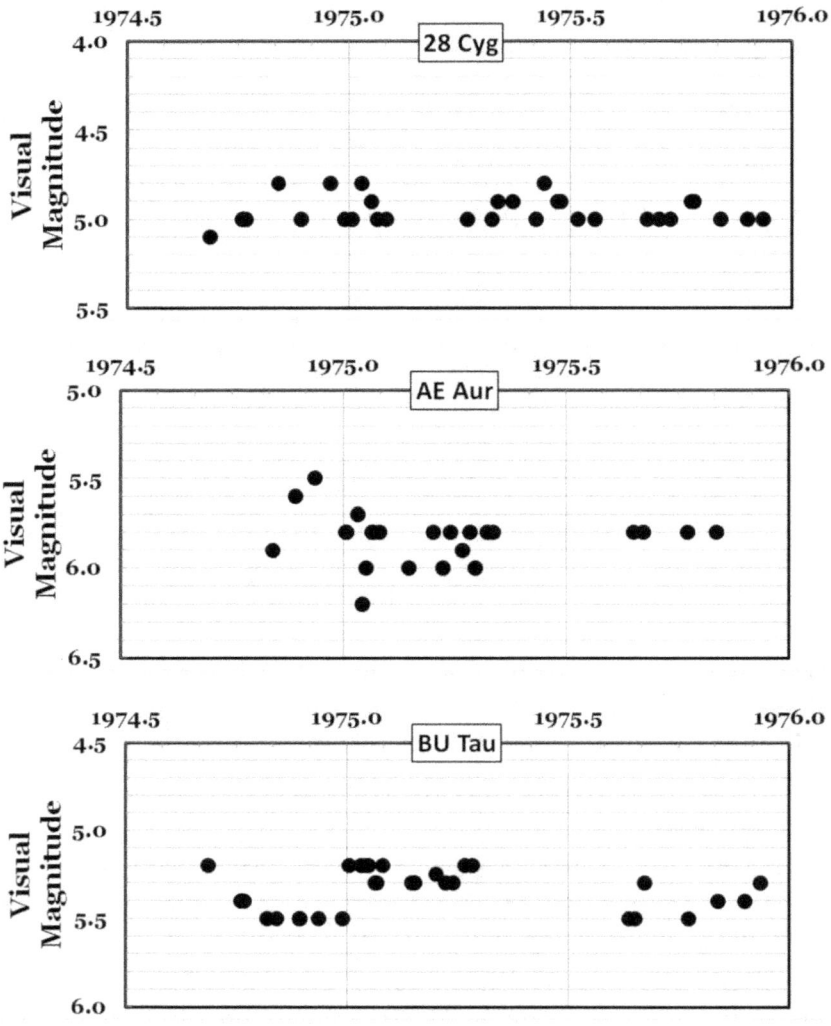

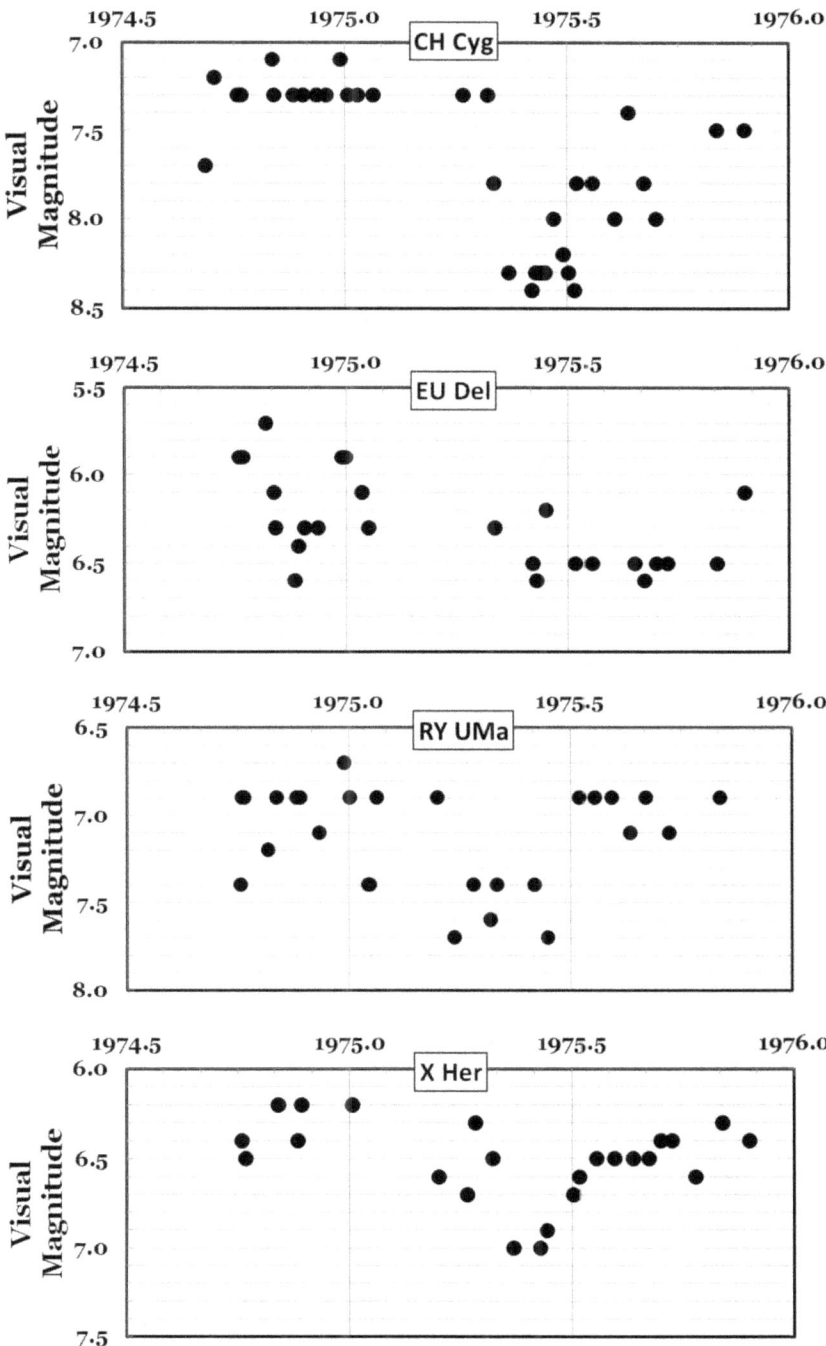

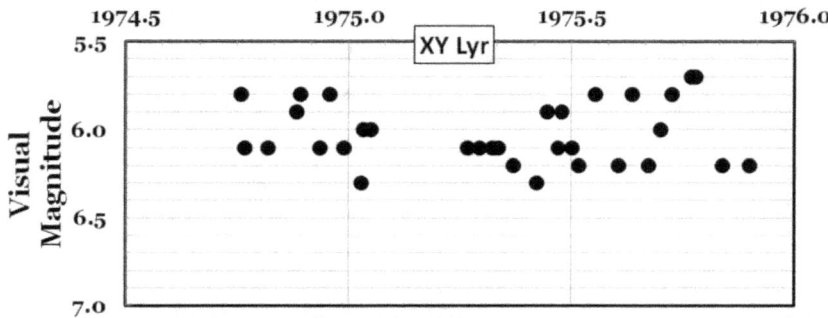

We see that some stars, such as 28 Cyg, did not appear to have varied by a significant amount over the period of observation.

In the charts above the stars CH Cyg, RY UMa and X Her genuinely seem to have varied because the changes I recorded are larger than the random error inherent in the visual observations.

Other stars, such as BU Tau and XY Lyr, may have been constant during the period I was observing them but there is a greater scatter in my measurements which confuses the analysis.

Why should the scatter be different?

One reason that has caused quite a lot of headaches for the analysists is that one or more of the comparison stars may actually be unrecognized variables.

I did quite a lot of work in the 1970s checking star charts and I found that some comparison stars on BAA charts were actually variable stars whilst others were constant brightness but were clearly not the visual magnitude stated.

There is a greater scatter of results for red stars due to an optical effect known as the Purkinje effect [293]. This causes red stars to appear to brighten as you stare at them.

That's why it's important to keep your eye moving back and forth between the variable and any comparison stars to prevent this effect kicking in.

Red stars can also look brighter than comparison stars depending on how they are oriented in the field of view.

If a red star is oriented above or below a white comparison star, the red star will sometimes appear brighter than if the two are oriented directly left or right of each other.

In addition light pollution, twilight, or moonlight, can also make red stars appear brighter.

These optical effects in the eye and brain are well known to experienced variable star observers and they do their best to avoid the Purkinje effect devaluing their estimates.

Eclipsing Binary Stars

The term 'Eclipsing Binary Star' appears to be universally used for star systems whose combined brightness varies due to eclipses even when there are several stars involved, as for Algol.

The BAA has published an excellent guide to observing eclipsing stars and this can be freely downloaded [294].

Observing eclipsing star systems can be exciting and useful.

For example, in June 1996 Mike Collins was checking through hundreds of Nova Patrol photographs that he had taken when he spotted that a star was invisible on four pictures. It was suggested that this might be an eclipsing variable and I was asked to check on this star.

I took on this task and observed the suspected star every possible night with my large Dobsonian telescope. I was able to confirm that it was indeed an eclipsing binary. The graph shows my results.

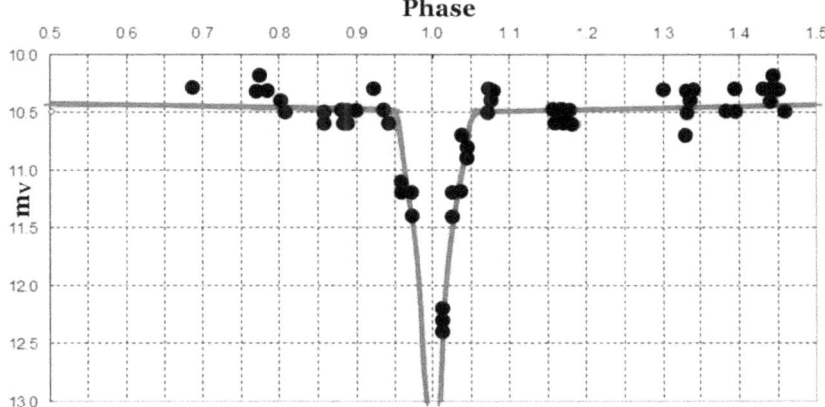

As you can see, something dramatic happens with this star and I was possibly the first human in the universe to *see* it happening visually.

In the above diagram the full curve is a mathematical model of the two stars' size, shape, spectral details, luminosity as well as their orbital characteristics derived by a professional astronomer at the Rutherford Appleton Laboratory.

Partly as a result of my observations and partly through surveys of old photographic plates and more recent CCD measurements the period was determined to be 10.352336 days.

This is a remarkable accuracy and is possible because the observations were made over a long period of time covering hundreds of rotations of the binary star system.

This star needs a fairly large telescope to observe.

I recommend that you start by observing two easier stars. These are Algol, an unaided eye eclipsing binary and RZ Cassiopeia which requires binoculars or a small telescope.

Observing Algol

Beta Persei - known as Algol - is perhaps the most famous eclipsing binary star system because of its recognition as variable in pre-telescopic times [295].

Indeed, it was announced in 2015 that the variability of Algol may have been recognized as long as three millennia ago by the Egyptians. An ancient Egyptian Calendar of Lucky and Unlucky Days composed some 3200 years ago is the oldest historical document referring to Algol as a regularly variable star.

In addition, the period of Algol's light changes was given as 2.85 days which compares with 2.867 days in modern times [296].

English astronomer John Goodricke was the first to suggest that the regular variations of Algol were either due to

a. a dimmer companion star in an orbit about the brighter star whose plane was close to the line of sight to the Earth, or

b. there was a single rotating star with a dark mark on its surface.

Goodricke's observations ruled out the second hypothesis because the dimming was over a short period of time relative to the star's rotation period.

Goodricke was deaf and dumb but he overcame these twin disabilities to become highly educated and industrious.

He presented his findings regarding Algol to the Royal Society in May 1783, and for this work, the Society awarded him the Copley

Medal.

He was elected a Fellow of the Royal Society on 16 April 1786.

However, he never learned of this honour because he died four days later from pneumonia.

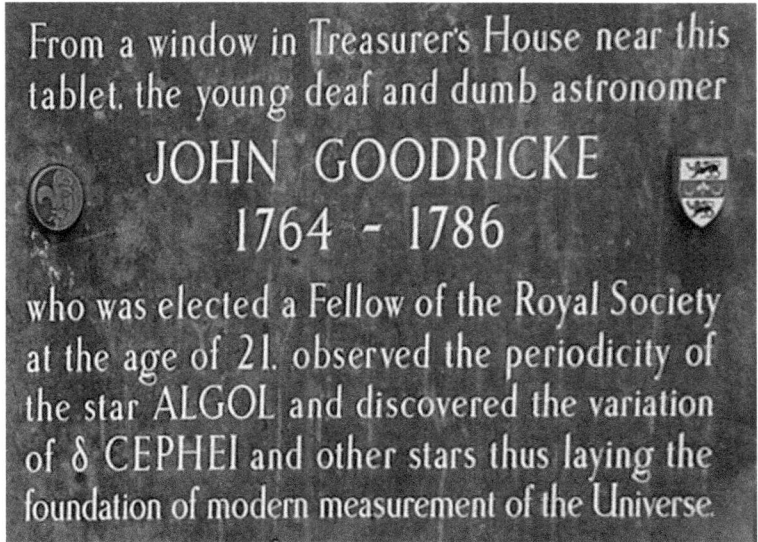

A memorial plaque to John Goodricke erected in York, England.

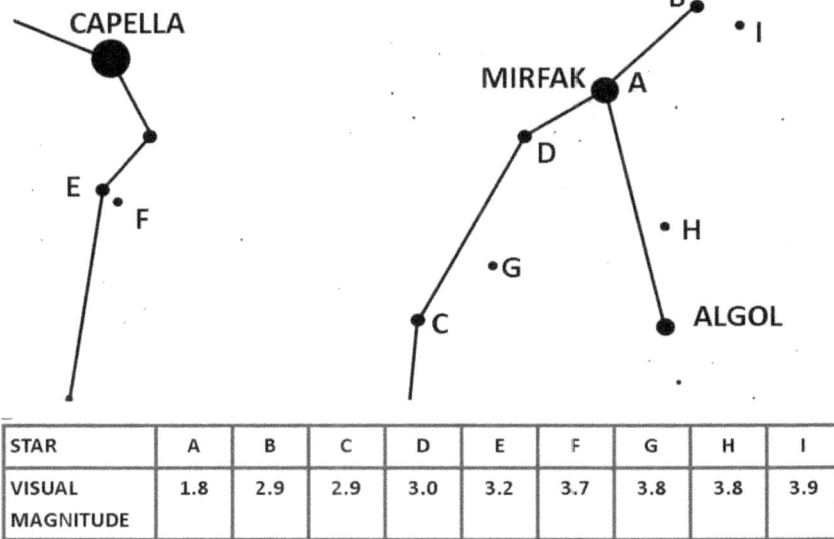

STAR	A	B	C	D	E	F	G	H	I
VISUAL MAGNITUDE	1.8	2.9	2.9	3.0	3.2	3.7	3.8	3.8	3.9

The above chart will help you to find Algol and estimate its brightness relative to nearby stars as it goes through its eclipses.

Capella is the brightest star in the constellation of Auriga and Mirfak is the brightest in Perseus.

The period between eclipses is currently about 2.867339 days [297] although this can vary very slightly due to interactions between the components in the multiple Algol star system.

The effect of these variations [298] is to make the predictions of mid-eclipse times be up to 4 hours early or late. So, you will need to start checking the brightness of Algol several hours before the predicted time of mid-eclipse.

Predictions for the times of mid-eclipse can be found in the monthly 'Astronomy Now' UK magazine. However, these predictions only are suitable for UK observers.

Predictions that can be used worldwide can be seen by scrolling down the webpage of the Krakow Observatory [299].

These predictions currently differ from those in 'Astronomy Now' by about 30 minutes. Also these predictions are only for a few weeks from the date of viewing the website page.

I have computed the predicted times of Algol's eclipse minima based on the Krakow elements [300] and these are tabulated on my website at [301].

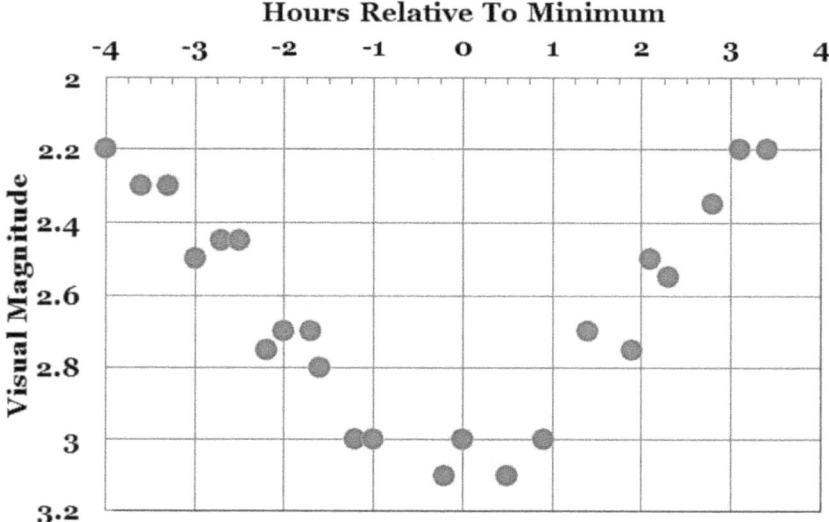

Above is one of my typical results from 2015 for unaided eye observations of Algol using the chart on the previous page.

It can be seen that the observation period needs to be about eight hours at most or four hours at least to adequately cover a single eclipse.

This means observing Algol in December, January or February when the star is high in the Northern Hemisphere skies around local midnight and the nights are long.

Observing RZ Cassiopeia

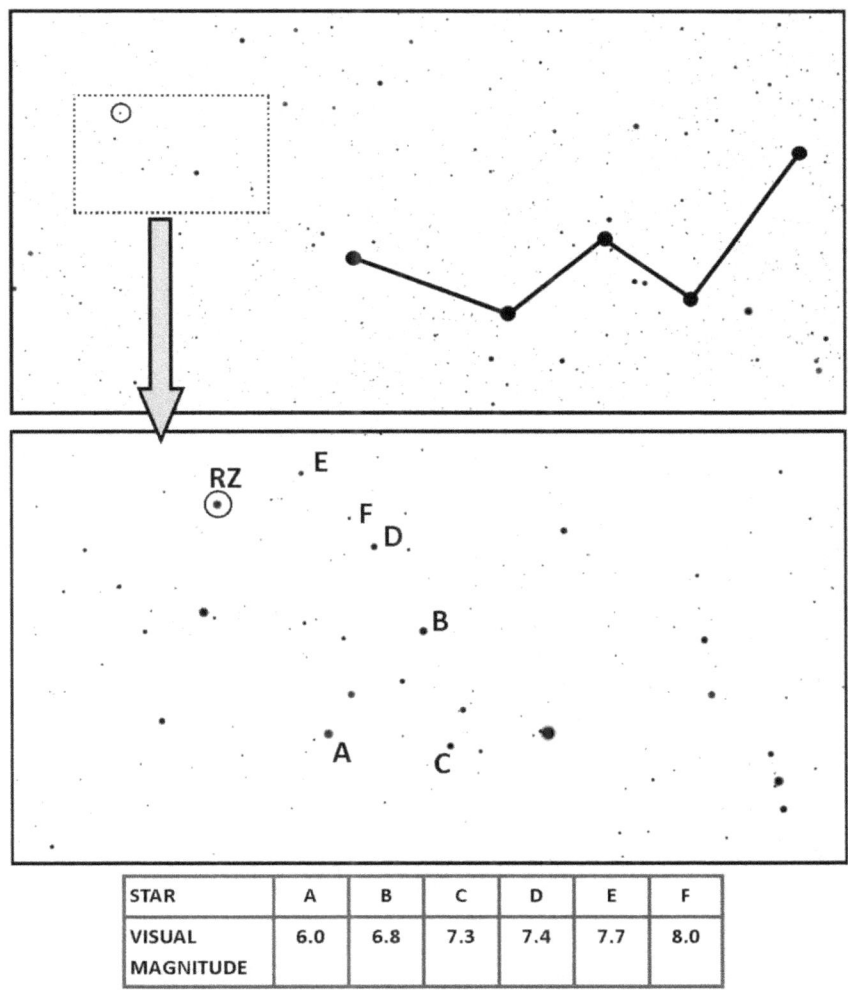

STAR	A	B	C	D	E	F
VISUAL MAGNITUDE	6.0	6.8	7.3	7.4	7.7	8.0

RZ Cassiopeia is easy to find as it is off one of the 'arms' of Queen Cassiopeia's chair as shown in the chart above.

It varies rapidly through each eclipse with a large brightness change - at least 1.6 magnitudes and sometimes more. The whole eclipse can be seen in one night which is not possible with many eclipsing binary stars.

An excellent short article on observing this variable star can be found at [302].

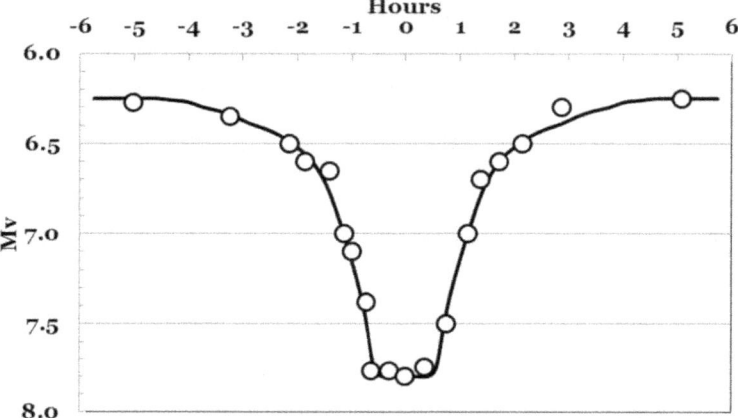

The main changes shown above occur within about five hours.

The above data show my very first night of observations of RZ Cassiopeia on 25th January 1975.

The full curve is a fit by eye to the variation of brightness with time.

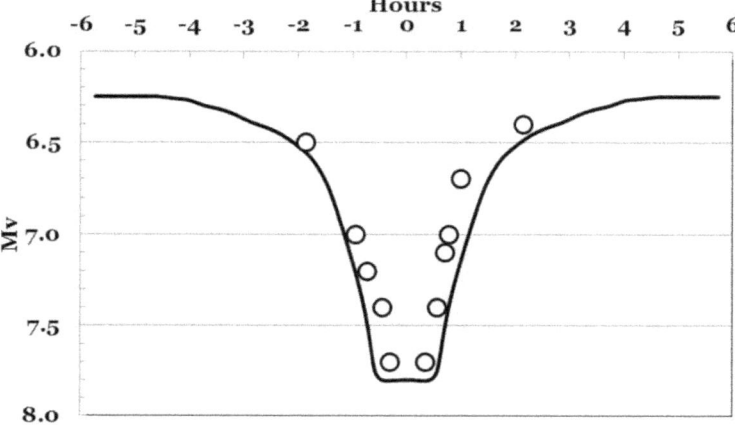

Above is another eclipse observed by me later in the 1970s.

The full line is the same curve fitted by eye to the earlier observations plotted above.

We see from the second example that the dip is narrower so that the star spends less time faint.

This is a real effect and is due to the two stars being so close that they are exchanging matter. This changes the combined brightness and the time between eclipses.

The eclipses occur, on average, at intervals of 1.19524700 days so the favourable dates for observing this binary star occur approximately every six days.

These effects can be seen by amateur astronomers using no more than a pair of binoculars.

As an inspiring example, amateur Wayne Lowder has made over 20,000 observations of RZ Cassiopeia's brightness and found complex changes in the behaviour of the pair of stars which were of great interest to professional astronomers [303].

The brightness of the binary when an eclipse is not underway varies between visual magnitudes 5.7 and 6.4 so that there is obviously something going on which is causing flares.

Sometimes there is a flat part to the curve when the two stars are in deep eclipse but, at other times, there is no flat part indicating that the star system is complex and changing over time.

There may be an accretion disc which could be contributing to the irregular changes in luminosity.

This star is not only easy to observe with binoculars or a small telescope but the characteristics of its light curve changes with time making it of particular interest.

Predictions for the times of mid-eclipse can be found in the monthly 'Astronomy Now' UK magazine. However, these predictions only are suitable for UK observers and differ slightly from the predictions I have generated.

Predictions that can be used worldwide can be seen by scrolling down the webpage of the Krakow Observatory [304]. However, these predictions are for only a short time into the future. The predictions on my website [305] are for twelve months.

I strongly recommend observing variable stars as a way of enjoying astronomy. Although you observe on your own you will be a welcomed member of an international fraternity.

15. Tracking Artificial Satellites

Sputnik I was launched in the month that I started university and within days of my eighteenth birthday in 1957.

It caused great excitement and panic. Suddenly the human race was probing out into space for the first time. The worlds of Dan Dare, the Treens and the Mekon were no longer beyond reach.

Humans had put a small ball in space that bleeped as it circled the Earth every ninety minutes. People stood in the streets to watch the spot of light passing over the sky.

Politicians went crazy - those 'Commie Bastards' were way ahead of the so-called 'Free World' and at any time nuclear bombs could rain down from space! The fears were widespread and the Cold War was approaching its height. By 1962 the West was close to war over the Cuban Missile Crisis and I recall listening to every radio news broadcast expecting hostilities to break out.

Sputnik started the space age and it also greatly accelerated Cold War fever.

With the threat of nuclear war escalating an even greater

disaster struck the entire civilised world threatening total annihilation. The Soviets put a dog into orbit! Laika was sent up with no means of being brought down again.

The outrage was unimaginable!

Dog owners besieged the Soviet Embassy. Newspapers screamed hysterical headlines.

Tracking Artificial Satellites

A small hard core of amateurs got involved in tracking artificial satellites from the earliest days.

I got involved in the late 1970s through the BAA.

The scheme operating in the late 1970s was for new observers to 'try their hand' at tracking a few bright objects whose orbits were well known. Their results would be checked against those from more experienced observers.

If they achieved a reliable and accurate performance they were accepted into the 'inner circle' of observers. I was soon accepted into this group and, for several years in the 1970s, I tracked satellites visually using binoculars whilst on a sun lounger.

The idea of observing satellites was - and still is - to fix their positions in space and time as seen from a given location on the Earth's surface.

The way in which I did this was to get a huge pile of predictions sent through the post from Slough. These arrived in a parcel every Saturday. They showed the predicted tracks of hundreds of satellites across the sky as seen from my back garden.

A small sample from one of the weekly documents is shown on the next page.

These predictions were obviously expensive to produce and post so only keen observers like me got them after serving their 'apprenticeship'.

I could update these predictions by telephoning Aston University which operated a recorded message service. I then plotted the predicted tracks on an old copy of Norton's Star Atlas from which the atlas sheets had been cut out and laminated in plastic. Armed with these marked-up charts, a stopwatch and a pair of binoculars I would wait in the garden huddled up on a sun lounger watching for the first satellites to appear.

```
SATELLITE 8111901    Orbit Age 202d   DATE 1987  19  LST at 19d 24h (UT) 11 38.2
SAT.LONG AT EPOCH 55   VIS.INT  0.0   NEXT TRANSIT   0.0 DAYS AFTER EPOCH
SATELLITE 8304701    Orbit Age -6d    DATE 1987  19  LST at 19d 24h (UT) 11 38.2

18  O O V  12   4  24.4  -7.39   0.0  0.0   17.7  30.1  200.5  35790  -0.01  18.52
19  O O V  12   5  24.6  -7.34   0.0  0.0   17.7  30.1  200.5  35792  -0.01  18.52
20  O O V  12   6  24.7  -7.29   0.0  0.0   17.8  30.1  200.5  35794  -0.01  18.52
21  O O V  12   7  24.8  -7.23   0.0  0.0   17.8  30.1  200.5  35795  -0.01  18.53
22  O O V  12   8  25.0  -7.19   0.0  0.0   17.8  30.1  200.5  35796  -0.01  18.54
23  O O V  12   9  25.1  -7.15   0.0  0.0   17.8  30.1  200.5  35796  -0.01  18.54
 0  O O V  12  10  25.2  -7.12   0.0  0.0   17.8  30.1  200.5  35795  -0.01  18.55
 1  O O S  12  11  25.3  -7.10   0.0  0.0   17.8  30.1  200.5  35793  -0.00  18.56

SATELLITE 8402301    Orbit Age       DATE 1987  19  LST at 19d 24h (UT) 11 38.2
SAT.LONG AT EPOCH 182  VIS.INT   0.0  NEXT TRANSIT   0.0 DAYS AFTER EPOCH
SATELLITE 8502501    Orbit Age -5d    DATE 1987  19  LST at 19d 24h (UT) 11 38.2

18  O O V  11   3  58.7  -7.36   0.0  0.0   24.2  28.5  207.5  35778   0.02  24.36
19  O O V  11   4  58.9  -7.32   0.0  0.0   24.2  28.5  207.5  35783   0.01  24.36
20  O O V  11   5  59.1  -7.27   0.0  0.0   24.2  28.5  207.5  35788  -0.00  24.35
21  O O V  11   6  59.2  -7.23   0.0  0.0   24.2  28.4  207.5  35793  -0.01  24.36
22  O O V  11   7  59.4  -7.19   0.0  0.0   24.2  28.4  207.5  35798  -0.02  24.36
23  O O V  11   8  59.5  -7.15   0.0  0.0   24.2  28.4  207.5  35802  -0.03  24.37
 0  O O V  11   9  59.5  -7.13   0.0  0.0   24.2  28.4  207.5  35805  -0.03  24.39
 1  O O V  11  10  59.6  -7.11   0.0  0.0   24.2  28.4  207.5  35806  -0.04  24.40
```

It was certainly never boring because satellites can only be seen when illuminated by the Sun and in a dark sky. This means that they can only be seen for a certain period after twilight has ended. (High satellites are visible all night but they were usually too faint or of no scientific interest.)

Thus, I was never expected to stay up more than about two hours after the sky was dark and I was then free to get on with other topics. In fact, the longest period of visibility is in summer when there is not much else to do astronomically.

My success rate for spotting satellites was low - only about 10% were picked up by me and yielded a good positional fix. This was because the satellite positions might be well in error compared with the computer predictions. This occurred because the satellites of interest were in low orbits and affected by air drag.

This drag is affected by stratospheric 'weather' and the solar sunspot activity - indeed, the appearance of a large sunspot group would cause low satellites to plummet out of orbit like flies in winter.

The reason I observed low satellites was mainly to find out how they are affected by air drag and how that is affected by sunspots.

The purpose of amateur observations was to get a visual fix in space and time from a known spot on the Earth's surface. This was done by following a satellite across the sky using binoculars until it passed close to a star or, better still, between two close stars. The position of the satellite was estimated relative to the star background and a time obtained using a stopwatch.

The satellite would be followed in my binoculars until it passed close to a pair of stars. Hopefully it would pass between the two stars as shown in the diagram below.

On other occasions it would pass outside the pair as also shown below.

In each case the ratio AC/AB would be estimated as accurately as possible.

The pair of stars would be identified and the time on my stopwatch would be noted.

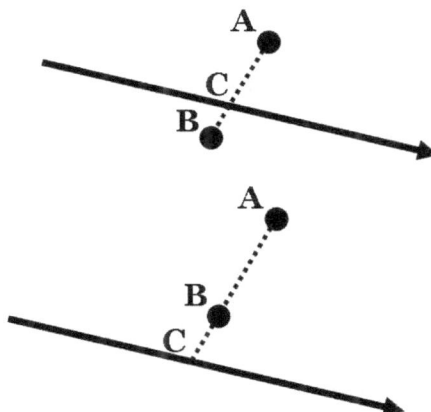

I made several hundreds of successful observations in this manner and my results were used to correct many satellite orbits.

To be useful, the observations had to be accurate to about 100 arcseconds in space and about 0.2 seconds in time. This was not too difficult with practice using binoculars.

Some satellites deviated from their predicted positions by several degrees or as much as a minute in time so visual observations were very useful as they could greatly improve the knowledge of the satellites' orbits.

Professional equipment was also available to make measurements.

However, radar systems were not then as accurate as amateur visual observations. Radar at that time could only fix a satellite's position in range to high accuracy. The radar position across the line of sight was only accurate to 10 arc minutes or so.

The Hewitt Satellite Tracking Cameras

One alternative method of satellite tracking capable of high accuracy - better than my back garden results - was the Hewitt camera. There is a picture of one of these large and very accurate cameras when it was operated by the University of Aston [306]. It was donated by the Ordnance Survey for use in satellite tracking.

I visited this camera in 1980 whilst at a conference, see page 365.

Only about half a dozen of these cameras were ever operational and they could not keep up the rate of observation achievable by an enthusiastic and large band of amateurs.

The history of the camera at the Greenwich Observatory can be read at [307].

In a one-second exposure, the cameras could photograph satellites down to about magnitude 13 against the star background. Satellite positions were derived to an accuracy of about 1 arcsecond and timing to within a few milliseconds. This positional accuracy corresponds to about ten metres in the satellite position in space.

This meant that the camera positions had to be known to an accuracy of one metre or better on the Earth's surface. In fact, very accurate surveying of the cameras fixed the latitude and longitude of the centre of the camera optics to within a centimetre or so. The results produced by the Hewitt Cameras enabled satellite orbits to be computed very accurately and the shape of the Earth to be deduced to within a few centimetres.

These results eventually allowed the Global Positioning System to be created.

In terms of numbers of satellite 'fixes' achieved human amateurs far exceeded the output of professional cameras and radar systems combined.

After the Hewitt cameras were decommissioned the archive of photographs was put into the safe keeping of the Crayford Manor House Astronomical Society [308].

My Study of Artificial Satellite Tracking Accuracy

I did some research into the accuracy of visual observations which I wrote specifically for the BAA Journal in its style and appropriate, useful content - or so I thought.

It took me months to write but eventually it was done.

I proudly sent it off to Burlington House.

Many months later the manuscript came back, mangled and dirty with a pencilled, barely legible scribbled note saying that the BAA would not be publishing it.

I was very upset!

It is, of course, every journal editor's prerogative to select what to print and I accept this. However, no publisher usually rejects a contribution without some explanation.

I sent the manuscript to Dr Desmond King-Hele who was then the leading UK expert on artificial satellites and their orbits. He wrote back saying that it was excellent and a very important contribution to satellite observing.

He arranged for it to be published in the Quarterly Journal of the Royal Astronomical Society - a far more prestigious and professional outlet than the BAA Journal.

My paper is reproduced at the reference [309].

In his excellent book *'Observing Artificial Satellites'* published in 1983, Dr Desmond King-Hele devoted eight pages to discussing my work and wrote of my paper [310]

"In discussing this subject, I am much indebted to an excellent paper by Geoffrey Kirby published in the Quarterly Journal Of The Royal Astronomical Society."

Many thanks Desmond!

Over 20,000 satellite positional fixes were being obtained every year in the 1970s with Russell Eberst soon accumulating over 100,000 positional fixes on artificial satellites and becoming by far the most prolific observer in the world.

The excellent book *'A Tapestry of Orbits'* by D. King-Hele [311] provides an entertaining and informative account of the activities of amateur artificial satellite observers and is well worth reading.

The Birmingham Meeting 1980

I got to know several of the personalities in the world of artificial satellites. My practical interest was in observing and reporting positions. However I also got interested in the mathematical aspects to the extent that I started computing satellite tracks across the sky seen from my house, given the mathematical elements of the orbits.

Because I was getting well known in the satellite spotting world, I was invited to attend - and give a lecture to - a meeting of UK artificial satellite scientists in Birmingham in September 1980.

For my lecture I spoke about my paper published in the Quarterly Journal of the Royal Astronomical Society.

It was during this weekend conference that I visited the local

Hewitt satellite tracking camera.

Dr. Desmond King-Hele

A personality at this meeting was Dr Desmond King-Hele who was a leading light in the satellite tracking world.

Desmond had written a biography of Erasmus Darwin (Charles Darwin's grandfather) who was a great polymath but mainly a brilliant scientist subsequently overshadowed by his equally brilliant - and controversial - grandson.

Desmond had also written an appreciation of Byron and had published two books of poems [312].

Desmond was a brilliant mathematician and juggled Legendre Functions like the rest of us juggle simple addition. I met and corresponded with him many times.

Dr. Pierre Neirinck

Pierre Neirinck was an interesting and amusing fellow.

In the early days of satellite observing, the prediction service was run from Slough. The person there who kept observers in touch and issued predictions was Pierre. Obviously French and with a thick Gallic accent, he was great fun. He issued newsletters with updates on what was happening in the world of satellites.

The US equivalent was a massive organisation called NORAD which consisted of a vast number of scientists and military personnel buried deep inside the granite Cheyenne Mountains.

It collects observations of everything in space and tracks it so that attacking Soviet missiles could be identified. It tracks every bit of crap in space - literally. They are still - reputably - tracking a bag of faeces ejected by a manned spacecraft in the 1970s!

Anything over the size of a glove is claimed to be catalogued and tracked.

Well, Pierre was rather scornful of this megabucks operation. He tried to do the same for Britain and did it very well.

He once mockingly wrote that NORAD had reported a certain space probe had re-entered the Earth's atmosphere and burnt up. In fact this was a Mariner probe sent to Venus! Pierre reported solemnly that the Venusians had captured the probe, refuelled it and shot it back to Earth. This was incontrovertible proof of life on Venus!

On another occasion he reported the re-entry of a space probe over France. A large lump of metal had whizzed over the heads of some farm workers and buried itself in a field. They described it as *'comme une marmite"* according to Pierre.

In a later newsletter he apologised. He was unaware that in Britain Marmite was a savoury spread. In France it is a large iron cooking pot!

Pierre died on 3rd January 2016 at the age of 89. There is an obituary which is well worth reading at [313].

The Slough service eventually faded out and, after a period of remarkable lack of interest by the Government, the prediction and collation of observations were passed to Aston University.

Visual observations of artificial satellites still continues but, since about 1982, without the benefit of my observations.

SKYLAB Crashes To Earth

An exciting example of the joys of tracking artificial satellites was the crash to Earth of SKYLAB in 1979.

As its orbit got lower it became more affected by the thin air it was pushing through. It became impossible to predict how it would behave from day to day unless the air drag was known.

Every day, NORAD computed predictions for the next day based on different assumptions of air drag. These were telephoned to active observers like me.

I would then rush out and watch for it to pass over. It was brilliant in the last few days - about as bright as Jupiter - so it was easy to see. As soon as a 'fix' had been obtained I telephoned this to the experts who relayed the information to NORAD deep in the Cheyenne Mountains. They then produced new predictions.

Meanwhile, NASA was trying to predict from all these observations where SKYLAB would crash. It was very important not to kill anyone. They manoeuvred it round so that the solar panels were at different angles to the direction of flight. This varied the drag and controlled - to some extent - its descent.

Eventually I was getting messages that it was due to crash at a certain time. I went out after that time - and it was still up there! I rushed to the telephone and reported my observation.

One night, several days after NASA and NORAD had predicted

its demise, the television news reported that it had come down in the Indian Ocean scattering fragments over the sea and Australia. A two tonne lead safe was reported to have been found in the desert and many more pieces were later recovered [314].

It was also subsequently reported that a large steel ring had hit the Australian desert and bounced for fifty kms before rolling another thirty kms. That sounds implausible but might possibly be true.

So, why were the observations of SKYLAB by my fellow amateur observers and me so important? The satellite was so low that it could only be seen from a small area of the Earth's surface at any time and radars were usually unable to contact it. Even then, visual observers were usually more accurate than the 1970s radars.

The advantage of the visual observers was that there would be many observations giving an overall high accuracy compared with a single radar observation of less accuracy.

Also, amateurs cost nothing!

Tracking Geosynchronous Satellites

Once I had my large Dobsonian telescope (see page 204) inside an observatory I tried to pick up geosynchronous satellites - often and incorrectly called geo-stationary. They are certainly not stationary but they are synchronous to the Earth's rotation.

Because my large homemade Dobsonian was inherently on an altaz (altitude and azimuth movement) mount I had to work out the coordinates of the track across the sky within which the satellite would be visible.

In the course of these calculations I discovered that geosynchronous satellites appear in a very narrow band which follows a line of constant declination.

This declination is a function of the observer's latitude as shown in the chart below.

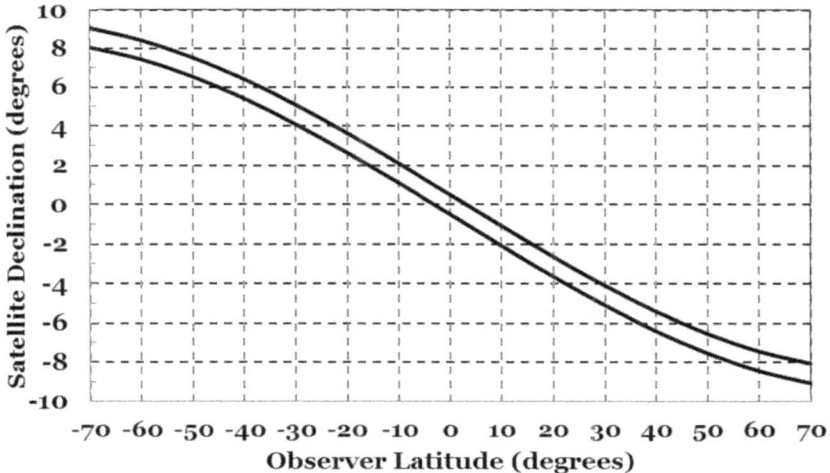

This feature is a boon to observers using an equatorially mounted large telescope.

Simply look up the range of declinations within which the satellites will be found and very slowly track along that narrow band of sky.

Remember that the geosynchronous satellite will be stationary in the sky whilst the stars drift from East to West.

A scheme for finding these satellites is to set the telescope on a star within the appropriate band for your latitude, as shown in the chart above, and set the telescope to motor drive or be hand driven to follow that star.

Scan the telescope a degree up and down either side of the star and watch out for a speck of light drifting through the star field.

That's a geosynchronous satellite!

Geosynchronous satellites all appear from my home town of Weymouth, UK to be within ± 0.25° of -7.2° declination for all hour angles. Because my Dobsonian telescope was not equatorially mounted I had to work out the Altitude and Azimuth of the band within which the satellites should be found.

I scanned slowly along that line of declination trying to pick one up.

There in the centre of my field of view was a tiny point of light, stationary in the field of view of my stationary telescope as the stars drifted by. The visual magnitude was about 13 - not far off the limit for my telescope from my urban back garden.

The position of this satellite was worked out, from the setting circles on the telescope, to be AZ = 193.9°; Alt = 31.5°. I reckoned that this was the satellite designated 1980-49A which was a communications satellite.

The satellite started fading at 00h 21m 20s on the 20th March and was invisible after about 20 seconds. These eclipses by the Earth's shadow only occur around the time of the equinoxes.

I picked up the satellite again the next day and timed twenty-seven very close passes to stars as they drifted past the image.

One close pass was with a star previously passed two days previously.

This was 7m 52.19s earlier after two days.

This was very close to the time expected based on a sidereal rotation period for the Earth of 23h 56m 04.09s.

I soon found a couple more geosynchronous satellites.

I decided to concentrate on observing these satellites and measure, as accurately as possible, their positions in space.

My scheme was to set up an illuminated graticule in the eyepiece (taken from a cheap microscope kit) which had North-South markings at about 10 arc second intervals. I then set the graticule on the satellite and noted the times and North-South relative values of the declination of stars as they drifted by.

One night I got over fifty measurements on stars drifting past one satellite.

From this, I was able to work out to about 5 arcseconds accuracy the position of the satellite. The accurate star positions were taken from the Smithsonian Astrophysical Observatory catalogue. Today I would use the HIPPARCOS catalogue data provided, for example, by the Stellarium planetarium software [315].

On April 20th 1987 I picked up another geosynchronous satellite. I made twenty-five timings of this satellite as stars drifted past.

On the following evening I picked up a different geosynchronous satellite at about Azimuth 207° and Altitude 28.3°. I made fifty-nine timings of this satellite against stars. Very accurate positions could be worked out.

Some geosynchronous satellites move in declination because they are not in an orbit which is perfectly aligned with the Earth's equator. Other satellites are moved around in order to optimise their performance. Sometimes I would see a satellite for days in the same spot in the sky and then - it would be gone, moved to a different location.

The random errors in my measurements corresponded to a positional accuracy of about 200 metres in space which, at that time, was more accurate than the professionals could track with radar.

This meant that my observations were of genuine value to the professionals responsible for maintaining and operating these satellites.

Iridium Flares

A popular and easily accessible way of observing artificial satellites with the unaided eye is to watch out for Iridium flares.

The 'constellation' of Iridium satellites started to be launched in the 1990s and, at the time of writing, there are 66 active satellites and six spares.

Originally there were intended to be 77 satellites and the system was named after the chemical element Iridium which has an atomic number of 77.

On 10th February 2009 Iridium 33 collided with the defunct Russian satellite Kosmos 2251. This was the first time two intact satellites had collided. Iridium 33 was in active service when the accident took place but was one of the oldest satellites in the constellation, having been launched in 1997.

The satellites collided at roughly 35,000 km/h (22,000 miles per hour).

This event of two intact satellites colliding was considered to have an extremely low probability and shook the satellite operating authorities whilst greatly raising the profile of the problem of 'space junk' and its threat to space craft operations.

The reason why the Iridium satellites are of interest to unaided eye observers is because the solar panels reflect a narrow beam of sunlight back to Earth and this results in a slow 'flash' which can reach magnitude -8. This is about forty times brighter than

Venus when the planet is at its brightest.

To see predictions of Iridium flares from your location go to the excellent 'Heavens Above' website [316].

The apparent brightness of a flare can vary considerably over a short distance on the Earth's surface.

As the satellite approaches it casts a beam of sunlight along the Earth's surface.

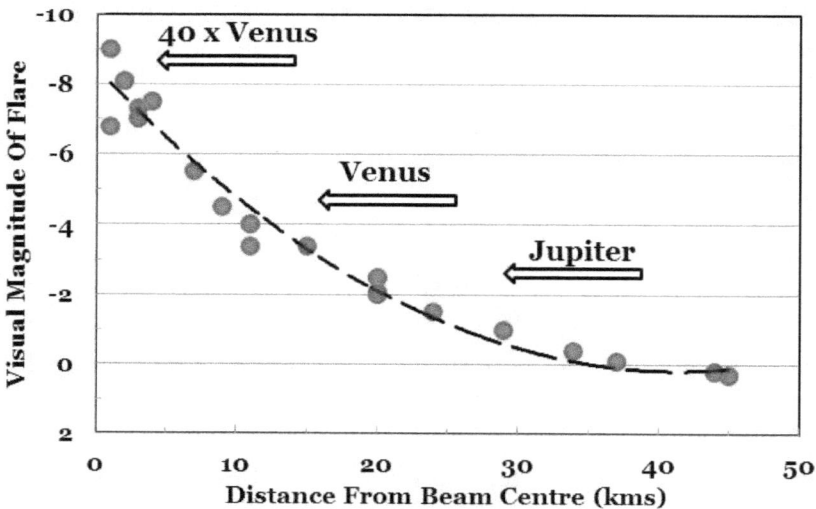

The above chart shows the variation of peak brightness achieved during a flare event as a function of distance from the centreline of the beam. I produced these figures using predictions from the 'Heavens Above' website for my home location.

It can be seen that an Iridium flare can be forty times brighter than Venus at its brightest if the observer is within a about three kilometres of the centre of the beam track.

Jupiter at its brightest is 400 times fainter than an Iridium flare at its brightest.

So, when obtaining predictions for these events try to put your location (latitude and longitude) into the website to at least an accuracy of one kilometre.

Observing a flare is really quite exciting and is a great opportunity to introduce those who scarcely ever look up into the sky to take an interest in the night sky.

As you watch the sky in the direction indicated by the 'Heavens Above' website you will eventually see a faint moving light. As you watch it gradually brightens, slowly at first, then the pace of brightening increases rapidly.

For about a second or two the flash occurs and it never fails to impress and excite me.

If this is a -8 magnitude flash then I guarantee that novices to this 'sport' of spotting Iridium flares will gasp with amazement and then watch the satellite fade to invisibility.

A flare of magnitude -8 can be seen during broad daylight.

Venus can be easily seen when at about magnitude -4 so a daylight Iridium flare should be easily visible although I have not yet seen one.

International Space Station (ISS)

As I write this the 'first' British Astronaut Commander Tim Peake is on-board and there is great enthusiasm for observing the ISS passing overhead.

At its brightest it can achieve a visual magnitude of -6 which is significantly brighter than Venus at its peak brightness.

Predictions for the ISS are given in the 'Heavens Above' website [317].

Let's not overlook the other seven British born astronauts who have been largely written out of history by the euphoria of Tim Peake's mission.

They are

Helen Sharman: The scientist from Sheffield who became the first Briton in space in 1991.

Michael Foale: The British-born astronaut who spent 375 days in orbit between 1992 and 2004

Mark Shuttleworth: South African/British entrepreneur became the second ever space tourist in 2002.

Piers Sellers: He completed three space missions, 2002-2010.

Nicholas Patrick: He spent over 300 hours in space, 2006-2010.

Richard Garriott: A self-funded space tourist in 2008 who went to the ISS.

Gregory Johnson: Piloted two spaceflights in 2008 and 2011.

ISS Passengers

At the time of writing there have been 221 individuals who have visited the International Space Station with Commander Tim Peake being the 221st to arrive.

Of these 33 were women (14%).

Those individuals, including seven self-funded space tourists, have performed 376 missions.

Yuri Malenchenko has made five spaceflights to the ISS, while five people made four, 28 people made three and 80 people made two.

The ISS occupants have come from 18 countries.

None of the 141 United States travellers to the ISS have been of African-American origin.

16. Observing The Sun

Observing Sunspots

A good reason for observing the Sun is that it can be done in the daytime and this is more sociable than observing late at night.

Of course, this may be a disadvantage for those in full time employment.

However, the late and much missed Bill Smith, the only person awarded Life Membership of the Wessex Astronomical Society in Dorset UK, had kept a careful watch on the Sun for over thirty years. He took a small telescope to work and he observed by projecting the Sun's image through his open office window.

Such a long record of observations by one person using the same telescope throughout is particularly valuable because the count of sunspots varies with telescope aperture and the observer. Here we have a long-term self-consistent series of observations.

The principle for making simple observations is very easy.

Pam Spence, then Editor of 'Astronomy Now' magazine, gave a fascinating talk on this subject at the Wessex Astronomical Society in 1994.

She used a simple undriven telescope and projected the image onto a plain sheet of paper inside a cardboard box.

Her talk inspired me into starting to make solar observations.

With the Sun under continuous observation by large numbers of observatories and space telescopes there may seem to be no reason for amateurs to observe the Sun.

You must remember that astronomy is a hobby that you should do primarily for enjoyment. If your observations happen to be useful to a professional astronomer then that's a bonus.

Above all, enjoy the heavens!

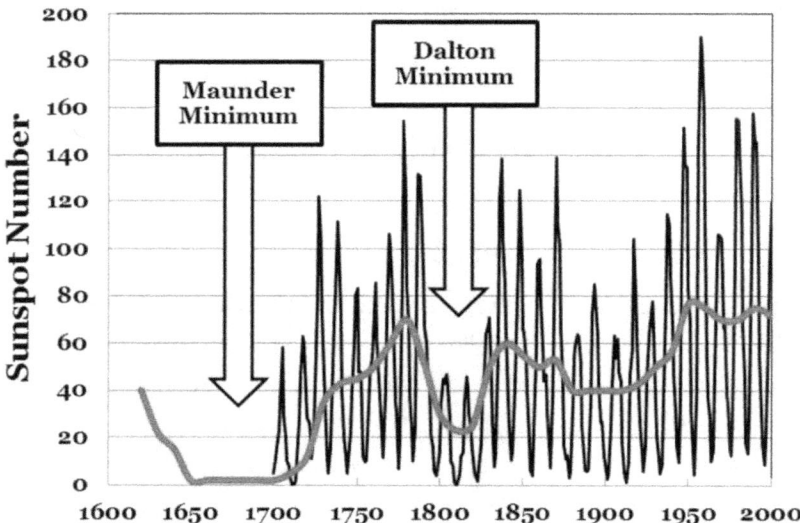

The annual averaged number of sunspots shows a periodic variation of about 11 years - more or less. This is shown by the chart above.

What you need to remember is that the overwhelming bulk of the observations used to compile this chart were collected by an unknowable number of amateur astronomers using simple telescopes.

In recent decades there has a remarkable reduction in the number of sunspots.

The sunspot maximum of 1990 peaked with a sunspot mean count of 220. The following sunspot maximum in about 2001 peaked at 165 mean counts and the 2014 peak only produced an average of 100 counts.

An extrapolation of these results suggests that the solar minimum due around 2030 could be followed a period of very few, if any, sunspots; a new 'Maunder Minimum' might then be underway with possibly serious consequences on our climate.

Of course this type of extrapolation is naive and only time will tell whether a new cold period for the Earth is coming for the middle of the 21st millennium.

Sunspots And Climate Change

The cycle of sunspots appears to have a measurable effect on the Earth's climate although some researchers believe this is probably coincidental because of the lack of a theoretical mechanism whereby sunspots could significantly affect the climate.

During the years of the Maunder Minimum (broadly the second half of the 17th century) Europe suffered a period of very severe winters and cool summers. The Thames froze over and starvation stalked Europe [318].

Other sunspot minima have been identified through measurements of the carbon isotope C^{14}. One of these was the Spörer Minimum between 1460 and 1565 [319].

Like the later Maunder Minimum, the Spörer Minimum coincided with a time when Earth's climate was colder than average.

One tentative theory is the modification of the Arctic Ocean Oscillation and North Atlantic Oscillation due to a change in solar output [320].

Another theory is that cosmic ray intensities arriving in our atmosphere are modulated by sunspots and the associated magnetic field variations of the Sun.

These cosmic rays may seed clouds whose presence is known to affect the Earth's overall climate through albedo changes [321].

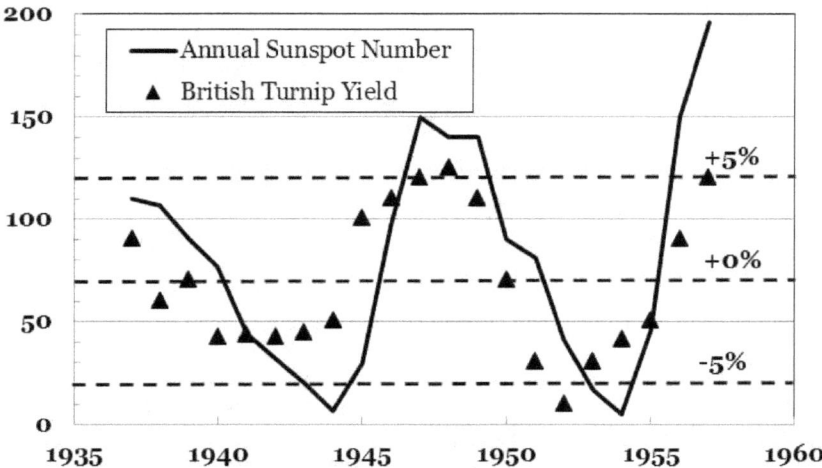

Many correlations have been produced purporting to show a relationship between sunspot numbers and climate.

Above I have plotted just one apparent correlation that has appeared in the literature. This suggests that the yield of turnips in Britain varies up and down more or less in line with the changes in number of sunspots.

There is an oddity in this chart. Can you work out what it is?

The answer is at the end of this chapter.

There are similar correlations between sunspot numbers and potato yields [322], the average temperature in central England and Canada [323] and even the frequency of lightning strikes in the UK [324].

Thunderstorms are 30% more frequent in Toronto at sunspot minimum compared with sunspot maximum [325].

Forecasting the Earth's climate over the next critical century needs climate models that are reliable. That reliability can only be assessed by looking back and forecasting past climate behaviour.

Explaining the effects of sunspots on past climate is vital to the way we prepare for and mitigate the effects of climate change on our grandchildren's lives.

Amateur Sun observers were the gatherers of this essential data on sunspot frequency for nearly four hundred years.

My First Solar Observations

When I built my first observatory in the early 1970s I had an equatorially mounted Newtonian reflector in it. I mounted a refractor piggy-back onto the reflector. This enabled me to track the Sun and keep the projected image steady on the screen.

Initially I just counted sunspots.

This involved counting the number of separate umbrae (the inner, black area of sunspots) and the number of groups. Umbrae within about 10 degrees of each other on the Sun's surface in solar latitude and longitude counted as a single group.

The number of umbrae seen depends on the quality and aperture of the telescope as well as the visual acuity of the observer.

This is why results vary from observer to observer and telescope to telescope.

My observations of the numbers of umbrae and number of groups were reported to the BAA. These figures were corrected by a factor which attempts to take into account the effects of telescope type, aperture and individual observers' biases.

The final figure was called the Mean Daily Figure (MDF).

My Solar Telescope In The 1990s

The arrangement I used later in the 1990s is shown above and is a 'bit of a lash up'.

Indeed, Heath Robinson [326] would have been proud of me!

The achromatic lens was mounted in a rigid wooden square tube which includes a projection screen. In the above photograph, the Sun's image can be seen on the screen in the viewing box.

A measuring grid was fixed on a wooden disc which could be rotated. This meant that the grid could be accurately aligned with the apparent drift of the Sun's image. With this arrangement I was able to make very accurate measurements of sunspot position.

These positions were referred to as heliographic (solar) latitude and longitude. These are equivalent to the latitude and longitude measured on the Earth. However, because the Sun does not have a solid surface, it is not possible to assign a reference zero longitude on the Sun's surface in the same way that Greenwich marked the pre-GPS zero longitude on Earth [327].

The longitude on the Sun was originally defined as the centre of the Sun facing the Earth at an arbitrary instant late on November 9, 1853 by the great solar observer Richard Carrington and a rotation rate of 25.38 days relative to the distant stars was - and still is - assumed.

In fact, the rotation period of the Sun varies with latitude so the assignment of a longitude and period of rotation is somewhat arbitrary. At the equator the rotation period is about 25 days whilst at a latitude of 30° north or south of the Sun's equator the rotation rate is about 26.3 days.

Eccentricity Of The Earth's Orbit

One spin-off from my measurements was a record of changes in the apparent diameter of the Sun. This came about because, for every observation, I measured the apparent north-south diameter of the Sun on the observing screen.

This showed that the Earth is indeed closest to the Sun in early January, furthest in early July and I also got a good figure for the eccentricity of the Earth's orbit.

The plot below is from measurements made between 1996 and 1998.

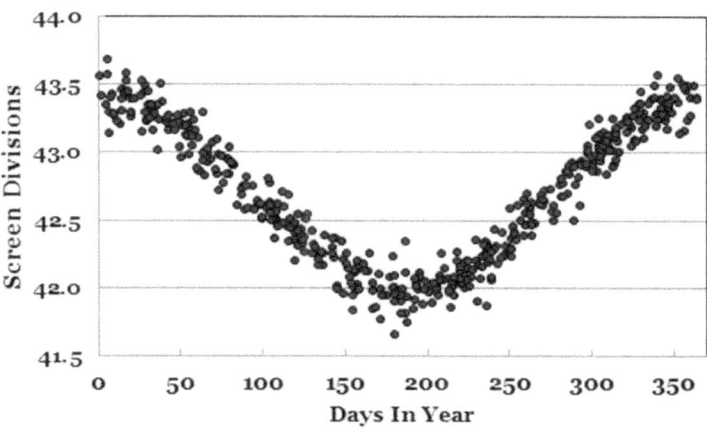

From these data the eccentricity of the Earth's orbit was calculated as 0.0175 which compares well with the actual value of 0.0167.

Recording Sunspots

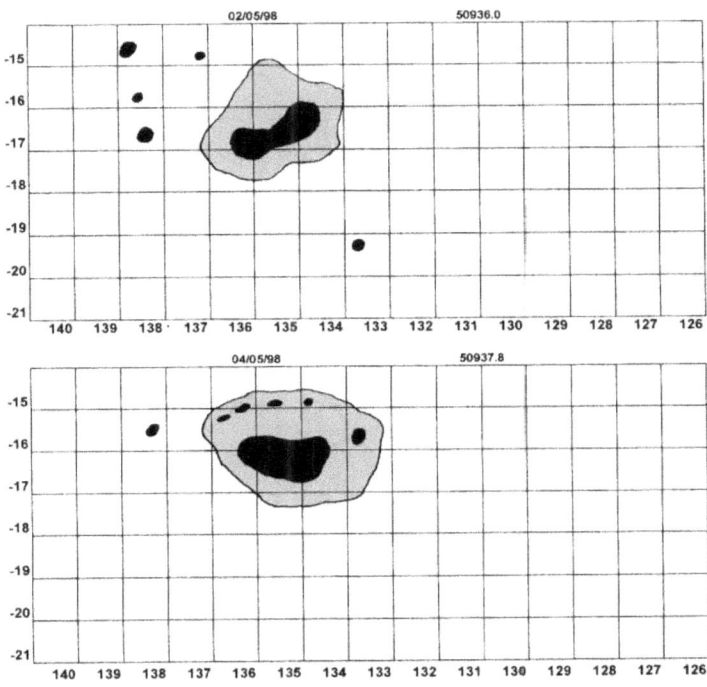

For several years I meticulously recorded the appearance of sunspot groups and their positions on the solar photosphere.

The above diagram shows a typical set of observations of a sunspot group as drawn by me on the 2nd and 4th of May 1998. The details are plotted on the solar latitude and longitude positioning system. On this scale the Earth would be about the size of each square on the grid.

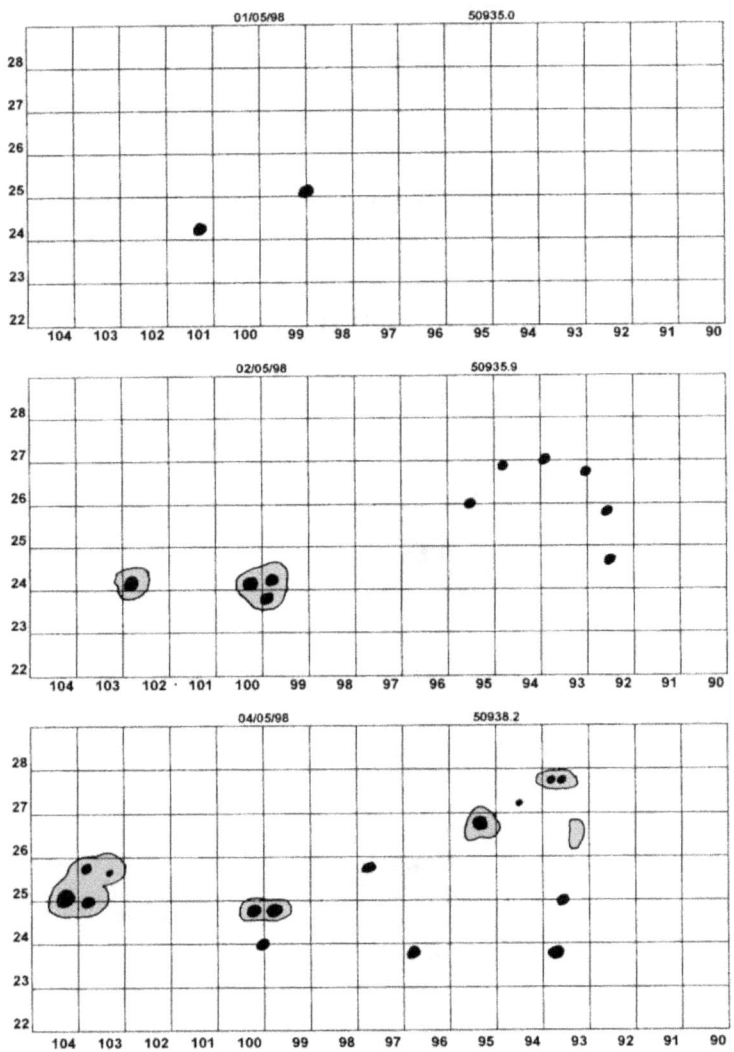

The above drawings from 1st, 2nd and 4th May 1998 show the complexity of sunspot groups and also the extent to which they change in size and position from day to day.

In some cases the smaller spots could almost be seen moving relative to each other and certainly changes were apparent over a period as short as five minutes.

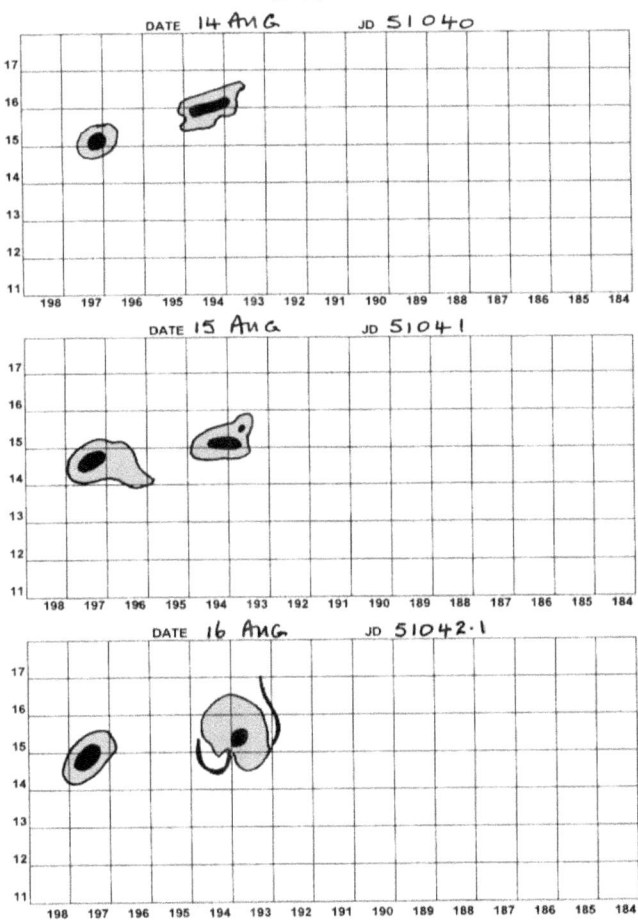

This third sample of my sunspot drawings was made on 14th, 15th and 16th August 1998.

This shows in the lower drawing a feature that I only ever saw once. These were two thin dark lines superimposed on the sunspot and were prominences arching above the solar surface and visible in white light to the eye.

Until this observation I had not realised that prominences could be seen visually with a small amateur telescope on the face of the Sun.

It was just that sort of surprising and exciting observation that made sunspot observations interesting.

Rotation Rate Of The Sun

My reason for plotting the sunspots accurately in solar latitude and longitude coordinates was to measure the rotation rate of the Sun as a function of solar latitude.

The solar rotation rate is actually well understood because it has been measured to high accuracy using fine resolution spectroscopic analysis of the Doppler shifts of the photosphere near to the Sun's limb.

However, I thought it would be an interesting project to come up with my own determination using a small telescope from my back garden.

There were problems in determining the Sun's rotation rate.

I needed to track sunspots that made at least one - preferably more - full rotations around the Sun. The individual umbrae (the dark centres of sunspots) move around relative to other sunspots in a group. Only after a month or so, when the spot group has been right around the Sun can a reasonably reliable rotation rate be determined.

I only managed to find a few umbrae that could be tracked in longitude for a full solar rotation as shown in the chart below.

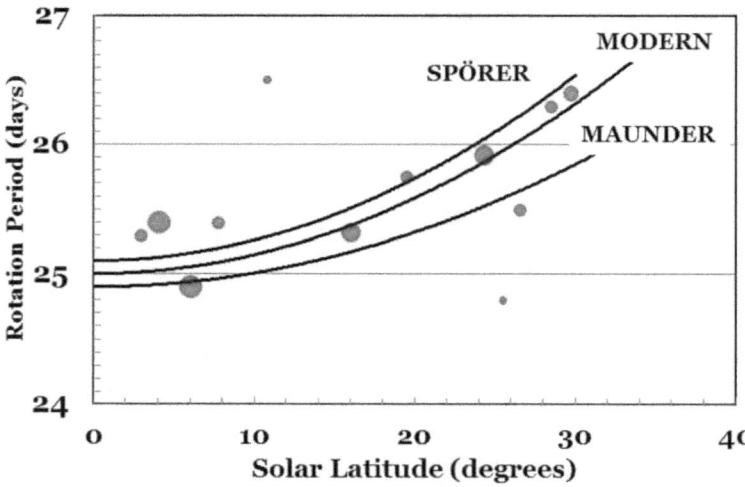

Three professional determinations of the variations of the solar rotation rate with latitude have been included.

The size of the plotted points is an indication of the time span over which individual sunspots were tracked with the longest timespan being three rotations and the shortest less than one rotation.

The Solar Event Of June 1972

In June 1972 a particularly large sunspot group appeared on the Sun. It was huge and the detail changed within minutes. It was a fascinating object for me to watch.

I made detailed drawings over several days and then later read in *'New Scientist'* magazine that a white light solar flare had erupted just fifteen minutes after I stopped observing on one day. This would have been visible to me had I still been observing.

This is extremely rare and very few people have seen visible white light flares - Carrington was the first in 1859 to see one.

Just before noon on September 1st, Richard Carrington and Richard Hodgson made the first observations of a solar flare. The flare was the visible ejection of a huge coronal mass ejection (CME) that travelled directly toward Earth.

On September 1st and 2nd 1859, one of the largest recorded geomagnetic storms occurred.

Aurorae over the Rocky Mountains were so bright that the glow awoke gold miners, who began cooking breakfast because they thought it was morning.

People in the north-eastern US could read a newspaper by the aurora's light. The aurora was visible as far from the poles as Cuba and Hawaii.

Telegraph systems all over Europe and North America failed, in some cases giving telegraph operators electric shocks.

Telegraph pylons sparked.

Some telegraph systems continued to send and receive messages despite having been disconnected from their power supplies.

In North America it was reported that a voltage surge of 60 volts occurred on the telephone cable between Chicago and Nebraska which severely disrupted the telephone system.

It has been estimated that the cost of a similar event to the USA alone at up to $2.6 trillion if it were to occur in modern times.

It was calculated that the energy emitted by the 1972 white light flare (which I missed by minutes!) in one hour was more than the total power used by the USA in a century.

Another storm in March 1989 produced spectacular aurorae which I saw from Somerset in the south of England, see page 85.

This caused huge electrical blackouts over Canada [328].

However, I was not involved in solar observing at that time so missed seeing the sunspot group that produced this Coronal Mass Discharge.

I strongly recommend taking up the observation of sunspots.

It is genuinely exciting to watch the changing detail in groups of sunspots not just from day to day but even over periods as short as minutes.

Of course, with so many professional astronomers watching the Sun continuously you will not discover any new but - it is fun, easy, requires only a simple telescope and is cheap to do.

It also does not require the observer to go out on freezing winter nights which, for me, was of considerable advantage!

What Was Odd About That Chart?

The oddity is that the minima and maxima in turnip yields occur about two years before the minima and maxima in sunspot numbers.

This makes it look as though the turnips are predicting the changes in sunspots.

If this is a valid effect then we should all be using growth measurements of turnips to predict future sunspot numbers and possibly climate change.

Or...

...maybe not.

This chart shows the danger of accepting a similarity in two parameters and their variations without looking in detail at the consequences.

17. Touring The USA

I have visited the USA and the Canary Islands for astronomical tours. We start with the United States tours and then move on to the Canary Islands in the next chapter.

The US Naval Observatory Washington DC

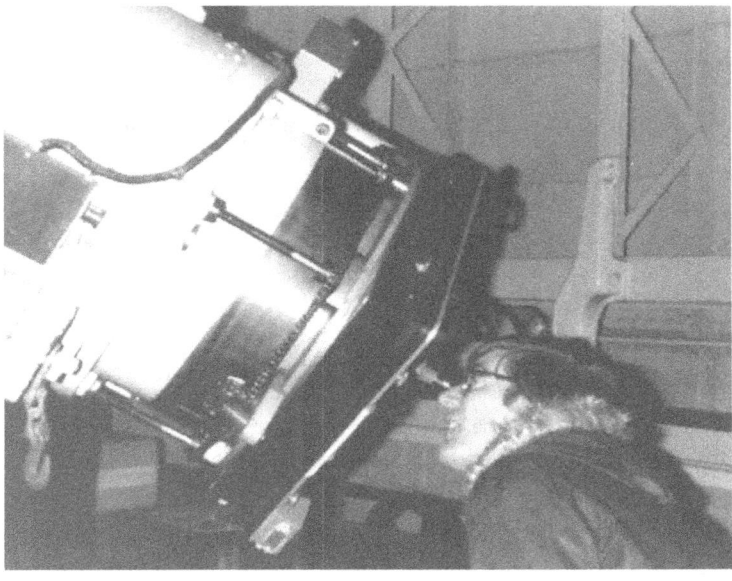

In 1979 I went on a business trip to Washington and managed to slip in an observing session on the magnificent 26 inch (660 mm) aperture refractor operated in Washington DC by the United States Naval Observatory [329].

I was very privileged because the telescope is in use on clear nights mainly being used to measure the orbits of binary stars and tracking the satellites of the planets in support of exploration of these bodies.

I was shown the Orion Nebula which was almost totally washed out by the sky glow of the nearby city.

The Moon was also on view (which did not help the sky brightness) but the view was no better than through my own telescope.

I also saw a few double stars.

Overall, I was not impressed by these views because of the very poor atmospheric conditions but the instrument itself was spectacular.

Asaph Hall discovered the two tiny moons of Mars; Deimos on 12 August 1877 and Phobos on 18 August 1877 using this telescope [330].

I also saw the Meridian Transit Circle [331] which was used for nearly a century to measure very precisely the positions of stars.

It was claimed that this device was so accurate that it was able to measure the continental drift which is separating the USA and Europe at the rate of about 20 mm per year.

Visit To Mount Palomar, 1983

In 1983 I had another business trip, this time to California and I was able to visit the 200 inch (5 metre) aperture telescope on Palomar Mountain.

Although this huge telescope was the largest accessible in the world at that time, it did not really impress me with its size.

True, it is big but somehow the scale was not evident. Perhaps it was because tourists were held inside a glass bubble.

A curious feature about this visit was a video loop playing for visitors describing the telescope. This was a copy of an old BBC TV Patrick Moore 'Sky at Night' programme! The authorities could not even put together a video of their own making!

There was a visitor's centre separated from the dome.

This was poor.

Tatty black and white photographs from the 1950s were displayed to illustrate the wonders of the heavens and the information was set out in a very old fashioned and boring fashion.

In fact, the whole thing was a disgrace.

Overall, the most interesting and impressive features on Palomar Mountain were the dome - which is huge - and the brightly coloured Blue Birds flying around.

"There'll be Blue Birds over the White Cliffs of Dover..."

What an inspiring song that was sung so emotionally by Vera Lynn in wartime Britain!

But, do you know that it was written by an American who never visited Britain and was so ignorant of the country that he thought there were the same 'Blue Birds' flying in England as in California? [332]

Also of interest were the warnings about deadly snakes in the area - unknown in the UK - and the property halfway up the mountain which was once owned by George Adamski of Flying Saucer fame - see [333].

Tour Of Western USA Astronomical Sites In 1991

My next trip to the USA was a holiday arranged with two friends in March and April 1991 who, like me, were astronomy enthusiasts. As a result, our trip had a strongly astronomical theme.

We spent three weeks driving around California, Arizona, New Mexico and Nevada. This was a mixture of visits to astronomical sites and to tourist sites.

These memoirs are about astronomy so I will whiz quickly over the non-astronomical aspects of our trip and get down to the astronomical nitty gritty.

My two friends and I flew to Los Angeles as this was the starting point for our tour of the astronomy world on the US West Coast and also because it had theme parks with scary rides - I never said this trip was purely one of scientific endeavour!

We were determined to avoid downtown LA. We wanted to survive our visit to the USA so we got a cab to the car hire office, got straight into our car and immediately struck off to the outskirts of LA where our hotel awaited.

The Griffiths Observatory, Pasadena

Off the next day to the Griffith Observatory - famous for its magnificent old buildings, the venue for many films and the place where we saw the 'HOLLYWOOD' sign in big white tatty letters on the hillside.

Now, if you have never been to LA you might think that the HOLLYWOOD sign can be seen from all over California but - no - it can only really be seen well from the grounds of the Griffith Observatory.

So that's where you must go!

Films where the HOLLYWOOD sign appears over the actors' shoulders are usually filmed in the observatory car-park.

The first barrier to getting in to the observatory was a sign at the pay desk stating.

'STRICTLY NO BILLS OVER $20 ACCEPTED'

Since we were fresh from the UK and carried nothing smaller than $100 bills this was a problem. However, I smiled my sweetest smile and apologised in my best cut-glass 'Hugh Grant' accent for the large denomination bill.

The lady smiled back and counted out my change.

One thing the Americans are good at is politeness and they generally having a high regard for the British and the British way of life.

I got invited to many parties during my frequent business trips to the USA. The other guests simply couldn't get enough of my vast repertoire of amusing British accents.

My 'Brummie' Birmingham, my 'Scouse' Liverpool and my Cockney 'Rhyming Slang' particularly made me the star of every party I attended.

At least that's what I believed.

Maybe they were angry inside for being stuck with the toffy-nosed Englishman making the embarrassing strange voices.

The exhibits in the Griffiths Observatory were interesting in their own way but rather old-fashioned in 1991 when I visited.

This is the plug removed from the centre of the Palomar 200 inch mirror; a respectable size in itself for an amateur mirror and this was just the bit they didn't want!

A solar telescope projected an image of the Sun onto a table for general viewing and we attended a lecture in the Planetarium on 'Dark Matter'.

During the show at the planetarium some idiot kept taking flash photographs of the images projected on the dome until they were thrown out.

From the Griffiths Observatory the centre of Los Angeles could be seen spread out before us in a vast and most impressive panorama. Apparently, the smog only lifts on a few days a year for this sight to be seen.

Over the next few days we expanded our Hedonistic senses at Universal Studios, Sea World, Disneyland, etc.

You don't want to hear all about that do you?

Of course not!

You, dear reader, are an astronomer - more interested in the pleasures of the skies than the pleasures of the flesh.

So, I'll just show two pictures to give you a flavour of what I got up to.

"Be gentle with me. It's my first time!"

"Frankly My Dear, with a face like yours I'm not surprised!"

I just cannot resist sticking my head into an oval cut out in a vertical board. It's a strange compulsion but usually amusing for onlookers.

I place myself in the service of Condom The Barbarian

Jet Propulsion Laboratory, Pasadena

On next to the Jet Propulsion Laboratory in Pasadena.

It is difficult to believe when visiting places like Pasadena that there is a square mile or so of murder and mayhem in the centre of Los Angeles. Pasadena exudes an ambience of gentility and esoteric studiousness.

We saw full-scale models of the many interplanetary probes and the discoveries JPL had made.

There were lots of handouts and, when we revealed that we were amateur astronomers from England, we were given large packs of glossy photographs of very high quality - not normally given to those outside the scientific community.

Here I am posing against a globe of Mars made up of space probe observations - remember that this was in 1991 before the high resolution surveys of Mars Explorer, etc.

By the time we had sated ourselves on the delights of Los Angeles the torrential rain had ceased and the landslides were cleared from the roads. We had been trapped in LA by floods, landslides and snow drifts. Not what we had expected in April!

We travelled east towards the astronomical delights that we had come to the USA to visit.

The Americans are interesting when it comes to names.

Look in any big city telephone book and you will see the weirdest concatenations of consonants and vowels imaginable. Look especially at the last entries where people vie to be last. I found the last name in the Los Angeles telephone book was Mr Zzzzyz.

A Town Called Zzyzx

We saw a Freeway sign directing motorists to 'Zzyzx Road' [334], [335]

And who else but Americans would call a city *'Truth or Consequences'* after a radio quiz show?

Personally I'll stick to sensible names like we have in Dorset -

such as Whitechurch Canonicorum, Winterbourne Zelston and Shitterton.

Shitterton is a hamlet near Bere Regis in Dorset.

According to Wikipedia [336]

"The unusual name of the hamlet dates back at least 1,000 years to Anglo-Saxon times. It was recorded in the Domesday Book of 1086 as Scatera or Scetra, a Norman French rendering of an Old English name derived from the word scite, meaning 'dung' - hence the word 'scatological'.

This word became schitte in Middle English and shit in modern English.

The name alludes to the stream that bisects the hamlet, which appears to have been called the Shiter or Shitter, or 'brook used as a privy'.

The place-name therefore means something along the lines of 'Farmstead on the stream used as an open sewer'".

My soulmate and partner once asked me to pose in front of Shitterton's village sign.

"A bit more to the left!" she repeatedly said until she clicked this.

So, I now know what her true feelings are for me.

Dorset has more than its fair share of amusing and embarrassing village names such as Happy Bottom, Scratchy Bottom, Squibb's Bottom and Shaggs.

The River Piddle runs through Dorset giving us Piddlehinton and Piddletrenthide.

There is the Piddle Brewery whose advertising tag line asks customers

"Do you fancy a Piddle?"

There are embarrassing genital problems alluded to in Sandy Balls, Droop, Dungy Head, Shatters Hill, Knacker's Hole and Knobcrook Road.

What goes on at Crumpet's Drive, Pistle Down and Belchawell?

Anyway, leaving behind scatological British place names we return to astronomy and my tour of the Western USA.

The Palomar 200 Inch Telescope.

On the way east from Los Angeles we decided to drive up to see the Palomar 200 inch telescope.

For me this would be a return visit.

We drove steadily up Palomar Mountain for hours - the snow getting deeper, the air colder and our ears popping like machine gun fire as tiny pellets of wax were ejected by the release of internal air pressure.

Just by the site of George Adamski's cafe - he of the excursions around the Solar System in a flying saucer [337] - we stopped to make a snowman.

Well, everyone else was doing it.

Curiously, some people had built their snowman on the roof of their car and were driving away. Would this snow sculpture last until they got home?

At last we arrived at the gates to the Observatory.

A sign greeted us

CLOSED

We couldn't believe it.

All that way up the mountain and no warning at the start of the climb!

The road was all perfectly dry and clear to the gates. The observatory staff had simply not bothered to clear the car-park for visitors. We were furious and took it out on each other with a snowball fight.

To fill in this part of my story in the absence of seeing the Palomar telescope on this trip, here is a quotation from a magazine I found.

"Hale (the founder of the Palomar Observatory) was a solar astronomer with a talent for extracting money from tycoons to build telescopes.

He was also a trifle odd.

In 1922, for instance, while he was Director of the prestigious Mount Wilson Observatory, he was arrested for challenging two traffic cops to a race through the streets of Pasadena on his three-wheeled motorcycle.

Hale's obsession was with building giant telescopes. His mind, imagining bigger and better instruments, often whirled out of control, spinning him to the edge of insanity.

It was in the midst of one such mental firestorm that an elf materialised before him. The elf gave Hale advice on how to run his life and continued to do so until his death."

If this had spread over the country it would become a *National 'Elf Service* (groan!)

Kitt Peak Observatory, Arizona

Having made many side visits to non-astronomical sites we arrived at Kitt Peak in Arizona.

We eagerly anticipated our trip to the top of the mountain.

When we arrived at the start of the road to the peak guess what we saw?

This time we were in rebellious mode.

These damn Americans were taking on three determined 'British Bulldogs'. The Dunkirk Spirit was aroused. With stiff upper lips, that had typically been borne by our forefathers at Agincourt,

Waterloo, the Battle of Britain... we turned to each other and said

"Wheear gging nnn!"

Damn those stiff upper lips - so difficult to understand what one is saying!

Probably Churchill had said

"Let's give in everyone - Hitler's obviously going to win"

but, because of his stiff upper lip, it sounded like

"Never, in the field of human conflict...."

And we British had carried on fighting despite the absence of our American Allies who, as in World War 1, turned up late.

And so at Kitt Peak we three British stalwarts drove our car around the barrier and on up the mountain.

The snow got thicker as we climbed.

Our thoughts turned to Tenzing Norgay and Edmund Hillary climbing up Everest - that wonderful British Achievement (by a New Zealander and a Nepalese) except that we were in a Mitsubishi Gallant with the heater full on.

As we arrived at the top we saw other cars in the car-park. Yes, the observatories were open to visitors!

The visit itself was really excellent.

The staff had gone to a lot of trouble to clear the car-park and walkways for visitors - unlike the Mount Palomar staff.

The observatory stands at 6,972 feet (2,125 metres) altitude - so we arrived somewhat breathless.

The buildings and instruments are really impressive.

See the website at [338] for a virtual tour.

We were invited to see the Mayall telescope - a magnificent instrument with a 4 metre aperture mirror [339].

Although this telescope weighs 250 tonnes it can be moved by a light touch of the hand. It can see the light of a match at 10,000 miles were it possible to light a match in space.

Personally I thought this telescope was very much more impressive than the larger 5 metre aperture Mount Palomar telescope.

The above picture shows the Mayall telescope dome. That's me on the left of the building.

There are several other telescopes on the peak.

These were all explained in very visitor-friendly brochures and in public lectures which were held hourly and given by professional astronomers.

The McMath Solar Telescope was very remarkable [340].

I recall an edition of 'The Sky At Night' during which Patrick Moore climbed up the shaft down which the Sun's light is directed and he emerged out of the telescope at the top like a large lumbering cave explorer.

Overall, this was a really excellent visit.

The staff were welcoming, a lot of effort had been put into making the visit an educational as well as an enjoyable trip.

Kitt Peak is sacred to the local Native Americans - the Tohono O'odham Nation - and a great deal of sensitive negotiation was undertaken before the tribal elders allowed the observatory to be built.

In return they got rights to work at the observatory and sell their trinkets in the site shop.

We then went on to visit Tombstone [341] *'The Town Too Tough To Die!'* and several other cowboy-related sites.

An unusual site we visited was a silver mining town called Calico which had been abandoned in the 1890s. It is now a 'ghost town' and was well worth a visit [342].

Our next astronomical stop was Tucson, Arizona.

Tucson is famous astronomically for its pioneering strict laws on reducing light pollution. All public lighting is 'full cut-off' so that the road is illuminated by patches of light instead of vast floodlights which send half their light into the sky.

It is strange walking in Tucson as the sky above is so black. Indeed, we were able to see the Milky Way from the town centre.

We visited the public observatory and enjoyed being shown deep sky objects from the centre of a city.

Biosphere 2, Arizona

Whilst in Arizona we visited the Biosphere 2 Research Facility [343].

At the time of our visit the huge biosphere containment greenhouse was being stocked with plants and animals ready to seal up the building with a team of humans inside.

Biosphere 2 was originally meant to explore the web of interactions within life systems in a structure with five areas based on biomes to study the interactions between humans, farming and technology with the rest of nature.

It also explored the use of closed biospheres in space colonization and allowed the study and manipulation of a biosphere without harming the Earth's ecology.

Unfortunately, in the years following our visit there were problems with the equipment, the team members who were shut away isolated from Earth and with the management.

The experiment did achieve results relevant to long-term manned space missions however despite its failure to meet the original objectives.

The 'Very Large Array' In Socorro New Mexico

We moved on to Socorro, New Mexico where the Very Large Array (VLA) operates - see next page.

Yes! That's me in the foreground. Everywhere I go I carry a large white hoop with which to identify me in photographs.

This facility is vast and has featured in many films such as the excellent 'Contact' [344].

Here, as at Kitt Peak but unlike Palomar, visitors are made very welcome.

Excellent educational facilities include a free 36 page booklet covering the history of radio astronomy and the role played by the VLA [345].

I was amused by the toilet (or the 'bathroom' as our American cousins like to coyly refer to this facility) where the designers had gone to the trouble of making astronomically themed tiles for the walls as shown below.

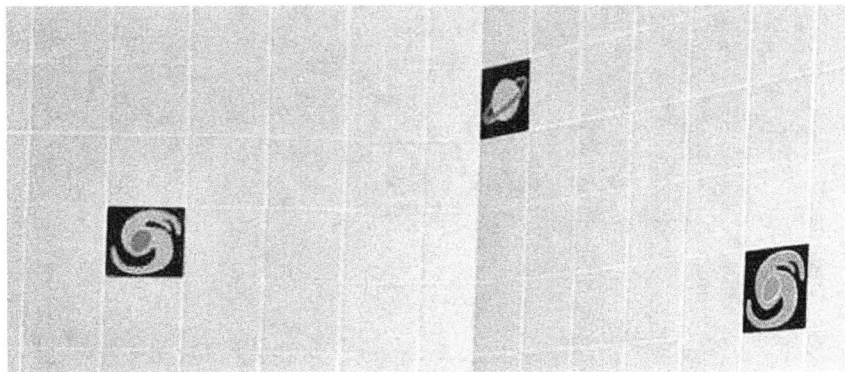

The VLA comprises twenty-seven 25-metre diameter radio telescopes in a Y-shaped array and functions as an interferometer.

This means that the outputs from the dishes are combined to give the resolution of a single huge telescope approximately equal to the distance between the furthest dishes.

Each of the massive dishes is mounted on double parallel railroad tracks, so the radius and density of the array can be transformed to focus on particular bands of wavelength.

Astronomers using the VLA have made key observations of black holes and protoplanetary discs around young stars, discovered magnetic filaments and traced complex gas motions at the Milky Way's centre, probed the Universe's cosmological parameters, and provided new knowledge about the physical mechanisms that produce radio emission.

The Great Continental Divide

After leaving Socorro we headed west and crossed the 'Continental Divide'.

I found the concept described on the information board shown below to be awesome. A drop of rain falling at this point has a 50:50 chance of going 400 miles to the Pacific or 1600 miles to the Gulf of Mexico. This large ratio of 4:1 shows how the Northern American tectonic plate tilts down from west to east.

Is it not amazing that the water finds its way over such a tortuous path to the sea?

I have worked out that the average gradient of the rivers flowing east is about 0.06 degrees. Allowing for meandering the actual mean gradient is about half this - and yet such a small gradient creates the mighty Mississippi and Missouri rivers.

We moved on to the Petrified Forest, The Grand Canyon, The Hoover Dam, Las Vegas, Escondido Animal Park and many more tourist attractions - all really wonderful in their own ways but not astronomical so they will be left out of this travelogue.

What I will mention is that I must be one of the few visitors to Las Vegas not to place a bet whilst there.

As a mathematical physicist I know that in the long term I will lose money gambling. How else can all those luxurious casinos make a profit?

Being a scientist sucks all the fun out of games of chance.

Meteor Crater, Arizona

Our next encounter was the impact of a small asteroidal body weighing a few million tonnes with the Arizona desert which was hit at about 16 km/second [346].

There are many photographs of this huge crater - a typical view can be seen above [347].

On the top rim in this picture is the winding road from the nearby highway and a building. Although this is dwarfed by the crater this building is a huge visitor centre with impressive facilities.

There is a large and very well presented exhibition hall with restaurant, etc. [348]

It is almost impossible to judge the size of this huge hole, blasted out of the desert.

The impact is believed to have occurred about 49,500 year ago.

It is estimated that about 300 million tonnes of rock were thrown out into space or onto the surrounding desert by the 43,000 miles per hour impacting body.

It is all very impressive.

My picture below was taken from the crater rim and shows snow on the southern walls.

I had not expected to see snow in Arizona in April - or indeed at any time of the year.

However, the crater is about 1,740 m (5,710 ft) above sea level so this accounts for its coolness in April.

The picture above shows a general view from the crater rim.

In the centre of the crater is a life-sized figure. It is just visible with binoculars from the rim but I had to use a 200 mm lens to get any picture at all.

The area within the rectangle above is blown up to give the picture below.

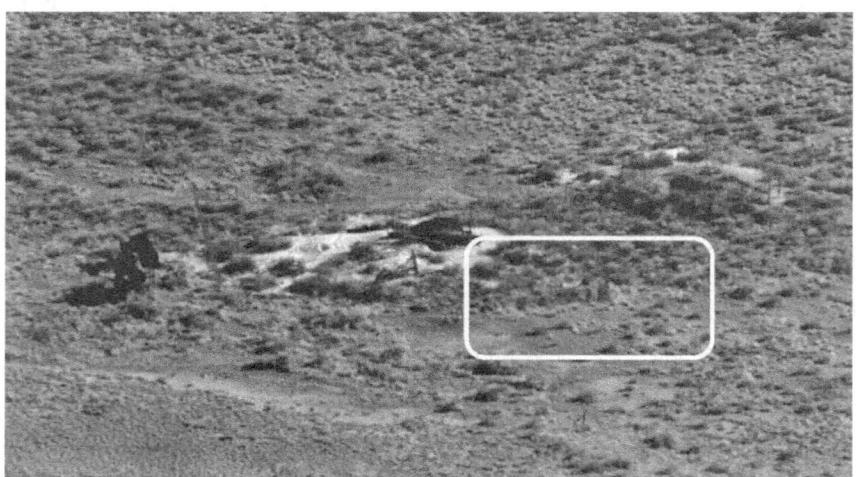

The area within that white rectangle is shown below.

Only at such magnification can the figure and the flag be seen.

We next tried to visit Flagstaff Observatory but we could not find it. This was long before the invention of Satellite Navigation system.

Eventually we spotted the domes as we were driving along the freeway out of the area.

Too late to turn back.

The Lowell Observatory

So, we carried on to the Lowell Observatory [349].

This is a quaint and friendly place. It exudes old fashioned astronomy from a long gone age.

We loved it!

We arrived after it had closed for the evening but we enjoyed the visit.

We saw the refractor with which the search for Planet X was conducted resulting in the discovery of the dwarf planet Pluto by Clyde Tombaugh in 1930 - see [350].

Although the observatory dome was closed, a glass door had been erected and an externally-mounted light switch so that visitors could illuminate and see the telescope after opening hours.

What a nice gesture - unlike the gesture given to us at Palomar!

Percival Lowell, who funded and built the observatory and then ran the programme of research, was an eccentric who resolutely believed in a civilisation living on Mars.

His observations of 'canals' convinced him that a dying race of aliens lived on that cold, inhospitable planet.

The non-existent 'canals' which he mapped so assiduously were believed by him to be engineered to move water from the ice caps as they melted towards the parched vegetation around the temperate latitudes.

Unfortunately - or perhaps fortunately for the human race - he was wrong.

There are no Martians.

He is currently resident in a mausoleum design by him - a very odd building indeed.

There are extracts from two of his writings on the plaques on each side of the entrance. These are reproduced below.

"Astronomy now demands bodily abstraction of its devotee...to see into the beyond...and securing it makes him perforce a hermit from his kind...He must abandon cities and forego plains...only in places raised above and aloof from men can he profitably pursue his search...he must learn to wait upon his opportunities and then no less to wait for mankind's acceptance of his results...For, in common with most explorers, he will

encounter on his return that final penalty of penetration the certainty at first of being disbelieved...

From **Mars and its Canals** by Percival Lowell

"Everything around the Earth we see is subject to the inevitable cycle of birth growth decay...Nothing begins that comes, at last, to end...Though our own lives are too busy to even mark the slow nearing to that eventual goal...today, what we already know is helping. To comprehension of another world...in a not distant future we shall be repaid with interest and what that other world shall have taught us will redound to a better knowledge of our own and of the cosmos of which the two form part."

From **The Evolution of Worlds** by Percival Lowell

Above is the Clark Telescope Dome on Mars Hill. This dome was built for Lowell by a local carpenter.

This contains the old fashioned but wonderful 610 mm (24 in) aperture Alvan Clark refractor used by Percival Lowell in his search for Planet X [351]. Planet X was never discovered but the much smaller Pluto was discovered in the course of this search.

In the grounds of the observatory buildings is a scale model of the Solar System with model planets in their relative places. This is not unique of course. A very similar one exists at the Jodrell Bank Space Science Centre outside Manchester.

However, the Lowell Observatory one is quaint - like everything on the site - a faint whiff of Victorian simplicity, naiveté and elegance pervades the place and gives a quiet ambiance.

It is a bit like thumbing through an old astronomy book. Comforting in the simplicity of human knowledge in times past without all the frenetic mega-science that has come from - for example - the Hubble Space Telescope.

I took the opportunity to do a little exercise with a giant snowball. Almost immediately afterwards, I offered up my prayers to Saint Hernia, the patron saint of large snowball lifting show-offs.

The altitude is 2,210 m (7,250 ft) and the air was noticeably thin.

Mount Wilson Observatories

Our final visit on this trip was to Mount Wilson outside Los Angeles. We were due to fly home that same evening so it was rather a rushed visit. The snow was deep but the weather was warm and sunny.

The area around the observatory buildings was crowded with children on skis and toboggans. Around the café humming birds hummed and sipped from feeders.

Although the brilliant white domes shone stunningly in the pristine snow, the idyllic effect was spoilt by the huge and very ugly array of massive aerials and transmitter masts. The concrete buildings were appalling but the demands of the military and the demands of civilians for zillions of TV channels overwhelm any sense of beauty on Mount Wilson!

The dome of the Hooker 100 inch aperture telescope had a forlorn message hanging on its door.

"CLOSED"

Nobody at that time was willing to take on the task of maintaining this venerable instrument which had been used to prove that the universe consists of billions of galaxies like our Milky Way and that our Milky Way is not the entirety of the universe [352].

Above we see the Hooker telescope [353].

For thirty-one years this was the biggest telescope in the world.

Such was the later rush to invest in gigantic new telescopes that an elderly telescope with a 100 inch mirror was dumped in the 1980s.

We were not allowed through the door to see this magnificent machine because of concerns that bits were falling off the decaying observatory building.

However, the telescope was revived after our visit and was fitted with adaptive optics to give it a new lease of productive life.

There is an excellent article about the Hooker telescope and the discoveries made with it at [354].

A small visitors' centre was really weird - like a time capsule of a past age of astronomy. Decaying exhibits covered in dust and cobwebs mournfully tried to teach us something about a past age.

Here I am with a polishing tool. Nothing to say what telescope the mirror was used for; it just stood there sadly exposed to graffiti and my embrace. Compare this with the polishing tool I used for my own mirror - see page 180.

Same principle, different size.

A poster explained what would be seen when Halley's Comet arrived in the inner Solar System - this was four years **after** the event!

All very sad and neglected - a lost opportunity to educate and excite some of those children tobogganing nearby - but not one of them or, indeed, any visitors other than the three of us were here.

A Solar Tower Telescope

There were other smaller facilities on the Mount Wilson site and the pictures above and below show a solar tower where the face of the Sun is scrutinised.

A rotating mirror (heliostat) at the top of the tower reflects the image of the Sun to an observing station at ground level.

The 'SNOW' Solar Telescope [355]

And so the three-week trip came to an end as we drove our rented car back to the airport in Los Angeles to catch our flight back to the United Kingdom.

We had driven 3,500 miles in our visits to the astronomical centres of the South-Western United States. We had also visited the fleshpots of Las Vegas, the grandeur of the Hoover Dam, the magnificence of the Grand Canyon with snow on its rim, Meteor Crater - and the hedonism of the theme parks.

Time to get back to England.

18. Touring The Canary Islands

Visiting Tenerife And La Palma In 1986

My other astronomical visits were to the Canary Islands and that's where we will go next in my story.

I made two trips to the Canary Islands to see the observatories and observe from high altitudes the celestial sights not visible from murky, sea level, northern latitude Weymouth in the United Kingdom.

The first trip was in April 1986 essentially to view Halley's Comet from a good site. In fact, the comet was a disappointment but the trip itself was interesting.

The whole family went along - this was the first time that my children - then aged 8, 10 and 13 - had flown. It was wonderful to share their excitement of flying for the first time when I had become blasé having flown many times in pursuit of my career.

The flight was mundane and punctuated only by the children's screams of laughter when I discovered a caterpillar crawling over my in-flight salad.

"Where's his friend, Dad?"

they mocked me.

I told them that this was a useful source of protein and then went to the toilet to secretly swallow several anti-nausea pills.

As we flew into darkness there was a glorious sunset with a brilliant Venus blazing out behind the aircraft's wing.

A 'James Bond' film was about to be shown and we were told to close all window blinds.

This was unfortunate as about one quarter of the occupants of the plane were fellow astronomers travelling with Trans-Solar Tours to see Halley's Comet. They, being nerds like me, wanted to see celestial stars and not terrestrial stars like Roger Moore. (I have often wondered whether he lived up to his name.)

Eventually, we negotiated with the cabin crew to have one window left uncovered - up by the galley - and we enthusiasts for the heavens queued to take a turn to look out of that window for about twenty seconds and then re-join the end of the queue.

We stayed in a vast skyscraper block hotel in the concrete jungle of Southern Tenerife - truly a ghastly place where people go to get sunburnt and drunk.

The hotel was not as bad as its name suggested!

My family stayed there most of the time and shivered in unexpectedly low temperatures and cloudy conditions. We came home almost as white as when we arrived - any difference being due to wind burn!

One day we hired a car and drove over Mount Teide which, at 3,718 metres, is the highest point in Spanish Territory.

Even though we were there in early summer the children were delighted and amazed to find snow at high altitude and enjoyed a snowball fight.

The last time Teide erupted was in 1909. We tried not to think about the possibility of another volcanic eruption whilst we were at the summit. When it does happen there could be little warning and the consequences are likely to include massive loss of life and a mega tsunami bringing death and widespread destruction to coastal inhabitants around the North Atlantic [356].

To The Summit Of Mount Teide

One evening coaches had been arranged to take the many amateur astronomers up to the summit of Mount Teide so that we could see the heavens by night and see Halley's Comet which was the main reason for coming.

At the arranged time - about 11 pm - we all gathered in the hotel lobby waiting for the coaches. We waited...and we waited...and we waited.

No coaches!

Eventually we all went to bed.

It turned out that the Spanish drivers got the date wrong. They turned up 24 hours later.

Mañana!

The Spanish lifestyle in a word.

The following night our coaches turned up. My wife and two of my children decided to stay at the hotel whilst I took my younger son up the mountain.

About thirty other coaches full of amateur astronomers had the same idea that night. The coaches queued to grind up the mountain tracks and they jostled for parking spaces in the middle of the otherwise bleak and lonely lava fields.

It took a long time before we could persuade all the coach drivers to turn off their lights. They wanted to read newspapers to pass the time.

With several hundred astronomers on a mountain top there will always be a few fools who want to do nothing but take flash photographs. After the last one was strangled and his body buried in the loose lava we could all get down to some serious observing.

We were totally overwhelmed by the brilliance of the sky and wandered around going "OOOOH" and "ARRRH" at the multitude of stars in a totally black sky. We thought there was a long cloud stretching over the sky but it suddenly occurred to me that it was the Milky Way brilliantly gleaming from one horizon to the other.

What a sight!

Behold Halley's Comet! Where?

But, where was Halley's Comet that we had paid a fortune to see? Some people were pointing in one direction - others pointing elsewhere. All these helpful indications turned out to be Messier objects blazing down - not the comet.

Eventually, I identified it in my binoculars near the stars of Corvus. Was that really it? It turned out that the tail was pointing towards the Earth and so it was relatively insignificant.

Despite this disappointment my son and I hunted in the dark for a flat piece of lava on which to set up my camera mount.

There wasn't any.

The whole area off the immediate cinder track consisted of razor sharp lava fragments.

We could feel them tearing at the soles of our trainers and there was no way any contact could be permitted with clothes or - God forbid - bare skin! I set up the camera as best I could on what appeared to be the edge of a precipice. I was nervous that, in the dark, I might tumble over some unseen edge and plummet to a grisly death.

I took pictures with my 'Barn Door' camera mount - see page 208 for a description of this device. This was a lightweight version that could be carried easily in pieces in a suitcase.

After use it was thrown away in the hotel waste bin - it cost less than £1 to make.

The disc on which five second intervals are marked has small nails sticking out. I used a Walkman (a forerunner of the iPod for you youngsters) with a tape of five-second ticks. At every tick I moved the rotating arm to the next nail. I called this my 'Braille Barn Door Mount'.

It was very difficult setting the polar axis up because of the fear of putting my cheek against some lava and being scarred for life.

However, my son and I got it set up and I took several pictures.

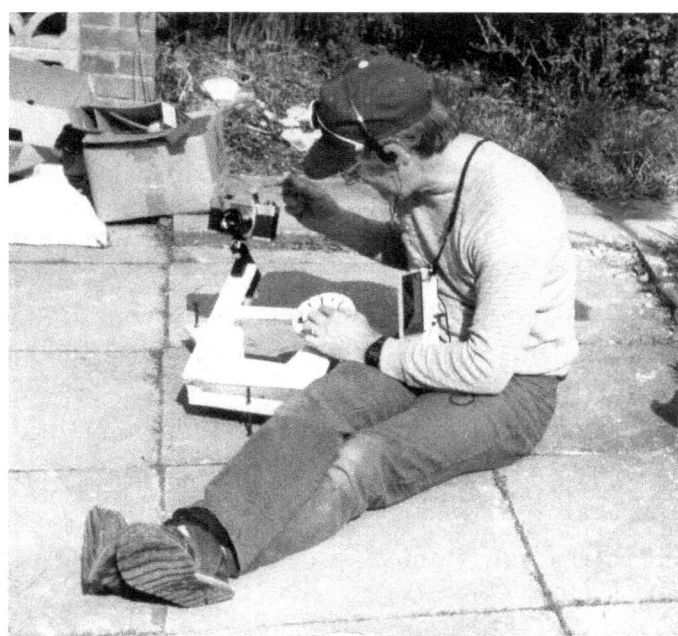

Above I am rehearsing using the mount at home.

The picture above shows Halley's Comet in the centre of the field.

We had travelled to see a magnificent tail stretching across the heavens but, because of the poor timing, we saw just an amorphous blob.

Notice how there is no apparent dimming of the stars down to the horizon. One star is just on the horizon but is as sharp and bright as when higher.

Superb conditions!

I had trouble getting the precious roll of slide film home and processed. At the airport some officious guard insisted that I put the film through the X-Ray machine.

Why?

It was, after all, an obvious small tube of film.

I refused. He insisted.

I raised my voice. He shouted.

I did not want my ISO 1000 film fogged by his crude X-Ray equipment. I pleaded with the crowd gathering around the customs area

"Does anyone speak Spanish?"

Nobody did - or if they did, they did not want to get involved.

It appeared that I would either have to put my tiny tube of film through the X-Ray machine or forfeit it. I put it through the machine whilst muttering insulting oaths and smiling in the hope that he might think I was paying him a complement.

I have heard that the test for new recruits to the Spanish immigration service is to tell them to walk towards a horizontal board suspended 1.5 metres above the ground. If he walks under the board the candidate is too short. If he ducks under the board he is too intelligent. If he walks into the board and hits his face he passes the test.

When I returned home with my precious roll of film I took it to Supasnaps a now defunct firm of high street film developers.

I explained that I did not want it cut.

The lady wrote on the envelope in huge letters "DO NOT CUT FILM".

I stuck a big note on the film cassette saying "DO NOT CUT FILM" and left the end of the film sticking out of the cassette with a third note on saying "DO NOT CUT FILM".

It came back cut!

Fortunately, I had taken some pictures of my children at the start of the film so that the cutting machine could be aligned.

Idiots!

After that diversion into the murky backwaters of human intelligence, back to the Canary Islands.

Visiting La Palma In 1986

My eldest son and I had booked for an excursion to the neighbouring island of La Palma to go up to see the Isaac Newton, William Herschel and other telescopes. So we flew with about thirty others and booked into a hotel.

We then were taken, in daylight, to the top of the mountain where the observatories were located.

The top of the mountain was very impressive. The road was especially built by the Spanish Government and is about 35 kms long.

We counted 330 hairpin bends, i.e. about one bend every 100 metres. It was not a comfortable journey but the road was a real engineering feat.

The first thing we saw on this trip to the top of La Palma was the decaying remains of the staging and seating for the opening ceremony some years previous. Nobody had bothered to clear it away.

What a mess!

Originally it must have been impressive.

The Heads of State from many nations had arrived and been welcomed. King Carlos of Spain and other monarchs had gathered here.

I believe that Britain had been represented by Mrs Honaria Gromit, a cleaning lady from Surbiton who had won the trip in a competition organised by a British national newspaper.

Seriously though, there had been a bit of a scandal about the UK not matching the rank of the representatives from other nations.

The crumbling remains of the wheel-like emblem had been the representation of the (possible) flags of alien nations who would have been here if they could have made it.

Alas, nobody turned up from another planet. They didn't even send their apologies for non-attendance.

How rude!

The trip to La Palma was like the curate's egg. (Incidentally what sort of eggs does a curate lay?)

My eldest son soaked in the whole experience and I had him running up and down the hill to experience the effects of altitude on breathing capability. The observatories are at 2,400 metres (about 8,000 feet) altitude.

Unfortunately, we were not allowed anywhere near the observatory domes.

In fact, the mountain top was heaving with coach loads of astronomers attracted by the prospects of viewing Halley's Comet from a dark sky location.

The domes were closed to all visitors.

We heard that barricades had been set up on Tenerife to stop visitors getting anywhere near the observatories on that island.

There were two interesting and amusing aspects of this trip.

The photograph above shows me standing by a bus stop indicating that I am at East Grinstead and not at the top of La Palma. This sign had been stolen from the bus company local to Herstmonceux Observatory in Sussex and erected outside the Isaac Newton dome.

Some remarkable damage caused by hailstones to the buildings of the Isaac Newton telescope.

Had anyone been out in the open during that storm I imagine they would have been killed!

The achievement on this trip was that we got to see the outsides of several domes before we were rounded up and taken down to sea level in the procession of coaches that bought us up.

Visiting La Palma in 1987

I was to return on a much more extensive trip the following year - in 1987.

For that trip I travelled to La Palma and Tenerife for a couple of weeks with two friends. We three intrepid travellers went straight to La Palma and stayed in an apartment.

It was very basic.

Ants marched everywhere in orderly lines and I had to sleep on a convertible bed - although what it converted into I never discovered.

It was cloudy the entire time over our apartment and we came home without a speck of suntan on us.

The island was dominated by German tourists because direct flights from Berlin to La Palma had recently been started.

We found these visitors generally overbearing, rude and demanding.

When they shouted at us in German because our hire car was in the 'wrong' parking space we waved, smiled and wished them a good holiday in exaggerated 'cut glass' English accents.

We did consider doing the Basil Fawlty walk [357] up and down the terrace just to upset them - after all they did start two World Wars and commit the most appalling genocide.

On our trip we usually slept by day and went up the mountain at night to observe the stars and tour the observatories. This meant - luckily - that we scarcely saw our German neighbours.

One day we trekked into the centre of La Palma. After taking our hire car as far as we could into the ancient volcanic caldera.

The mountain track stopped where a landslide had fallen a few days previously.

The island is a volcano with a huge caldera in the centre. We decided to backpack into this wilderness and spent the whole day walking, fording streams and getting into the most inaccessible parts of the island.

La Palma appears from time to time in the scarier sections of the media because of a threat to inundate the Eastern coast of the USA and the Western coast of Europe by a huge tsunami [358].

It is claimed that a large flank of La Palma could slide into the Atlantic Ocean and create a wave height of between 650 metres and 950 metres.

This wave would be about 50 metres high on the western Atlantic Ocean and would devastate coastal regions up to 25 kms from the coast on the eastern seaboard of the USA, Canada and other coastal areas.

Tens of millions of US and Mexican citizens would be killed.

However these claims are refuted in the website at [359].

Never-the-less, a tsunami created about 70,000 years ago when a part of Cape Verde's Fogo volcano collapsed is estimated to have created a wave 170 metres high. When this hit the coast of Santiago Island boulders weighing over 700 tonnes were lifted up over 200 metres and thrown well inland [360].

My home is 300 metres from the English Channel and 12 metres above sea level.

Best I put a small boat in my back garden in case La Palma

collapses.

With our sturdy hire car we went many times up the 300+ hairpin bends to the observatories near the summit and down those 300+ hairpin bends.

Only on one occasion did I vomit. That was on a Sunday and we had failed to find a café or restaurant open.

We were starving. All we could find was a roadside stall selling weird things from the sea.

Being ravenous, we bought anything available.

I ended up with octopus tentacles in olive oil. The suckers were all present and accounted for as I closed my eyes and tried not to think about what I was eating. Later that evening, the whole lot came up around hairpin bend number 205 - or was it 206?

Luckily I got out of the car in time.

The surrounding countryside changed slowly but surely as we climbed the tortuous road. At sea level, La Palma is lush and wet. The deciduous trees thin out with altitude to be replaced by conifers which themselves die out to reveal vast stretches of red coloured lava and a few red rocks.

The astronomers had given names to geological features on the boring drive up and down the mountain.

This block of volcanic rock shown above was called '*The Submarine*'.

This barren landscape continues for many kilometres until suddenly - there are the gleaming observatory domes.

Every time we reached this point after ninety minutes of driving it was exciting. Here were a band of esoteric hermits leading a life of intellectual conquest and celibacy.

Some astronomers lived up the mountain in dormitories. These were not popular but avoided the long drive up and down the mountain.

Other astronomers did the drive each day just to share in the cosmopolitan wild living of La Palma. By the way, I am being sarcastic!

Life up the mountain was typified for me by the games room.

The astronomers had asked the Spanish authorities to provide a snooker table. Eventually it arrived. A snooker table - with no pockets and three balls.

No rules were provided.

I think that we three visitors were the only people to have played on this table - and we had no idea what we were supposed to be doing!

One feature of the mountain top of La Palma which makes it great for astronomy are the clouds.

This seems a strange thing to say but the clouds rise up in the daytime - and often rise well above the mountain top - but at night they compact down below the peak.

They are then so dense that all light from the inhabited region of La Palma is trapped below the cloud level.

Indeed, the lights of the capital city, Santa Cruz de la Palma, can just faintly be seen as a soft glow on the top of the clouds when looking down from the observatories.

The two pictures below show the clouds in the morning and in the late evening. (The pictures were taken from different places on the peak so that the mountain peaks do not quite line up.)

These show the vertical movement of clouds during the day that allows light pollution to be trapped at night.

The potential of the summit of La Palma as a good astronomical site had been checked by an astronomer who trekked up to the top with a few donkeys loaded with instruments, food, water and a tent.

For a year he measured the clarity and steadiness of the skies using a small telescope pointed at the star Polaris.

The improvised mounting on which his telescope had been

propped survives as seen in the photograph below.

This picture shows the moment of sunset. The tapering shape is the shadow of the island blending into the Earth's Shadow. The peak's shadow reached to Tenerife itself some 110 kms distant (visible by eye but not in the above picture) - a sign of the phenomenally clear skies.

I explore the horizon using a pair of 20 x 100 giant binoculars.

These gave superb views of the skies at night.

Below we see me posing at the highest point, above the observatories which gleam brilliant white in the setting sunlight.

In the above view we see the William Herschel dome at left, the Swedish solar tower telescope next to it, the Isaac Newton dome almost hidden and the Dutch Jakobus Kaptyn dome.

The latter contains a telescope with a one metre diameter mirror which is small by modern standards [361].

Since 2014, the Jakobus Kaptyn telescope has been owned by the Instituto de Astrofísica de Canarias (IAC) and operated by the South-Eastern Association for Research in Astronomy (SARA).

The latter has worked to revamp the facility as a remotely operated observatory.

Increasingly over the past two decades the telescopes have been converted to remote operations.

At the time of our visit the summit of La Palma was populated by astronomers, engineers and support staff. By 2016 the astronomers were mostly back in their home bases and the telescopes move and collect data under the control of internet-based remote control schemes.

A few engineers and support staff are on hand on La Palma to keep the instruments working correctly.

This photograph from the time of our visit shows the 'Nordic Telescope' under construction. This is a joint venture by

Denmark, Sweden, Norway, Finland, and (since 1997) Iceland. It became operational in 1990 [362].

This was the first major telescope to use active optics to correct the shape of a thin primary mirror supported on electro-mechanical actuators.

The main mirror has a diameter of 2.56 metres (101 in).

Some local astronomers we spoke to were convinced it would have a poor performance due to the location hanging on the edge of the mountain side but subsequent tests show it to have an excellent performance.

The Carlsberg Meridian telescope was sponsored by the brewers of what is '*Probably The Best Lager In The World*'.

This instrument swung very precisely in a North-South plane and was fully automated to measure precise positions and brightness of stars, planets and asteroids.

In the past, such instruments were widely used for measuring star positions but the requirement is less now thanks to the

HIPPARCOS, Tycho and similar missions.

The asteroid work appeared to be the main function for this instrument when we visited.

We knocked on the observatory door and asked the rather surprised technician if we could look around - having established our credentials as informed amateurs. The Danish technician seemed to be delighted to have some company.

He explained that he had very little to do.

A task list of objects was programmed into the computer memory every day and the machine then ran itself - as we saw it simply nodded up and down clicking away.

The instrument was decommissioned in 2013 and moved to a museum.

Because one of my companions on this trip was, until a short time previous to our visit, a computer software engineer at the observatories, we were able to go freely into all the domes and meet the professionals. In fact, many seemed to think he still worked there!

We got to look over the instruments in daylight and then see them in operation at night.

Right at the peak is a shrine, possibly erected by the astronomers to appease the gods and bring fine weather and research funds to the observatory site.

One rather curious relic we found by the roadside near the peak was the cylindrical box in which the Isaac Newton 2.5 metre (100 inch) mirror had been transported from Britain.

This was bought on the deck of a container ship.

During the journey, a Harrier jet aircraft pilot from one of the Royal Navy's aircraft carriers got into trouble far from his ship. The pilot searched around and found a large container ship.

He put the aircraft down on the top of one of the containers until the ship could reach port and the aircraft could be rescued.

The container on which he landed held inside the Isaac Newton telescope mirror.

In view of the risk to the expensive mirror it might have been cheaper to lose the aircraft and save the mirror!

In addition to surprizing the astronomers by knocking on observatory doors and inviting ourselves in, we went out at night to just gaze at the incredible night sky.

The La Palma sky was truly awesome.

We watched the clouds descend and the Sun sink to the horizon.

As the last speck of Sun disappeared it turned an emerald green - the elusive Green Flash! I had a 200 mm lens attached to my camera and was waiting for the flash to take a picture but it was over too fast and took me by surprize.

As darkness fell we cursed the light pollution from the island's coastal strip but then, as we traced the shape of the glow, we saw that it was a cone inclined along the ecliptic and that it was setting in the direction of the Sun. It was the Zodiacal Light rising like a shaft in the west [363] and not light pollution.

As an aside, in 2007 Brian May of 'Queen' completed his PhD thesis 'A Survey of Radial Velocities in the Zodiacal Dust Cloud' 36 years after he started, and abandoned it, to pursue a career in music [364].

He was able to submit it only because of the minimal amount of research on the topic undertaken during the intervening years. His original observations were carried out at the observatory on top of Mount Teide in Tenerife [365].

We had borrowed a pair of 20 x 100 giant binoculars for this trip. With these we spent hours tracing out Messier objects and just sweeping along the Milky Way.

It was overwhelming - especially when I swept up Omega Centauri - without doubt the finest deep sky object of them all - a huge globular cluster sadly not visible from the United Kingdom.

There is a helicopter landing pad at the peak - the only area free of razor-sharp lava apart from the roadways.

We relaxed on our backs on this concrete pad and stared up at the stars. We seemed to be floating up and away from the Earth. It was truly a weird and wonderful feeling - breaking free of gravity and floating off to the stars.

Overall, I think the two most impressive things about these days and nights spent on the peak of La Palma were the sky and the immense William Herschel telescope.

With its altazimuth mounting and its huge size it was much more impressive than the Hale 5 metre (200 inch) telescope on Mount Palomar even though the William Herschel mirror is 'only' 3.75 metres (150 inches) diameter.

This was the last telescope built by the Grubb Parsons Company based in Newcastle, England. Since 1833 this company had built some of the finest large telescopes in the world [366].

These include the 28-inch (710 mm) refractor at the Royal Observatory, Greenwich - the UK's largest refractor. This was built in 1893.

The William Herschel Telescope Dome [367].

We also explored the rest of the island.

La Palma is a volcanic island with a festering volcano in the south. Everyone thought it was dormant until 1971 when it erupted as shown at [368].

Is it wise to set up a load of expensive telescopes on an earthquake zone and volcanic island?

As we drove around the island we came across huge lava flows; some of which had poured over main road and blocked communications across the island.

Above we see my eldest son climbing out of the volcanic vent in 1986 and, below, some tourists were frying eggs on the rocks to show how hot the whole area still is.

We returned to Tenerife to explore the observatories on Mount Teide. Many of the observatories were for observing the Sun.

We knocked on the door of a German solar observatory but they declined to let us in; the only time we were not welcomed into an observatory. Not wishing to have a 'Basil Fawlty Moment' we didn't mention the war.

One site on the mountain top was like a concrete pyramid.

This housed the Global Oscillation Network Group (GONG) project instruments [369]. The name is a good pun because the GONG observatories continuously observe the Sun and measure the acoustic waves on its surface. Typically these have periods of a few minutes and can give an insight into the internal structure of the Sun.

Above is a general view of the solar observatories on Tenerife.

The dull appearance is due to Saharan sand blown in on a strong easterly wind. This blots out any chance of making observations and leaves a film of red dust over the domes as well as filling the air with a fog of grit.

Another instrument operating at the time of our visit was a device for measuring differences in the Cosmic Microwave Background.

Although COBE came along later and made sensational observations of these differences, the Manchester University experiment was claimed to be about four orders of magnitude cheaper than COBE and gave more accurate results.

It was simply a microwave detector with an oscillating plate which reflected the cosmic noise from two adjacent patches of sky into the detector.

We also took the opportunity to explore the lava fields of Mount Teide.

These were extensive and fascinating. In the distant past, rivers of red hot lava had poured over the countryside and scoured out tubes underground which were up to a kilometre long.

We climbed over the lava and into some of these lava tubes. It was a truly eerie experience but we didn't go too far underground in case we got trapped in the dark.

Also, the lava was like a field of upturned razor blades.

Similar volcanic activity created the lunar rilles, such as Schröter's Valley [370] (also known as the 'Cobra Head') which is seen below [371].

After having explored briefly the 'lunar landscape' of Tenerife - the part that the vast majority of tourists never venture into - our trip came to an end.

19. Light Pollution

Above is one of many pictures taken in and around my home town of Weymouth in Dorset as part of my Open University degree course in Environmental Studies in 2006.

This shows the Safeways (now Morrisons) supermarket flooding the skies with wasted light energy from globe lamps in the carpark and from a transparent roof.

As an amateur astronomer I naturally have had a long held interest in the curse of light pollution.

Back in the early 1960s when I was starting to make serious observations there was little thought that flooding the skies with light was a bad practice.

This was before the public were thinking much about energy saving schemes, climate change, the effects of light pollution on wildlife and the blanking out of our views of the heavens.

Predicting Light Pollution

As part of my Open University studies I created a map of Dorset which predicted light pollution levels using the simple 'Walker Formula' [372].

The formula states that the relative level of light pollution from a community as seen within 45 degrees of the horizon in a direction facing the source is proportional to $Pd^{-2.5}$ where 'P' is the community population and 'd' is the distance of the community from the observer [373].

This simple formula predicts the relative brightness of a sky as a function of the distribution of local communities.

I generated the map below showing predicted light pollution levels in my county of Dorset.

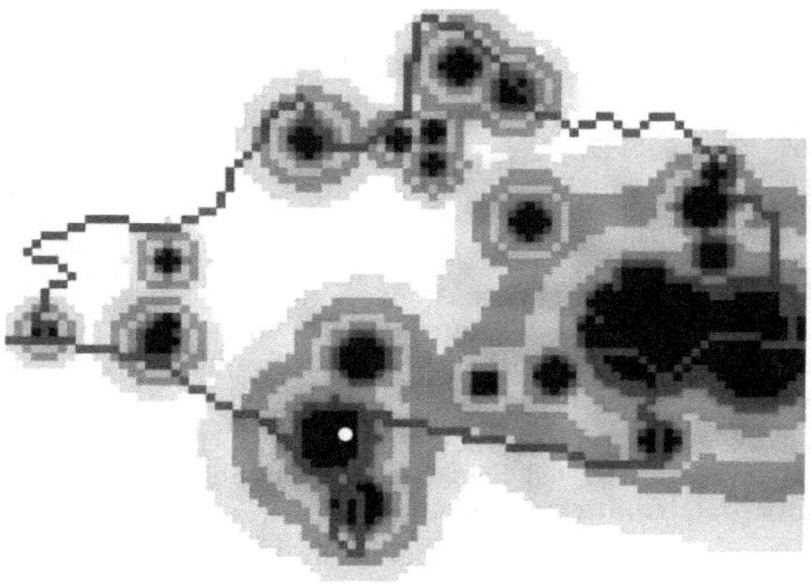

My home is where the white marker is located in Weymouth.

I used the MS EXCEL™ spreadsheet software and computed the predicted light pollution levels to a geographical resolution of one kilometre.

I included all communities with a population exceeding one thousand in Dorset and over the border into Hampshire, Somerset and Devon; light pollution does not recognize county boundaries.

I limited the calculations to a community size exceeding one thousand because smaller communities in Dorset do not general have street lighting.

The influence of the main towns can be seen; Weymouth, Dorchester and Portland in the South, Lyme Regis, Beaminster and Bridport in the west, Bournemouth and Poole in the east as well as Yeovil and Sherborne to the north.

It can be seen that there is a 'dark skies' area to the north west of Dorset around the area of the Marshwood Vale.

My map has since been confirmed by the Campaign to Protect Rural England (CPRE) which has generated a light pollution map for England using data collected by space probes [374].

Measuring Light Pollution

How light polluted are your night skies?

There are several methods of assessing the quality of your celestial environment ranging from buying a complex electronic metering device to using a method that uses only the unaided eye.

The simplest assessment is to find a patch of sky and see the faintest star that can be detected with the unaided eye. Then check the brightness of that star using a planetarium program such as Stellarium [375]. (Other planetarium programs are available.)

A method that avoids having to check individual star visual magnitudes is to count the number of stars seen in an easily defined and unambiguous area of sky.

There are several such areas that are popular.

One is the Square of Pegasus.

Count how many stars you can see inside the Square of Pegasus (not including the four corners of the square) and check your limiting visual magnitude on the chart below which has been constructed using data from the excellent freestarcharts.com website [376].

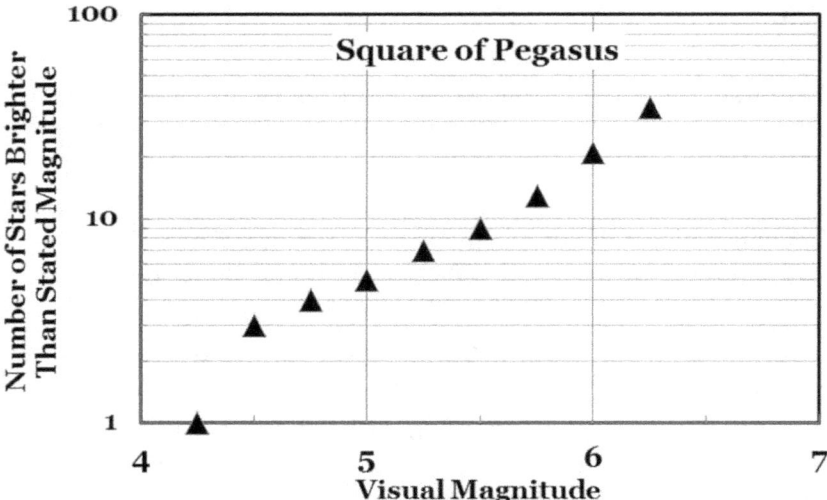

There are several other areas of sky, e.g. the 'body' of Orion, where these star counts are used to give an indication of sky quality.

Some attempts to use this method of star counting can be inaccurate. Many years ago Patrick Moore set up an experiment on the 'Sky at Night' TV programme. He asked viewers to count how many stars they could see in the Pleiades star cluster.

Responses ranged from plausible numbers around a dozen up to over one hundred. Clearly people were counting stars well outside the star cluster showing the need for a well defined definition of the test area.

Using areas like the above have serious flaws which make such star counts of very limited value.

These problems are

1. The star count areas are not visible in dark skies for long periods of the year.

2. The star count areas can be at widely different altitudes above the horizon when measurements are made.

It is known that the visual limiting magnitude of stars can vary enormously with altitude above the horizon due to mist, atmospheric absorption, local light pollution in certain directions, etc

What I propose is for observers in the northern hemisphere to count stars in an area centred on the star Polaris using a simple device for defining the same area of sky for all observers. Observers in the southern hemisphere should use the southern polar region although the lack of a bright star near the southern celestial pole is a disadvantage.

Because such an area is at constant altitude - or at least far more nearly constant than the Square of Pegasus or Orion - a consistent set of measurements of sky conditions can be obtained for any geographical location.

To define the area of sky to be measured we take the cardboard tube from inside a toilet roll and cut it so that the length is 2.5 times the diameter as shown below.

This creates a view of view of approximately 20 degrees in diameter when looked through.

Put the tube to your eye, place the star Polaris or the southern celestial pole in the centre of the field of view and then count how many stars you see.

Light Pollution Monitor Mark 1

We can see Patrick Moore using such a device for measuring light pollution at [377].

I have constructed the charts below using data from the Yale Bright Star Catalogue [378].

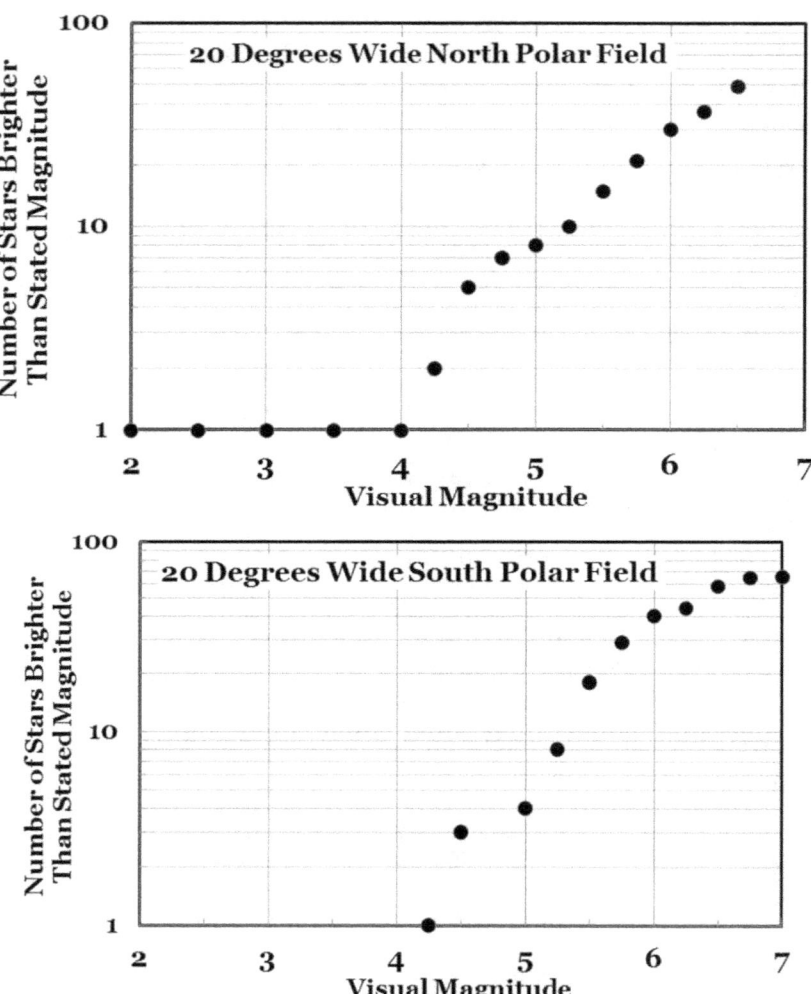

Using this simple technique allows the condition of the sky to be monitored in a way that is consistent over time and date.

If light pollution is so bad that stars around visual magnitude 4.0 cannot be seen then another simple technique must be used.

Look towards a wider region around the celestial pole (north or south) and find the faintest star visible closest to the pole. Look up this star's visual magnitude using Stellarium or similar source of information and use that as your limiting magnitude.

To determine the limiting magnitude using binoculars I propose a smaller field of stars close to Polaris.

The area chosen is the 'bowl' of Ursa Minor - the 'Little Bear'.

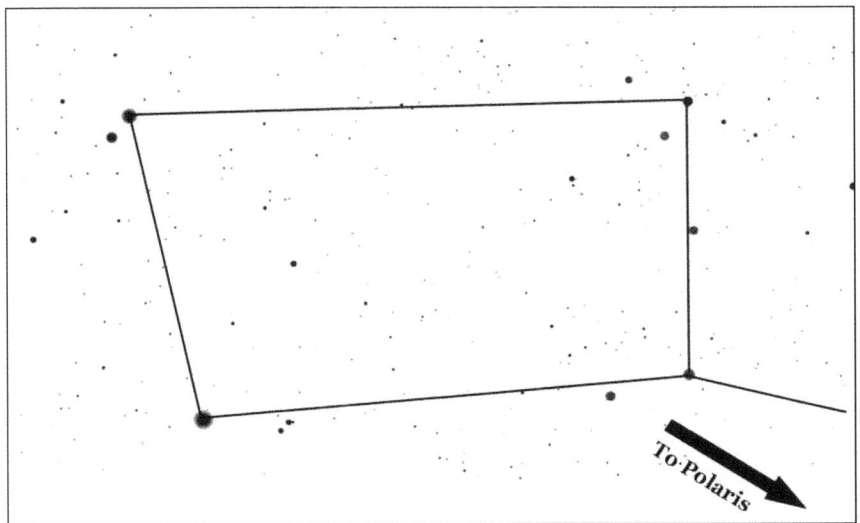

The area is shown above.

Simply count the number of stars within the 'bowl' of Ursa Minor as shown by the lines on the star map. Do not count the corner stars. Then look up the limiting magnitude on the chart below.

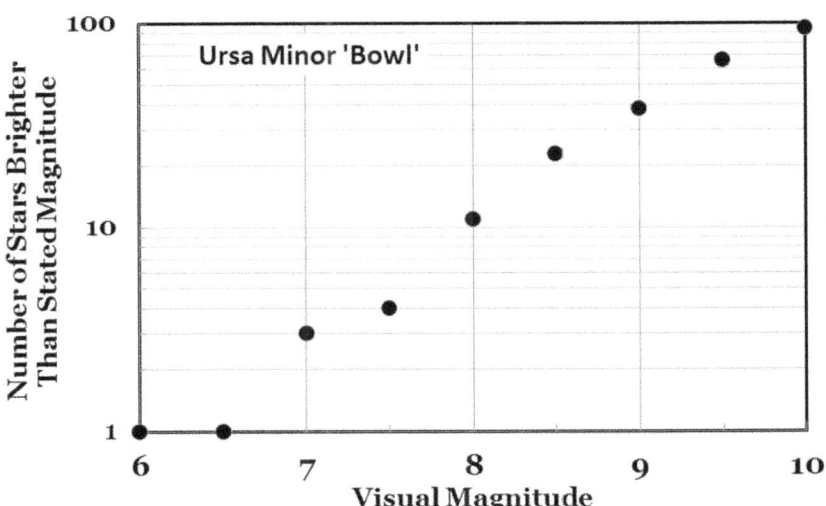

The BAA's Commission for Dark Skies (CfDS)

As the 1980s progressed, a ground swell of opposition to unnecessary lighting was growing and organisations such as the BAA's Campaign for Dark Skies (CfDS) - now renamed the Commission for Dark Skies - were set up.

The CfDS was founded in 1989 [379] and made its voice heard, for example, through its work with the House of Commons Science and Technology Committee on legislating against light pollution.

This resulted in the government including provisions in their Clean Neighbourhoods and Environment Bill [380] in 2005.

There is now international recognition of the problems created by stray lighting and the forming of the International Dark Sky Association has given great impetus to solving the light pollution problem [381].

It was a long struggle to get the problems of light pollution recognized and legislated against. Many people believed that opponents of light pollution were only a small number of campaigning amateur astronomers whose interests were not important.

This could have been understandable were it true but it was not.

Every lamp sending unnecessary energy into areas where it is not needed is wasting money, discharging unnecessary carbon dioxide into the atmosphere, causing nuisance, affecting wildlife and possibly increasing crime and illness.

One of the UK's 'leading lights' in the battle against light pollution had been my friend Bob Mizon who has been a tireless committee member of the CfDS and an author of several milestone books as well as newspaper and magazine articles on mitigating light pollution, for example [382], [383], [384].

The BAA Commission for Dark Skies makes the following important points on its website

- Light intruding into bedrooms at night can cause problems involving the human immune system and cancer [385]. In general, excessive exposure to artificial light, for example by shift workers working at night in bright workplaces, correlates with a greater incidence of breast cancer [386].

- Lights have caused highway fatalities by blinding car drivers [387].

- Inefficient lighting wastes over £1 billion per annum in the UK alone [388].

- Less than 10% of the UK population can see the beauty of a natural night sky full of stars [389].
- Lighting can help criminals see what they are doing. For example, in Saffron Walden, Essex, the police reported that night-time crime almost halved after the street-lighting was turned off at night [390].

I will give no further general warnings about the evils of light pollution because these are expertly described in the websites of the CfDS and the International Dark Sky Association.

What follows in the rest of this chapter is my account of my personal involvement in fighting light pollution which resulted in me being given the award shown below.

My Local Campaigns For Darker Skies

What follows are some of the many letters I have had published in the media and some of the responses.

On 2nd January 1996 I had the following letter published in the Dorset Echo.

"I was delighted to read your excellent editorial plea for sensible street lighting that does not throw energy wastefully into the sky.

We all pay for the electricity wasted in this way. Our councillors cannot be allowed to squander our Council Tax lighting up the sky.

Other major sources of light pollution are security lights.

Studies have proved that increased lighting actually increases crime. Where brighter street lights and security lights have been installed, crime increases.

The Government agrees. Home Office Crime Prevention Unit Papers 28 and 29 state:

'Better lighting, by itself does not reduce crime. There is no scope for reducing crime by investing in better lighting.'

The worst security lights are those which flood the building and surrounding countryside with brilliant light. Burglars love them.

Floodlights identify property worth raiding. There are plenty of deep shadows in which to hide and onlookers are dazzled so that intruders cannot be spotted. In any case, three out of four witnesses to a burglary pass by and do not report it.

If you really must fit security lights - and there is really little reason to do so - forget what some so-called 'experts' tell you. They are more interested in selling you the most expensive and brightest lights available - known as 'Rottweiler Lamps'.

Fit lamps not brighter than 25 watts and aim these downwards. These do not dazzle onlookers, do not provide harsh shadows and provide insufficient light for burglary, forcing burglars to use torches - which are evidence of a burglary in progress.

Some farmers in remote areas have suggested floodlighting their farms and have even been encouraged in this by the Dorset Police Crime Prevention Officer (Echo, February 2, 1993).

How ridiculous!

Which would a burglar prefer - to stumble around a pitch-black farmyard thick with dung or to have farm machinery nicely illuminated so that it can be selected and easily stolen?

Why floodlight valuable equipment if nobody is there to see the burglar at work?

Farm dogs on the loose would be far more effective than useless floodlighting!

Lights out, please!

G. J. KIRBY"

On 15th January 1996 I wrote in the Dorset Echo

"No astronomers want to plunge the country into the sort of darkness that existed in the war even though my life-long interest in astronomy was fired by the magnificent starry skies I saw as a child during the London Blitz.

All we want is the reduction of unnecessary lighting which goes straight out into space and wastes energy.

As an example, there are many very attractive globe lights in Hope Square (Weymouth) and elsewhere. However, half their light goes out into space - wasting energy and money. Fitting a small reflective cap on the top would stop this waste and send more light down onto the pavements where it is needed. The lights would still be highly attractive.

A few years ago I was in Tucson, Arizona, where all lighting is controlled and efficient. The city is very pretty with well lit streets and decorated buildings but I was able to see the Milky Way shimmering overhead from the city centre. In Britain you can only see that magnificent sight at least 40 miles from a town and that, I'm sorry to say, rules out much of Dorset.

G. J. KIRBY"

A report in the Dorset Echo dated 20th January 2007 stated that the Dorchester Town Crime Prevention Panel wanted to spend as much as £1 million floodlighting local car parks.

The Panel Chairman stated that lighting levels had to be increased to *"make people feel safer"* and went on to state that

"Increased lighting levels in the car parks are in the interests of public safety"

which, of course, is not borne out by crime statistical evidence.

I responded with a letter as shown below dated 24th January 2007.

"The story 'Let there be light in parks at night' (Dorset Echo, January 20) is based on the wholly inaccurate assumption that increased lighting in public places, such as car parks, will reduce crime.

Indeed, Jon Halewood, the deputy chairman of the panel investigating the use of more lighting, is quoted as saying 'Increased lighting levels in the car parks are in the interests of public safety'.

Amazingly, he goes on to admit

'Obviously, we are not lighting engineers but we would be able to use our personal judgment to see how we feel about lighting...'

Please do not allow decisions on crime prevention to be made by poorly-informed and misguided amateurs when there are well-informed experts on hand - especially when more than

£1 million of West Dorset District Council's taxpayers' money is about to be squandered.

There is no evidence that increasing lighting levels reduces crime - in fact, the available evidence shows that crime levels increase when lighting is increased in public spaces.

After all, the majority of muggings occur in daylight or in brightly-lit shopping centres for the obvious reason that muggers need to identify their victims and not attack a policeman in mistake for a pensioner.

Members of the panel deliberating on this issue must read the Home Office Crime Prevention Unit papers 28 and 29: 'The Effect Of Better Street Lighting On Crime And Fear - A Review'.

They will see that increasing lighting will reduce the fear of crime while - paradoxically - increasing the risk of crime.

Please, keep lighting levels down for the sake of the vulnerable members of our society and for the environment.

Geoff Kirby"

On 22nd August 2011 this false link between lower lighting levels and crime increase was again aired in the pages of the Dorset Echo when a letter writer stated amongst other myths and misleading statements

"Will Wyke Regis be provided with extra police patrols to deter those who like to indulge in a little late night crime made all the easier with Wyke Regis being in pitch darkness?"

What this writer, like so many others, failed to grasp is that in pitch darkness it is much more difficult to commit a crime than in a well-lit area. After all, most burglaries take place in daytime.

On 24th August 2011 I had the following response published in the Dorset Echo.

"Dear Sirs,

It is wrong to assume that switching off certain Wyke Regis street lights overnight will increase crime. Indeed, crime should decrease when the lights are out.

The Home Office commissioned a study by Southampton University (Home Office Crime Prevention Unit Papers 28, 29) which showed that increased lighting does not reduce crime - indeed, the opposite is true.

The reasons are obvious with a little thought.

The majority of home burglaries, muggings and vandalism occur in broad daylight, not in the dark night hours.

Burglars prefer to see what they are doing rather than fumbling around in the dark waving a torch which gives them away to the public. Also houses are more likely to be empty in the daytime than overnight.

Home security lighting can actually increase the risk of burglary because the lighting dazzles passers-by, provides light for the burglar to see what he is doing and allows him to see anyone watching him hidden safely in the deep shadows cast by the lights.

Muggers need to be sure that the person they are eyeing up is a teenager with an expensive mobile phone and not an off-duty policeman or an 18 stone rugby player. They do their mugging in daylight or in well lit areas at night for obvious reasons.

Vandals cannot spray graffiti in the dark.

The only effect of increased street or building lighting is to make people feel safer. Sadly, this is an illusion due to our primeval fear of the dark.

It is the light that actually exposes us to crime.

G J Kirby"

These were a small sample of the many letters I have written on the topic of light pollution.

For my Open University Degree project I conducted a study of light pollution in South Dorset.

As part of this project I took the following pair of pictures outside my house with my partner standing outside pretending to be a burglar breaking into the house.

This well illustrated the futility of 'security' lights; if anything they attract burglars and assist them to break in under cover of the light's glare.

Above we have my security light switched off and below the security light is switched on.

A conclusion of my Open University Project was that a complete conversion of all street lighting in my home town of Weymouth to new, environmentally friendly lamps would save about £100,000 per year in electricity costs (plus maintenance and capital depreciation costs not considered here), and save about 1,480 tonnes per year of CO_2.

Eventually the County and Local Councils 'saw the light!' and every street light in the County of Dorset was replaced by new low energy and low maintenance lamps starting in 2011.

Also, street lights were turned off after midnight on many housing estates.

Of course there was a huge barrage of complaints at the latter campaign with people saying that they were so afraid of being attacked that they never left their houses after midnight.

However, in a report in the Dorset Echo dated 30th July 2015 it was confirmed that £300,000 has been saved every year by these measures and that crime had fallen in areas where street lighting had been switched off after midnight.

At least on a local level the battle to reduce energy waste and light pollution is being won.

But there is still a long way to go especially to convince people that criminals prefer better lit areas in which to commit crime.

I Am Given A CfDS Award

For my support for keeping our skies dark I was awarded a certificate of appreciation by The Campaign For Dark Skies;

It was Bob Mizon who presented me with my award and that made me very proud indeed.

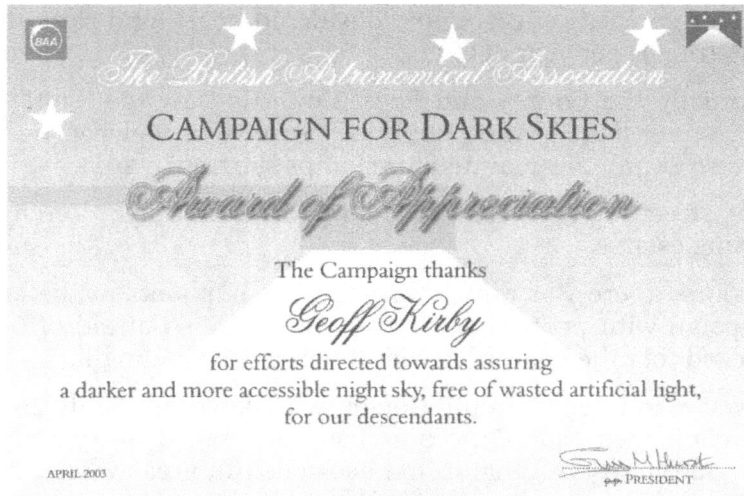

The British Astronomical Association

CAMPAIGN FOR DARK SKIES

Award of Appreciation

The Campaign thanks

Geoff Kirby

for efforts directed towards assuring
a darker and more accessible night sky, free of wasted artificial light,
for our descendants.

APRIL 2003

PRESIDENT

Why Floodlight Churches?

The floodlighting of churches has long been a source of aggressive light pollution [391].

Why anyone should think it a good idea to floodlight churches is beyond my understanding. Many churches so illuminated are not of interest.

In many cases the floodlights are aimed from ground level and many miss the building to go straight into space.

"The Heavens Declare The Glory Of God And The Firmament Showeth His Handiwork!"

Psalms 19.1

Unless you live near a floodlit church that it!

The picture below shows a church just 300 metres from my house. No wonder the unaided eye limiting magnitude from my back garden is only +3.5 on a 'good' night.

Exterior church floodlighting schemes were encouraged by the both the Millennium Commission and the Church Floodlighting Trust, using money from both the public and from interested parties to urge churches to install exterior lights.

20. Astrology And Other Belief Systems

I must say at the outset of this chapter that I don't believe in astrology because I am a Libran and we Librans are very sceptical.

Astrology may seem out of place in a collection of essays on astronomy.

However, my Google search for 'astrology' bought up 145 million entries whilst typing 'astronomy' bought up 115 million results.

I am not sure what that proves except that the two subjects share comparable interest.

Astrology Or Astronomy?

It always surprises me how many people muddle up astronomy and astrology. I have encountered many people who refer to me as an astrologer and, in many minds, the two names are interchangeable.

Even some people who are supposed to be well educated and intelligent do not know the difference.

As an example, at the 40th anniversary party for the *'Sky At Night'* TV programme, BBC1 controller Alan Yentob [392] congratulated Patrick Moore on his great services to *'astrology'* and praised the programme for cultivating public interest in 'astrology'.

Martin Mobberley has a wonderful photograph in his biography of Patrick Moore taken at this party [393]. Yentob is making a fool of himself raising a glass of champagne at Patrick who is looking totally appalled by Yentob's ignorance of the difference between astrology and astronomy. However, in Yentob's favour he did successfully launch the CBBC and CBeebies children's channels.

The science of astronomy was born out of, was nurtured by and was strongly related to, the developing art of astrology. As they grew apart astronomers went on the attack. The ensuing battle has, at times, been ferocious.

Evidence Against Astrology Being Of Value

This embattled and embittered feud has often - and sadly - shown both sides to be unreasonable and unscientific. I'm sorry to say that the professional scientists sometimes come out worst because of their arrogant manner.

The problem with astrology is that it is an ancient belief system - like religion - and, in both cases, supporters do not consider there to be a need for it to be examined in a rigid scientific fashion.

That it works is self-evident to believers.

Like religion, no *specific* predictions are ever made that would be a crucial test of the value of astrology. It can never be proven to be of value or to be worthless as a system for explaining and forecasting.

That is not to say that astrology has never been tested. It has, but always by sceptics who demand a high level of reliability.

Scientists are happy to accept rather flimsy evidence to 'prove' what they want to believe or what they expect to be true - see my article on Pathological Science in Chapter 10.

Astrologers do the same - choosing the occasional randomly generated successes and ignoring the many failures.

What is the evidence against astrology?

Look at the Astrological Compatibility charts below which have been produced using data collected by James Randi for his book 'Flim Flam' [394].

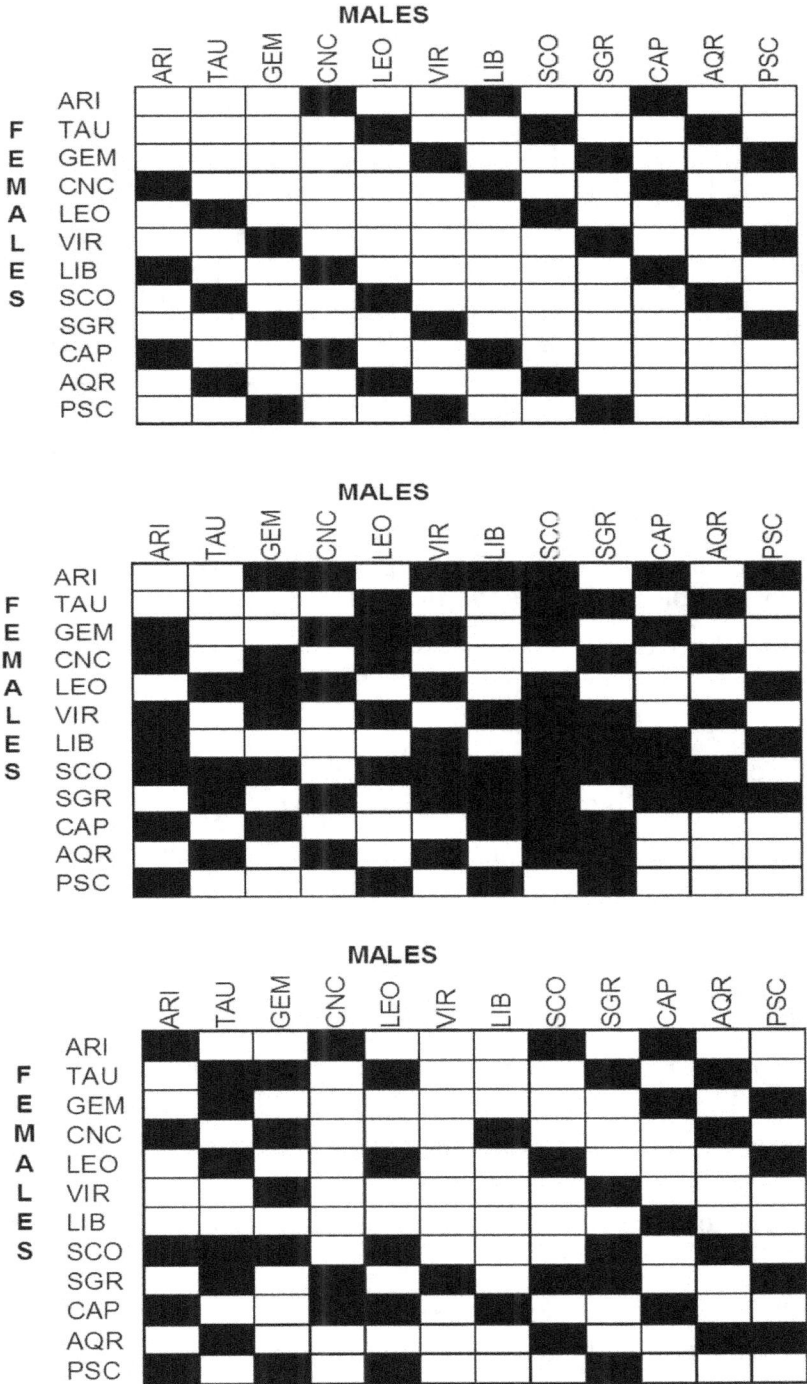

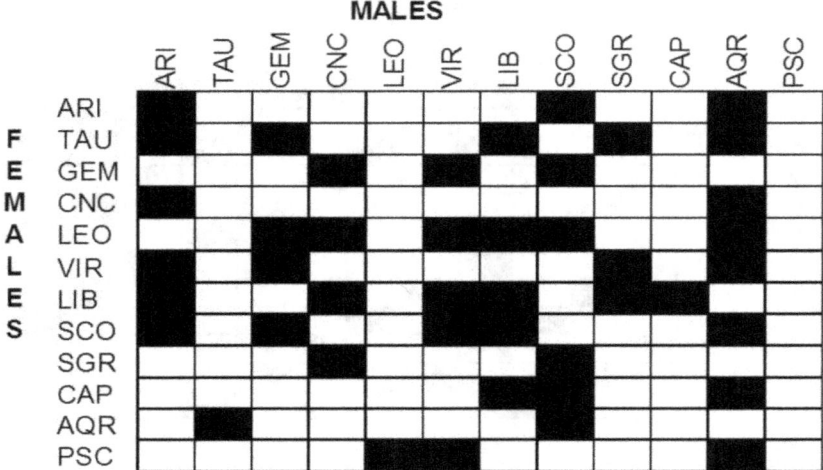

These predictions were produced by four top professional US astrologers. They show what each astrologer reckoned were the incompatible birth signs for males and females.

Notice how these four professional astrologers totally disagree with each other in their assessment of birth sign compatibility.

How can this be?

Astrology is supposed to be a 'science' with 'degrees' in astrology being awarded by many 'astrological universities'.

This inconsistency is damning enough but there are many other tests that have been conducted - some scientifically rigorous and others more showmanship but just as illuminating.

A large scientific investigation was carried out with the collaboration of several professional astrologers. They were invited to set up experiments to test - and prove - that astrology really works. By that we mean that it comes up with successful and reliable predictions - like any good science.

The experiments were evaluated and the results published in the prestigious science journal Nature [395].

The results were highly complex and the statistical evaluations lengthy.

However, the results were unambiguous.

Astrologers could not predict star signs better than random chance.

One might have thought that this would have put paid to all further astrological predictions. However, the paper in Nature has been analysed in great detail by astrologers and many papers have been published attempting to demolish the conclusions with some commentators even claiming that the paper in Nature proves that astrology works! [396]

What this confirms is the old adage that there lies, damn lies and statistics.

We can make a comparison here with religious belief. If the need to believe is strong then whether astrology or religion actually works is irrelevant to the followers of the belief system.

Attacking astrology is a waste of time. No believer will ever be convinced that it does not work. I know because I have tried.

Other investigations have been less thorough than the investigation published in Nature but they have been just as spectacularly damning and have reached a wider audience.

For example, an American TV network offered a prize of $100,000 to any person who could assign twelve people to their astrological star signs after interviewing them.

When this challenge was taken up by a professional astrologer, he was presented with twelve persons each having different star signs.

After extensive interviews with each person the astrologer announced their star signs.

Every one was wrong.

Michel Gauquelin's 'Mars Effect' Study

In 1955 Michel Gauquelin published his book 'L'influence des Astres' ('The Influence of the Stars') in which he claimed to have found that a statistically significant number of sports champions were born just after the planet Mars rises or culminates (is due south at the place and time of birth) [397].

This caused a considerable stir in both the astrological world and in the world of sceptics who believe that astrology is bunkum.

This controversy has rumbled on ever since with further tests being carried out with varying degrees of success and failure [398].

Much of the argument revolves around the accuracy and

selection of the data.

The doubts over accuracy mostly are due to an accurate time of birth being required because the so-called Mars Effect correlates sporting excellence with the position of Mars in the sky relative to the local horizon at the place and time of birth.

Michel Gauquelin's data set contained the precise dates, times and locations of the births of over two thousand athletes.

That such data could be regarded as accurate for such a large number of athletes seems highly implausible to me!

There must surely be biases in the data due to poor recording and baby delivery practices.

For example, my first and third children were both born (head appeared) at 4.30 pm to an accuracy of five minutes either way. I was there with a stopwatch timing everything to I know that to be true!

This was not a coincidence.

During the 1970s in many British hospitals women in labour were given intravenous drugs to control the labour process so that babies were born in the afternoon because this fitted in best with the availability of midwives and doctors.

This meant that midwives could hope to work a normal day shift and relax at night.

My second child was born at 2.30 am bucking the system. This happened because labour was proceeding slowly and my wife was given a sleeping drug to enable her to have a good night's sleep before the baby was delivered the following morning.

However, our son had other ideas and delivery started around midnight.

Astrologers are divided as to whether horoscopes can be cast for babies who are induced to be born early to fit hospital working patterns.

Astrologers say that the 'mystical rays' that flood down from the stars and planets to set up a baby's brain ready for later life enter the brain at the moment the head appears so it does not matter whether the birth is natural or induced.

However, the astrologer who cast the horoscopes for my three children - see page 482 - said that the horoscopes would not work for my first and third children who were artificially induced.

She still took my money though!

I have 'laboured' (sorry for the pun!) this point to show how data on birth times may be cluttered up with all sorts of confusion which would upset any astrological predictions of the type proposed in 'The Mars Effect'.

The second objection I have to Gauquelin's work is that very interesting correlations can be found in random data if you look hard enough.

People see the face of Jesus on a slice or toast [399] and [400] as well as faces in clouds, etc.

Gauquelin's data set apparently found an effect of Mars on sportsmen.

However, there are hundreds of career and personality traits that could have been tested and dozens of astronomical alignments against which to test those many traits.

Even if there were no effects, by testing enough combinations of traits against astronomical factors, some remarkable correlations will appear but only by random chance.

I can illustrate this by a mathematical study I have carried out.

I programmed my computer to assign a sign of the zodiac to each of one hundred imaginary individuals at random and plotted the distribution of numbers of people in each sign.

I repeated this many times which is the equivalent of Gauquelin sifting through his masses of datasets attempting to match a large number of traits (e.g., athletes, politicians, scientists, etc.,) against a huge number of possible astronomical configurations of the stars and planets.

On the next page is a result that caught my eye.

I will randomly say that the people in my sample were fishermen.

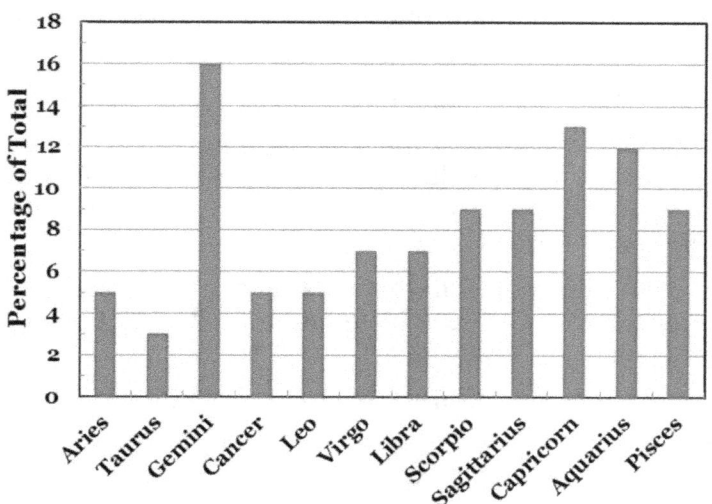

This chart shows the percentage of 'fishermen' whose zodiac sign is as shown at the horizontal axis.

An astrologer would excitedly state that Gemini is the dominant zodiac sign for 'fishermen'. Furthermore, there is an increasing influence of zodiacal signs as we move from Leo towards Capricorn.

Is this a wonderful astrological breakthrough?

No - because this chart was generated entirely using random numbers.

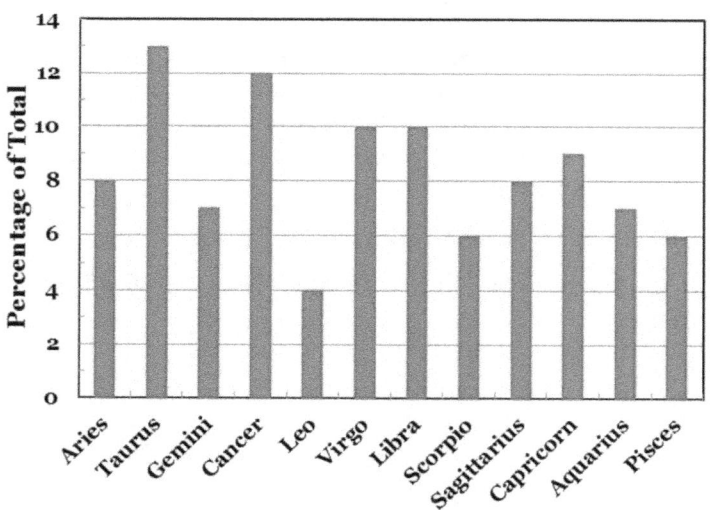

The second chart shown above was also generated by my computer using only random numbers.

This chart shows no particular bias and is typical of the majority of results I obtained.

What has happened here is that a wholly random set of numbers has produced an apparent correlation in the first chart but no such effect on the second chart.

In this way, by searching through random data spurious 'correlations' may be found and these could be published as 'evidence' that the star sign for 'fishermen' is Gemini.

I mentioned above a probable cause of bias in astrological tests where it has been the procedure to induce babies to be born during daytime working shift times. How would that affect the astrological reliability of horoscopes?

The book *'Outliers'* [401] by Malcolm Gladwell examines other forms of bias in searching for correlations.

For example, Gladwell observed that a disproportionate number of elite Canadian hockey players are born in the first few months of the calendar year.

Does this confirm Gauquelin's 'discovery'?

No - for two very good reasons.

Firstly, Gauquelin claimed that sport people were influenced by the position of Mars in the sky at the time of birth. This is wholly different from Gladwell's correlation.

Secondly, there is a very good non-astrological explanation for Gladwell's discovery.

This is that Canadian youth hockey leagues determine eligibility by calendar year. Children born on January 1 play in the same league as those born on December 31 in the same year.

Because children born earlier in the year are bigger and more mature than their younger competitors, they are often identified as better athletes. This leads to extra coaching and a higher likelihood of being selected for elite hockey leagues.

Being a Libran, I was always the oldest in my classes at school. This gave me the disadvantage that there were children up to a year younger than me in my class so that, by a given birthday, they had received up to one year's extra schooling than me.

So, there is a correlation between performance in athletic and educational performance and date of birth - and hence sign of the zodiac.

I believe that the so-called 'Mars Effect' and other perceived astrological relationships can similarly be explained by non-astrological mechanisms.

In the extremely unlikely event that *'The Mars Effect'* is eventually shown to be statistically water tight then what is the mechanism? What influence does the planet Mars 'use' to influence a new born baby to preferentially follow a life in sport?

Frankly, I cannot see anything that would do this.

For a start, it is claimed to be the position of Mars in the sky that affects the future athletic performance of a baby and not the distance to Mars which can vary between 56 million kms and 401 million kms so whatever is the influence of Mars on a new born baby it does not seem to be attenuated by the distance of the planet.

So, if Mars were to suddenly shoot off out of the solar system how far would it have to go to stop producing athletes?

All very profound questions that need answers from the astrologers!

I Debate With Four Astrologers

My local newspaper, the Dorset Echo, has a lively attitude to publishing readers' letters. I have had over 300 letters published since 1964 on a wide range of subjects.

I took on astrologers through the newspaper's letter pages after a local practitioner had had an article published saying that it was a science on a par with physics and chemistry.

I attacked by pointing out that there was no known mechanism by which the positions of the planets against the background of stars could affect our lives.

This was a mistake, as I realised later.

To argue that something cannot be true because no mechanism is known is the usual way of attacking astrology and it is fatally flawed.

Galileo knew that the Atlantic Ocean tides rise and fall in sympathy with the apparent motions of the Sun and Moon around the Earth. However, he dismissed this as a coincidence because he could think of no mechanism that could link these two heavenly bodies with the oceans.

Then along came Newton and provided that mechanism - Universal Gravitation - and the tides were soon being accurately predicted using Newton's Theories.

Astrologers quite rightly came back to me and said that since astrology works (their view) then it is up to science to seek out and study the mechanism. After all, if we had always said that things cannot be true because no mechanism is known then nothing new would ever be discovered.

I retreated to lick my wounds and wrote some more scathing attacks from different angles.

When the date and location of a certain person's birth were given to a professional astrologer who produced a horrorscope - sorry I meant to write horoscope!

He said that the birth date and location were for a kindly person who loved children, etc.

The birth details given for assessment were those of Adolph Hitler!

An astrologer then wrote that, because astrology is accurate, Hitler was in fact a kind and loving person who was bent toward evil late in his life by forces stronger than those exerted by the stars and planets at his birth.

I pressed on with my quizzing of astrologers.

Why had astrologers not predicted the existence of the planets Uranus, Neptune and Pluto before they were found by astronomers in 1781, 1846 and 1930 respectively? [402]

Why did the characteristics ascribed to the planets follow the names given to them by their discoverers?

The response was surprising.

I was invited to spend an evening with a group of four professional astrologers and debate their 'science'.

In response to their offer I agreed to allow them to ask me any questions about my life on condition that, at the end of the evening, they had to tell me what my star sign was.

It was all set up and the meeting took place at my house in 1990.

I was flattered that one astrologer had travelled all the way from Bath - over 100 kilometres away - to meet me and take up my challenge.

So, I started by asking why the planets Uranus, Neptune and Pluto had not been predicted before they were discovered in 1781, 1846 and 1930 respectively.

The answer given was that astrologers in the 18th, 19th and 20th centuries had realised that there was something wrong - although none had actually said anything at the time.

What a shame!

Modern astrological charts are now seen to be 'complete' with the three 'new' planets added.

(Since this meeting Pluto has been downgraded to a 'minor planet' and is no longer regarded as one of the main bodies orbiting around the Sun. Of course, astrologers have quickly taken all this in their stride.)

I pointed out that astrologers had missed a huge opportunity to predict the existence of the three outermost planets. Indeed, they should have been able to deduce their periods of revolution around the Sun by the periodicity of failures in the predictions using the known planets out to and including Saturn.

The astrologers agreed that their failure to predict Uranus, Neptune and Pluto was a shame.

However, a contrary view is widespread amongst many astrologers. They state that planets have no astrological influence on humans until discovered. That is why they were never predicted. Indeed, it is never possible to predict new planets by astrology for this reason.

They go further and say that their failure to predict the existence of the outer three planets is in agreement with astrological theories and therefore proves that astrology works.

Think carefully about that remarkable slice of logic!

Next I asked my four erudite astrologers why the influences of planets and stars are based broadly according to their names.

For example, the mythological figure of Pluto was the Roman god of the underworld and the judge of the souls of the dead. The astrological influence of Pluto is considered to be sombre, gloomy, dark by some - but not all - 'universities' of astrological thought.

I pointed out that the name was chosen in 1930 from a long list of submitted names and Pluto was chosen from a short list of three names; the other names being Cronos and Minerva.

In fact, the name 'Pluto' was proposed by Venetia Burney then aged 11 years [403].

So was it a coincidence that the chosen name Pluto aligned with the astrological properties given to the planet?

The answer given to me was that the planet itself had influenced the committee drawn up to choose a name to call the newly discovered planet Pluto.

This again proved that astrology works according to my four visiting astrologers.

Hang on a moment. Don't some astrologers say that a planet has no influence before it is discovered?

Yes, but there was a time gap between discovery and naming. The planet exerted its influence on the members of the naming committee in that time gap.

I tried again.

"William Herschel had originally wanted to name his planet after King George III - 'Georgium Sidus'. If this name had stuck would the astrological influence of the planet have reflected the characteristics of the king?

Would those born under the influence of 'George's Star' have had an overwhelming desire to go to Weymouth in Dorset for his holidays, bathe naked in the sea and be raving mad?" [404]

The astrologers hinted that I was being facetious.

I asked about the precession of the equinoxes and the fact that there are now thirteen zodiacal constellations spread irregularly around the sky; Eridanus being a new interloper into the Zodiac.

The astrologers' reply was clear.

Astrology was based on twelve equal partitions of the zodiac in Babylonian times. The fact that the International Astronomical Union redefined the constellations and their boundaries in the 1920s is irrelevant.

You might as well say that old bridges will all fall down because we have now changed to the metric measurement system from imperial units.

OK, I reluctantly conceded that point.

However, the Babylonians invented astrology and there is no evidence that they systematically recorded the exact time of birth of thousands of people and followed their development throughout their lives in order to correlate all the subtle nuances of their lifetime personalities with the huge number of combinations of characteristics possible by the use of natal charts.

The four astrologers agreed that it was an amazing feat that these ancient people had been able to work out from all the possible positions of planets in Star Signs, Sun Signs, Houses, etc., that anything sensible could have been developed.

Many weird belief systems take their authority from the ancient people who are supposed to have invented them.

Some spiritualists 'channel' through ancient spirits such as Native Americans over 100,000 years old; never mind the inconvenient fact that Continental America was unoccupied by humans that long ago.

Astrology was invented by ancient peoples living in the area around Babylonia who, incidentally, not only believed that the positions of the stars and planets at birth determined a person's character but they also believed that women were able to give birth to elephants.

To discover what characteristics were implanted on a child by the combination of the ten planets plus the Sun, the twelve zodiacal constellations and the twelve houses would have required over one million children to have been followed into adult life and their characteristics correlated with the astrological circumstances of their birth.

I doubt that the near-Stone Age Babylonians ever did such a carefully controlled data gathering exercise.

However, astrologers continue to claim that because astrology is an ancient wisdom it therefore works.

After about five hours of questioning and cross-questioning I challenged the visiting professional astrologers to tell me my star sign. They said that this was very difficult to judge because I had given away no clues.

Of course, that is how astrology, fortune telling, spiritualism, etc., works.

Random guesses are put forward and the responses of the person - no matter how subtle - are noted. It's called cold reading.

I gave no clues as they discussed my possible Sun Sign. I sat Poker-faced and silent.

Eventually, they settled on either Leo or Sagittarius for my star sign.

Both wrong.

When I told them I was Libra they gasped.

"Yes! Of course! It is obvious now!"

They smiled and nodded to each other knowingly.

I am sure they went away thinking they had succeeded when they had actually totally failed.

My next tactic was to dismiss Sun Signs horoscopes as worthless nonsense fit only for tabloid newspaper.

On that we all agreed!

I Buy Three Horoscopes

Having stated earlier that we should not dismiss a belief system because no mechanism can be seen I have to admit to being an open-minded scientist - but not so open-minded as to have an empty brain!

I had recorded the times of birth of my three children to the nearest minute – at least, the time when the head emerged since astrologers say this is the significant timing.

I paid a lot of hard-earned cash to have professional horoscopes cast for all three of my children.

I got for my money several pages of charts, diagrams and text all computer generated for each child.

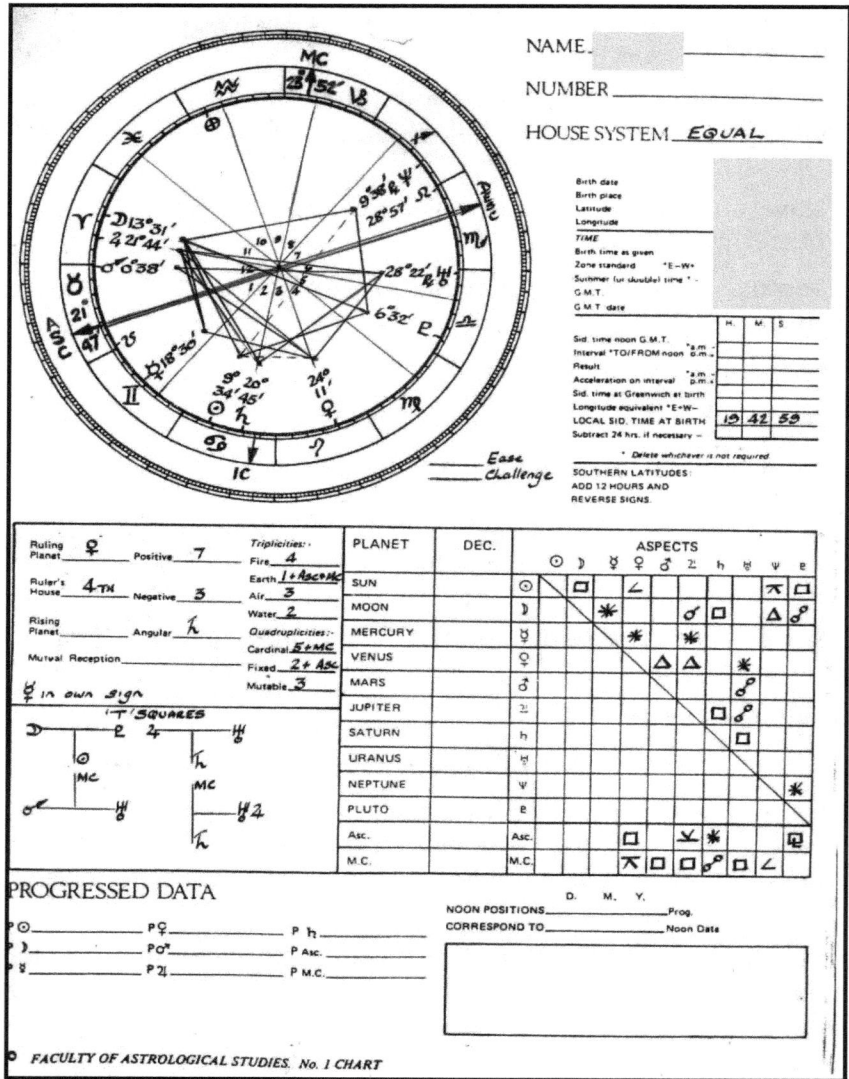

NAME

NUMBER

HOUSE SYSTEM _EQUAL_

FACULTY OF ASTROLOGICAL STUDIES. No. 1 CHART

Above is just one of the four pages for one of my children.

I have to confess that I got more for my money than I expected and, had I not been sceptical about astrology, I would have been very impressed.

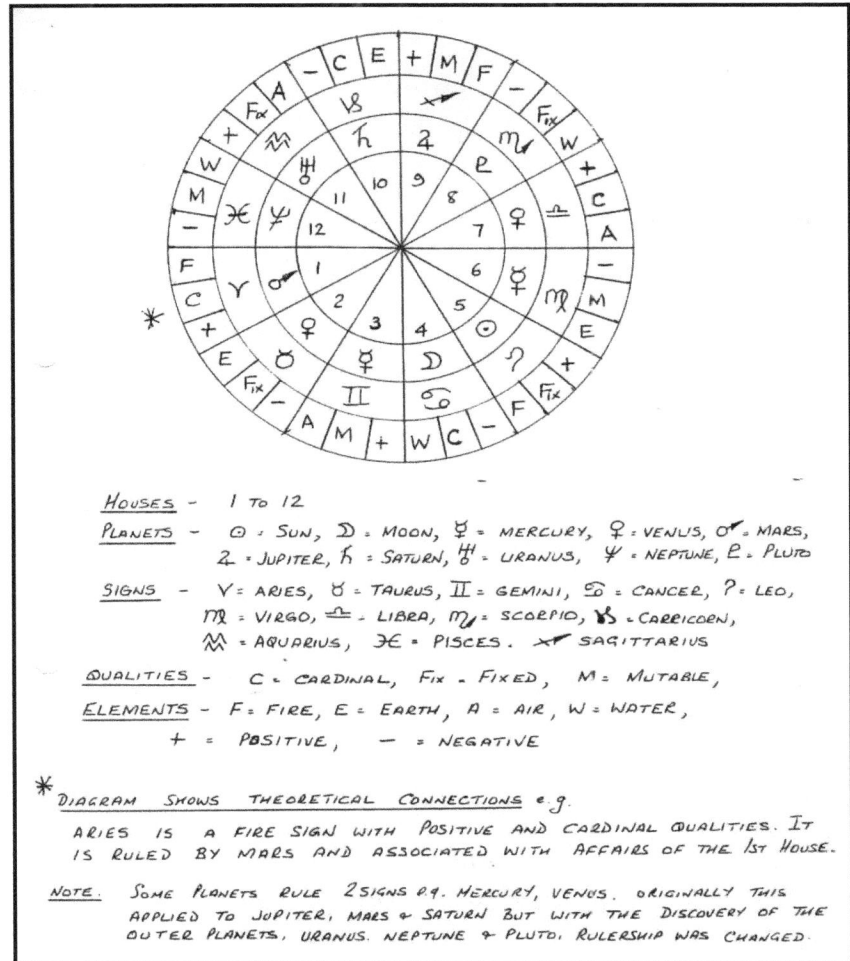

HOUSES - 1 TO 12

PLANETS -   ☉ = SUN, ☽ = MOON, ☿ = MERCURY, ♀ = VENUS, ♂ = MARS,
            ♃ = JUPITER, ♄ = SATURN, ♅ = URANUS, ♆ = NEPTUNE, ♇ = PLUTO

SIGNS -   ♈ = ARIES, ♉ = TAURUS, ♊ = GEMINI, ♋ = CANCER, ♌ = LEO,
           ♍ = VIRGO, ♎ = LIBRA, ♏ = SCORPIO, ♑ = CAPRICORN,
           ♒ = AQUARIUS, ♓ = PISCES. ♐ SAGITTARIUS

QUALITIES -   C = CARDINAL, Fix = FIXED, M = MUTABLE,

ELEMENTS - F = FIRE, E = EARTH, A = AIR, W = WATER,
           + = POSITIVE, — = NEGATIVE

* DIAGRAM SHOWS THEORETICAL CONNECTIONS e.g.
    ARIES IS A FIRE SIGN WITH POSITIVE AND CARDINAL QUALITIES. IT
    IS RULED BY MARS AND ASSOCIATED WITH AFFAIRS OF THE 1ST HOUSE.

NOTE. SOME PLANETS RULE 2 SIGNS e.g. MERCURY, VENUS. ORIGINALLY THIS
       APPLIED TO JUPITER, MARS & SATURN BUT WITH THE DISCOVERY OF THE
       OUTER PLANETS, URANUS. NEPTUNE & PLUTO. RULERSHIP WAS CHANGED.

Above is another of the four pages.

Let it be noted that the predictions about my child's life events and personality described on the two pages not reproduced here, to save my child's embarrassment, have turned out to be wildly untrue.

The Universal Horoscope

Finally on the topic of astrology, let me present a 'Universal Horoscope' invented by psychologist Bertram R. Forer as reproduced below [405].

It appears to be making predictions but these are actually rather general and vague statements.

In an experiment, Forer gave out identical copies to a large number of students but misled them into believing that they were getting individual astrological predictions based on their dates of birth.

Over 80% reported that the horoscopes were uncannily accurate and believed the horoscope was highly applicable to their own star sign.

This experiment has been repeated many times and the average score of reliability is about 84%.

The Universal Horoscope

You have a great need for other people to like and admire you.

You have a tendency to be critical of yourself.

You have a great deal of unused capacity which you have not turned to your advantage.

While you have some personality weaknesses, you are generally able to compensate for them.

Your sexual adjustment has presented problems for you.

Disciplined and self-controlled outside, you tend to be worrisome and insecure inside.

At times you have serious doubts as to whether you have made the right decision or done the right thing.

You prefer a certain amount of change and variety and become dissatisfied when hemmed in by restrictions and limitations.

You pride yourself as an independent thinker and do not accept others' statements without satisfactory proof.

You have found it unwise to be too frank in revealing yourself to others.

At times you are extroverted, affable, and sociable, while at other times you are introverted, wary, reserved.

Some of your aspirations tend to be pretty unrealistic.

Security is one of your major goals in life.

I have repeated this test several times on groups of people with much the same result; an agreement that this 'horoscope' was surprisingly accurate in describing the person reading it.

I recommend trying out this 'horoscope' for yourself. It is an interesting project for someone to try on an astronomy club membership.

This 'Universal Horoscope' shows how astrology works.

It never makes useful predictions of specific events – like the date of death of a famous person, devastating earthquakes, the location of missing aircraft, etc., because this would expose it as a sham.

All predictions are wrapped up cleverly to appear specific whilst actually being vague.

I Attend A Séance

I have included this short diversion on the flimsy argument that only one letter separates 'Heavens' from 'Heaven'.

If you do not accept that as justification for the story I am about to tell well...

...Bad Luck!

About forty years ago I started a 'Discussion Group' to provide some local community spirit on the housing estate where I was living at the time.

About a dozen people would join me in a room of our local Community Hall and discuss a wide range of subjects in a light-hearted manner, often with alcoholic nourishment.

I once led a discussion on religion provocatively entitled

"God Does Not Exist!"

The local curate refused to attend because he deemed it 'inappropriate' and I got quite a lot of hassle from locals who were very suspicious of an atheist.

I was used to antagonism because of my disbelief.

Whilst going through university in the late 1950s one of my uncles warned me that if I persisted in my refusal to believe in and worship the Christian God I would never get a job in the Civil Service.

The fact that my entire working life was as a Civil Servant debunks that bit of fatuous advice.

In the spirit of scientific investigation, I persuaded a self-styled Spiritualist to hold a séance at the Discussion Group.

He called up a spirit from the 'Other Side' and started telling a member of the Discussion Group lots of really quite impressive 'facts' about his life and relatives.

He said that the person's mother was happy in Heaven. In life she had owned a small black and white dog that she was fond of and it was now dead.

There were lots of impressive details being passed on and I have to say that the detail was such as make the hairs on the back of my neck bristle.

Could Spiritualism actually be a fact?

After all, Patrick Moore was a firm believer in some sort of 'after world'. He believed that his dead mother was looking down on him and that they would meet up again when he died [406].

When the séance was over I could scarcely hide my astonishment over the detailed 'facts' that had been passed on from the spiritualist!

I went up to speak to the member of the group who had been the receiver of all the wisdom.

"That was impressive" I enthused. *"What did you think?"*

"Not a lot!" he admitted.

"It was all rubbish. My mother is not dead, she never had a dog and, in fact, she hated dogs! It was all nonsense!"

I was appalled.

I asked why this person had not said at the time that it was rubbish.

"I didn't want to embarrass the guy. He seemed such a nice person."

And so it was that a spiritualist and his audience (except for two of us) went away believing that this had been a successful séance.

I recall a television programme about a group of Spiritualists sitting around a table listening to a blank audio cassette tape being played.

There was regular *"Pssst Ahhh.....Pssst Ahhh.....Pssst Ahhh..."* noise coming from the tape player.

The listeners believed this was a noise of a dead person

desperately trying to contact the group through the blank cassette tape.

To me it was the noise of the take up spool on the player rubbing on a piece of projecting plastic each time it rotated.

The desperate need to believe that a dead loved one can make contact is - sadly - very strong in some people.

More Belief Systems

Do have a look at a YouTube movie of someone speaking Venusian [407] and you will get the idea of what the rest of this chapter is about.

I have long had a fascination for 'Independent Thinkers' as Patrick Moore called them. These are people who reject conventional views on the way the universe and life within it works.

There is a full recording of the TV programme *'One Pair Of Eyes'* [408] which you simply must see. If you don't follow up any other reference in this book then you must watch this movie!

In 1972 Patrick Moore published a book entitled *'Can You Speak Venusian?'* [409] based on the TV programme.

Of all the books in my large collection I have re-read this more than any other.

It is brilliant!

Look at some of the chapter headings to get a flavour of this wonderful little book:

- Better and Flatter Earths

- Hollow Earths and Solid Skies

- The Frigid Sun

- Up Atlantis!

- Crockery From The Void

- Report From Mars, Sector 6

and so on!

So inspired was I by the weird and wacky ideas in this book that I spend years tracking down the source material for some of these alternative theories of the Universe, including some not included by Patrick in his book.

The Earth Has A Ring!

Frank G Back, Me. Sc. Dr. privately published his theory that the Earth has a ring in about 1960 [410].

He is well qualified to propose this theory because, unlike so many 'Independent Thinkers' he built a novel spectroscope and travelled into the stratosphere as a passenger in a US Air Force jet fighter to obtain measurement of the vicinity of the Sun during a total eclipse which occurred over The Philippines on 20th June 1955.

Beck described himself as a great friend of Albert Einstein and there are no fewer than six pictures of him with Einstein discussing his theory for a ring around the Earth.

One of these pictures shows him attending Albert Einstein's last birthday party.

Beck's book includes a facsimile of a letter sent to Beck by Einstein dated 16th March 1955 which (in translation from German) includes these words

"In any case, I believe that your discovery is of considerable interest. It is strange that so far nobody has thought of this fact."

Beck's theory that Earth has a ring has not yet been confirmed.

The Sun Is Cold!

I managed to track down a copy of *'The Temperate Sun'* by Rev. P H Francis MA. which was privately published in 1964.

His theory was that the Sun is cold but the feeling of heat that we get is due to rays from the Sun releasing heat on the Earth.

Francis was no fool and his MA degree in mathematics was earned at Cambridge University.

Francis's arguments in favour of a cool Sun were numerous and plausible - up to a point!

He pointed out that, as a mountaineer climbs a high mountain the temperature falls until snow is experienced. Even Kilimanjaro, which straddles the equator, has snow and glaciers on its summit.

The fact that the surrounding temperature in the environment falls as we move away from the Earth and towards the Sun shows that we are moving from a hot body towards a cold body - the Sun.

Another argument proposed by Francis was that heat cannot travel across a vacuum. This is why a vacuum flask keeps our tea hot all day.

The space between the Sun and the Earth is a better vacuum than can be found in any vacuum flask and so it follows that heat cannot be flowing from the Sun to the Earth.

The Rev. Francis MA was a good company because the great Sir William Herschel believed that, although the surface of the Sun was hot, beneath the surface was a cool solid surface upon which an alien race lived [411].

Above we see my best friend and partner enjoying the illusion of heat emanating from the Sun whilst reading Francis's book.

In 1952 Dr Gottfried Buren claimed that the inside of the Sun was cold and that he had seen vegetation and creatures through the umbrae of sunspots.

He put up a large prize to anyone who could prove him wrong.

A group of astronomers did just that but he refused to pay up.

They sued him in court and won.

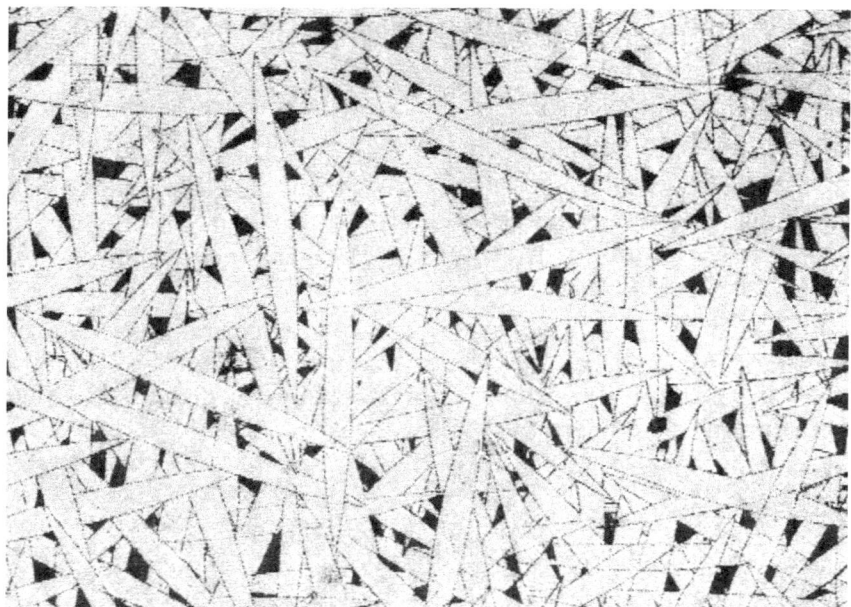

Incidentally, Sir William Herschel's son John believed that the granulation seen on the surface of the Sun, as drawn above by James Nasmyth, was made up of solid 'flakes or scales' and he regarded these

"...as organisms of some peculiar and amazing kind"

Wilbur Glenn Voliva was not so silly as to claim that the Sun was cold.

He argued that the Sun was only 32 miles in diameter and 3,000 miles away

"..in accordance with God's great wisdom and economy."

The famous Irish playwright and philosopher George Bernard Shaw supported this argument

Religion has also been used to justify the idea that the Earth is flat.

Martin Luther argued that on the day of judgement when the angels come to collect souls, the Earth would have to be flat to give everyone a fair chance of going to heaven.

If the Earth were round half the world's population would be on the 'wrong' side and miss out on the huge celebration.

The Earth Is Hollow!

The idea that the Earth is hollow has a long history stretching back to the ancient Egyptians [412].

In recent centuries the belief has persisted and been supported by famous well-educated people.

Edmond Halley in 1692 put forth the idea that the Earth consists of a hollow shell about 800 km thick, two inner concentric shells and an innermost core.

Atmospheres separate these shells and each shell has its own magnetic pole. The spheres rotate at different speeds.

Halley proposed this scheme in order to explain anomalous magnetic compass readings. He envisaged the atmosphere inside as luminous (and possibly inhabited) and speculated that escaping gas caused the Aurora Borealis.

Claimed by some to be the most accomplished mathematician of all times, Leonhard Euler [413] also proposed that the Earth contained an interior sun 1,000 km across to provide light to an advanced inner-Earth civilization.

Sir William Herschel also believed the Earth to be hollow with intelligent creatures living inside as on the Sun.

The Aurora was believed by him to be lights from their fires shining out of a hole at the North Pole.

In modern times a brand of Flying Saucer theorizing has suggested that the Earth is hollow and that UFOs and their occupants are located inside the Earth. They emerge to taunt us by using a hole into the Earth located close to the North Pole.

The Earth Is Inside Out!

A popular theory especially in the early years of the twentieth century in Germany was the theory that the Earth is inside out.

It was proposed that we live on the inside of a sphere. The Earth beneath our feet goes outwards for an unknown but very large distance whilst the whole of the universe is contained within the sphere [414].

In the 1930s this theory is said to have attracted Hitler to the idea that the universe is contained within a sphere approximately 12,000 kilometres in diameter and that we live on the inside surface of this hollow sphere with Australia almost directly above the United Kingdom.

In 1933 the Magdeburg City Council in Germany funded a rocket to be sent vertically up to see if it would carry on across the universe and land in Australia [415].

The rocket was not a success and never went anywhere near high enough to approach Australia hanging over our heads.

In the 1950s, as a typically nerdy activity I calculated what modifications would need to be made to the Laws of Physics to make an inside-out Earth appear the way the Earth and universe appears to us now.

I found that light would need to travel in circles passing through one's eye at one end of the arc and through 'infinity' (i.e., the centre of the hollow Earth) at the other end.

The speed of light would slow down as the ray got nearer to the centre of the sphere such that its speed would approach zero at 'infinity'.

Indeed, a simple mathematical transformation turns our current Earth and universe inside out and the two theories are mathematically indistinguishable!

Of course, Newton's simple laws of motion would be ruined by turning the Earth inside out.

The Sun, Moon and planets in an inside out Earth would follow complicated orbits and Newton's simple law of gravitational attraction would become horrifically complicated.

In 1992, the famous recreational mathematician Martin Gardner came to the same conclusion as I had in the 1950s noting that

"Most mathematicians believe that an inside-out universe, with properly adjusted physical laws, is empirically irrefutable".

Gardner rejected the concave hollow Earth hypothesis only on the basis of Occam's Razor [416].

It is not possible to prove the Earth is not inside out by observation of the universe above our heads.

21. Unidentified Flying Objects

In the late 1960s I became interested in the collation and study of reports involving Unidentified Flying Objects - UFOs.

I read up the current literature and decided that there was a phenomenon worthy of scientific study.

I was anxious not to dismiss something as worthless without a properly conducted study being carried out.

After all, if we dismiss everything we believe as untrue then progress in our understanding of the universe will be greatly limited.

I believed then, and still do now, that we should never fail to investigate a phenomenon even though it runs contrary to popular scientific understanding.

There seemed to be a substantial number of good eye witness accounts of UFOs such that the topic deserved study.

For example, if 'Cold Fusion' [417] had worked it would have revolutionised the generation of electrical power and solved the climate change problem. So, it was worth exploring even if it didn't have a fully validated scientific credibility.

Serendipity has always played a large role in scientific and engineering progress. No idea should be dismissed without a trial.

And so I became an investigator for the British Unidentified Flying Object Research Association (BUFORA).

My interviews with people who claimed to have seen UFOs in my area (Dorset) left me puzzled as to how some had seen the most amazing apparitions which were apparently invisible to others in the same area.

Witnesses were convinced that they had seen something extraordinary and yet these events had been unseen by others who should have seen the same things.

As an example, I received a report in the mid-1970s of a report of a UFO - indeed, a classical 'Flying Saucer' - having been seen over the area north of Bournemouth in Dorset in the vicinity of Hurn Airport; now renamed Bournemouth International Airport.

There were three witnesses; a man, a woman and their teenage daughter.

It was a Saturday lunchtime and the man was sitting in his dining room which was about 500 metres from Bournemouth Airport.

Suddenly he saw what he described as a flying disc coming towards his house from the direction of the airport. He called his wife and daughter and all three went into the garden.

The man said he saw a typical flying saucer with a dome or cupola and he counted three 'portholes'.

The object hovered over his house and he saw what he described as 'landing gear' beneath it.

The woman confirmed this story except that, by the time she went into the garden, the disc was almost overhead and she was unable to get as a good view of the dome as her husband.

The daughter was so terrified that she ran in doors and locked the door thus preventing her parents from getting into the house and getting a camera.

As the parents watched this object it hovered for a few minutes over their garden and then suddenly shot vertically upwards into the clear blue sky and vanished as a small dot.

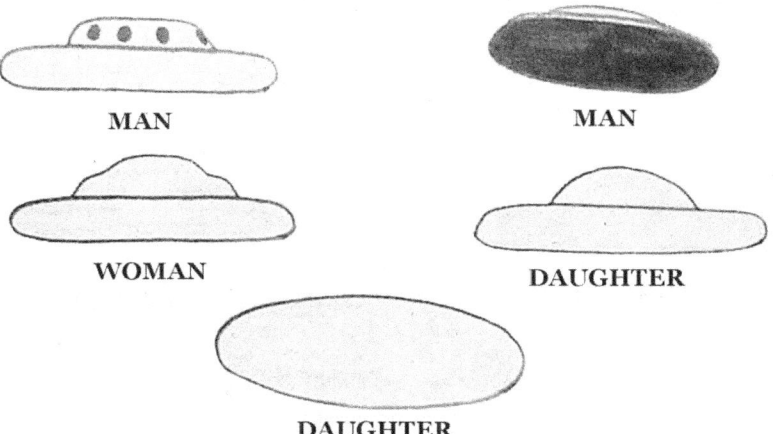

MAN

MAN

WOMAN

DAUGHTER

DAUGHTER

The drawings shown above were produced by the three witnesses independently when I interviewed them.

I followed this interview up by contacting the Air Traffic Controller at the airport who said that nothing unusual had been reported on that day. I also found out the wind direction at the time of the observation to see if this could have been a weather balloon.

The 'object' would have been travelling against the wind so a weather balloon was ruled out.

So what did these terrified people see?

I have no idea!

This was just one of hundreds of intriguing reports that arrived at BUFORA every year.

Another of my interviews was with a lady living on Portland in Dorset who told me that she had been sitting indoors watching evening television when she heard a noise outside her house like a very loud vacuum cleaner.

She went out and saw a huge triangular object silhouetted against the dark sky. It showed rows of flashing coloured lights on the rear edge and she remembered clearly the colours of the lights and the order in which they flashed.

The object 'flew' low over the witness's house and she watched it heading towards Weymouth.

The photograph below shows the view towards Weymouth from

the witness's house.

As the object approached the town of Weymouth seen here across the water, it turned right and 'flew off towards Lulworth Cove'.

This sighting might be accepted by some as a genuinely strange appearance of an alien spacecraft travelling over a highly populated area of South Dorset.

However, it was my brief as an investigator to be sceptical and my first consideration was that nobody else had reported seeing this huge and apparently noisy object travelling over the heavily populated area shown in the above photograph.

I appealed through the local newspaper for other witnesses but none came forward.

At the time of this report the Royal Naval Air Station was operational in the area centrally in the photograph above.

This station was used at that time for very frequent military helicopter flights so I would have assumed that the witness was confused by a low flying helicopter heading off towards Weymouth at night with its flashing coloured navigation lights and - possibly - a white searchlight on the nose.

Although the simplest explanation, the witness would have been well aware of these helicopter flights and she specifically told me that this was not a helicopter.

It may be relevant to mention that the witness told me that she had often observed ghosts in her house.

These two investigations were fairly typical of my encounters with UFO witnesses; people confident that they have seen something very unusual which I was unable to explain.

I also experienced the freakier side of UFO investigations.

For example, back in the 1970s I travelled with a couple of friends to Warminster in Wiltshire to experience standing all night on Cradle Hill which is *THE* centre for UFO spotting in the United Kingdom [418].

Warminster had been a focus for UFO spotters since the mid-1960s when Arthur Shuttlewood first publicised the UFO events allegedly occurring around the town [419].

Cradle Hill became the focus for enthusiasts and that's where two friends and I arrived one clear evening in the mid-1970s.

As the Sun set and the sky darkened dozens of people arrived carrying small telescopes, binoculars and boxes with wires and headphone attached. These latter devices were to alert the user to the imminent approach of a Flying Saucer by detecting and amplifying the 'Ethereal Emanations' from the alleged alien spacecraft.

As soon as the sky was dark enough to show the brighter stars the excitement increased amongst the groups of believers around us.

Suddenly a cry went up

"Here they come!!"

Everyone turned to look in the direction indicated.

A fairly bright point of white light was moving at a steady speed across the sky.

"They're coming!"

My friends and I knew this was one of several artificial satellites that we were to see that night.

Each appearance set up a growing frenzy of excitement.

My friends and I went into a huddle in the dark and muttered

"These are people are weird! Those are Earth satellites"

I cautiously approached an observer who was hopping from foot to foot with excitement at the presumption that there were Flying Saucers heading our way.

"Are you sure those are not artificial satellites? They are the right brightness, travelling at the appropriate speed and that one..."

I pointed in the dark sky with the beam of my torch

"...is SKYLAB"

The atmosphere went suddenly cold as I felt, even in the dark, many pairs of aggressive eyes turned my way.

"Are you a reporter or a Government dis-information spook?"

I tried to sound convincing as I explained that I was just an interested observer.

I was told that the brightly lit army base nearby was actually a TOP SECRET communication centre where the Government cooperated with aliens who gathered over Warminster for their daily chats.

Hmmm...

Shaking our heads at what we had seen and heard we got into my car and we headed back to the relative sanity of Dorset having experienced the power that a need to believe in alien visitors can exert on the mind.

In around 1986 I became obliquely embroiled in a famous UFO controversy.

In 1954 a book had been published by 'Cedric Allingham' [420] entitled *'Flying Saucer from Mars'*. This told of the author's encounter with a Flying Saucer and the alien occupants whilst on holiday in Scotland.

In the 1980s it came to be widely believed that this book had been written by Patrick Moore as a spoof to prove how gullible people were to believe in UFOs.

As evidence of his own disbelief he denigrated and mocked believers in many publications and in the media. One example can be seen in his book *'Can You Speak Venusian'* [421] in which there is a chapter entitled *'Crockery From The Void'*.

A well researched article in the magazine 'New Scientist' [422] concluded that Patrick Moore and Cedric Allingham were one and the same person.

Patrick Moore went ballistic!

Whenever he got annoyed with anything written by one of the residents in his 'Kingdom of Serpents' his voice would go up an octave and he would squeak

"I'll SUE! I'll SUE!!"

Well, I wrote a blog which referred to this 'New Scientist' denouement of Patrick Moore as Cedric Allingham and my writings were referred to in the UFO magazine 'Magonia'.

The prospect of being sued by Patrick Moore was enough to get me to retreat and avoid being hauled into court for stating what was self-evidently true - that Patrick Moore and Cedric Allingham were one and the same person.

In 1981 I was approached by the Editor of The Journal of Naval Science; a publication for sharing research within the Royal Navy science community in the United Kingdom.

The supply of articles was drying up and he was desperate - could I write up something quickly?

I suggested an article on UFOs. Although doubtful that this was a suitable topic for a serious journal the editor was willing to 'give it a go' as long as the article took a neutral and scientific approach.

My article was published in the December 1981 issue [423].

Despite the journal carrying a military classification of UK RESTRICTED, it soon leaked out into the open literature and has been frequently referenced since 1981, for example see [424].

The article is reproduced below.

Remember that the article was written in 1981.

UFOs - Evidence Of Alien Invasion?
Geoff Kirby B.Sc.

Abstract

Over 150 thousand UFO reports are now catalogued and of these several hundred include descriptions of alien intelligent creatures. These reports have led the public to believe that alien space travellers are visiting the Earth. The evidence becomes unconvincing under detailed analysis and no scientifically acceptable evidence exists to support this idea. However, a small number of well investigated multiple witness cases are provocative and suggest that a currently unrecognized phenomenon may be the stimulus for these sightings.

Introduction

In 1947 Mr Kenneth Arnold reported seeing a group of metallic discs 'like upturned saucers' skimming over the Mount Rainier area of the USA. This started the modern era of Ufology and coined the misnomer 'Flying Saucer'.

Over 150 thousand reports have now been catalogued worldwide and many people believe that this vast body of information represents a firm commitment to the notion that aliens are visiting us and, if some reports are to be believed, actually abducting people.

What does this wealth of information tell us?

Does it provide the evidence so often claimed by UFO buffs?

You can judge for yourself as I present a very condensed review of the evidence.

Before wielding our machetes through the jungle of hoaxes, misrepresentations and (possibly) genuine sightings we must first make a clearing using Occam's Razor from which to view the jungle and chart a course.

The name 'Unidentified Flying Object' is not an adequate definition to work to since we are not interested in aircraft of unknown nationality but genuinely strange sightings which defy analysis. Broadly speaking we are concerned with reports of sightings of apparently material flying or landed craft which do not correspond with terrestrial machines.

There is no universally accepted definition since we are always dealing with reports of objects and not examination of the physical objects that prompted the reports—if they exist.

The genuine sighting of an alien spacecraft on the Earth would be so miraculous that the evidence must be unimpeachable. Occam's Razor demands that we accept any rational explanation for UFO sightings unless the best rational explanation is much less probable than the 'Alien Visitor' explanation. This eliminates about 99.9% of all sightings since these do not have totally convincing evidence.

Most of the really interesting personalities vanish if we eliminate the least convincing cases.

Mr Valiant Thor, for example, arrived from Venus looking remarkably like an American earthling but writing and speaking in a strange language. Since Venus has an atmosphere of over 500°C temperature, a pressure of 91 atmospheres and a continual rain of sulphuric acid it seems that he adapted remarkably quickly to our conditions.

Dr King, who runs the Aetherius Society from Hollywood, is in regular contact with Emperors on Saturn and lesser dignitaries on other planets. His followers communicate with the pilots of flying saucers and there is a widely held belief in Aetherian circles that Jesus is alive, well and living on Venus.

Valiant Thor doesn't recall meeting him there however!

Belief in any of these and many other 'explanations' of UFO sightings requires a remarkable suspension of faith in established science and, whilst these theories have a great appeal to me, I find it difficult to accept any of them as a valid alternative description of the world I see around me!

The bookstalls abound in UFO literature but there is only one book which has my confidence and that is 'The UFO Handbook' written by Alan Hendry [425].

Hendry personally investigated all UFO sightings in the USA over a period of 15 months during which time he spoke to 1,300 witnesses. The remarkable fact that emerged was that not one sighting provided adequate evidence that an alien spacecraft had been sighted and, on the strength of his compilation of data, we can find no patterns emerging that suggest any explanation for the reports other than simple misinterpretations and hoaxes.

It is true to say that some genuinely strange sightings have been reported but these form a tiny minority of cases and, as a group, offer no consistent data for any explanation of the sightings.

In the rest of this article I will review the evidence and then leave the reader to form his or her own views. The presentation is biased here towards evidence of the material nature of UFOs.

The predominant view amongst UFO theorists is that the reports have a non-rational basis, i.e. the UFO sightings are either prompted by psychological causes or are manifestations of psychic phenomena. Jacques Vallée [426] has demonstrated a remarkable correspondence between some UFO events and the events associated with spiritualism and the perception of religious miracles.

It seems very likely that these experiences have a lot in common but I will restrict myself to the type of sighting that allows some degree of rational and scientific investigation.

No sightings earlier than about 1950 can be assessed because they have not been scientifically investigated and indeed we could draw a line at about 1970 without losing much useful data. Earlier cases were either not investigated or were investigated badly, particularly as in the latter case by a US Air Force team of investigators.

The table below shows Hendry's analysis of sightings for the month of March 1978.

Identification	US Cases in March 1978
Aircraft	21
Astronomical	31
Advertising Aircraft	12
Radio Mast	1
Road Sign	1
Satellite	1
Searchlight	1
Ground Lights	1
Total Identified	69
Unidentified	1

This table shows a fairly typical month's acquisition of reports.

The road sign was reported to have descended from the sky with lights flashing and seated itself so that cars passed between its legs!

The single unidentified sighting was strongly suspected to be a military aircraft.

Hendry's sample of reports is typical and has the advantage of a consistent set for analysis. I will be introducing other cases from earlier periods however where these illuminate the subject matter.

Hendry has shown that more than 95% of cases investigated in the USA during his study period could be attributed to misperceptions of rational material objects. Very few cases, less than 1%, were hoaxes.

The high proportion of sightings attributed to advertising aircraft is typical and illuminating. In the USA helicopters and specially built aircraft fly very slowly over towns and cities displaying messages on a programmed array of hundreds of lights.

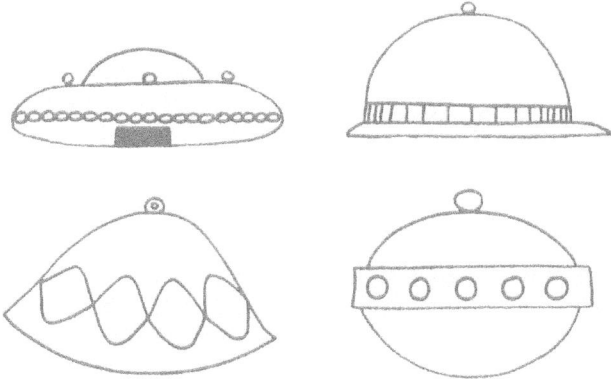

The above pictures show some typical witness drawings of these aircraft and it can be seen that a great deal of imagination has gone into their production. Such is the publicity given to UFOs that any strange object is often transformed, in a disturbed mind, to a saucer.

In one case a woman reported that she held a telepathic conversation with a UFO pilot and was mentally disturbed for weeks after the event. She had actually witnessed the flight of an advertising aircraft.

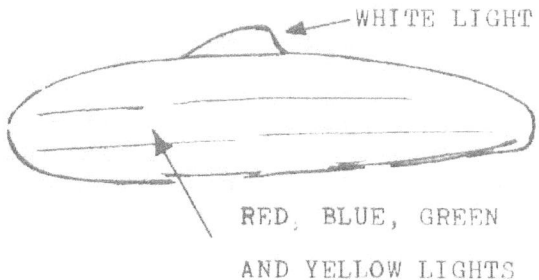

The above drawing shows a sketch of the star Antares at low altitude. Here again the expectation of seeing a flying saucer has transformed the star into such a craft.

The above drawing shows the planet Venus as seen by a witness.

These cases show how the great publicity - most of it nonsensical - given to the UFO phenomenon distorts the observation of perfectly normal objects.

Each of these cases required a great deal of investigation without which these would have been reported in newspapers or books as yet further sightings of alien spacecraft.

The emotional responses of witnesses have often been quoted as evidence for the validity of the sighting but consider the following quotations. The object observed is indicated.

"Extra-terrestrials? God? Russians?"

(Star watched for six hours)

"It's a sign, a premonition.....called my relatives to see if anyone was sick....."

(Advertising aircraft)

".....scared.....screamed for my husband to get back in the car....."

(Star)

"Made my hair stand on end....."

(Moon)

"I was torn between terror and curiosity....."

(Advertising aircraft)

"Oh my Lord, I'd better get down on my knees and pray....."

(Star watched for several hours)

These may be extreme examples of witness reactions but it must be remembered that over 95% of cases reported in the press and in books are not investigated in sufficient detail to reveal that the sources of such terror were common objects.

In this way the UFO phenomenon breeds hysterical responses by the dissemination of cases which have not been adequately investigated.

One woman committed herself to a mental hospital in the USA after spotting the planet Venus in the sky. What is more she was being extensively - and expensively - treated for this derangement!

It has been claimed that sightings by trained observers must be more reliable than sightings by others but the evidence is to the contrary.

Airline pilots and policemen come out badly in terms of the proportion of sightings attributable to common rational objects. The value of single witness sightings has been claimed to be acceptable since law courts accept witness sightings as evidence.

However, on the basis of the above experiences the argument should be reversed such that courts should only believe 5% of witnesses.

Despite the observation that at least 95% of UFO reports are stimulated by misperceptions of common objects there remains the small number of cases which might be stimulated by genuinely strange causes. It is to these cases that I now direct my attention.

Do these represent sightings of alien spacecraft or are these cases simply further mis-identifications which scientists cannot explain because the evidence is inadequately explicit?

Radar Evidence

There have been numerous cases where simultaneous radar and visual tracking has been reported of anomalous objects. Two of the 'classic' cases have been described in great technical detail and I will not therefore repeat the details here.

The Lakenhurst case involved visual sighting of an object both from ground level and from an aircraft—the object passed below the aircraft—simultaneously with radar tracking of an object travelling at speeds between 2,000 and 18,000 mph [427].

A detailed investigation of this case by a US Government funded research organisation resulted in the conclusions that [428]

'... this is the most puzzling case in the radar/visual files. The apparently rational, intelligent behaviour of the UFO suggests a mechanical device of unknown origin as the most probable explanation of this sighting.'

The second case of merit [429] involved visual sightings of a bright light at night time and radar tracking of the object by ground and airborne radars together with interception of radar-like transmissions from the object. These latter did not correspond to any existing airborne radar set at that time.

Although a large number of radar/visual sightings are on record most fail to give conviction that the radar track correlates with the visual sighting or that the visual sighting is other than a rational object such as a star.

It is relevant to enquire whether the extensive radar defensive chain around the NATO area detects anomalous tracks. The answer is certainly 'yes' - about 800 non-satellite tracks are recorded every day.

Most of these are meteors and the aurorae. The NORAD computers analyse all radar contacts and reject those that cannot be correlated with expected satellite or missile behaviour. To recover and analyse the tracks of objects which may be UFOs would involve enormous effort and cost and is unlikely to be performed.

Similarly, airport radars tend to track aircraft and reject objects which show anomalous Doppler frequency shift, no identification transponders (IFF), etc.

Thus, apart from a mere handful of cases there appears to be little evidence from radar tracks that material objects are travelling through our atmosphere with trajectories inconsistent with Earth technology.

Physical Traces

Several thousand cases exist where an aerial object has been reported to have landed or hovered at low altitude and left damaged vegetation.

A typical example is that of the destruction of a circular patch of soya plants by an object shaped like an inverted bowl with a shallow convex base and a circular red/orange band of light.

The typical ring imprinted on the ground is about 10 metres in diameter and all vegetation is killed and the ground stays sterile for several years afterwards. Very few of these rings show characteristics differing from fungus rings; the attack on a soya field [430] being rather exceptional.

Probably the most publicised ring originated in Delphos, USA.

Three witnesses reported the landing of a craft which left a three metre diameter ring. A detailed analysis of the ring by X-ray diffraction techniques, fertility tests, pH measurements, etc., showed this ring to have properties identical to those of fungi rings produced by Actinomycetacae Nocardia.

Some rings appear initially to be quite different. A ring of 'niobium globules' was reported by an aeronautical engineer but analysis showed these to be fungus spores.

Photography is a physical trace of another kind since the UFO leaves its impression on a film. However, hoaxes are almost impossible to rule out.

A photograph much reproduced in books and the media was of a Spanish disc which reportedly flew over fifty witnesses. On its base was a symbol very much like the astrological symbol for the planet Uranus. This created a cult which believed that UFOs were piloted by aliens from that planet and attempts were made over several decades to contact the 'Uranians' directly.

However, a computer analysis proved the picture to be of a model 'flying saucer' about 150 mm in diameter suspended by a thread in front of the camera lens!

A trio of pictures by the Jaroslaw brothers shows a finned disc turning beyond some trees. These photographs have been published as the

"Best UFO photographs ever taken".

After these pictures were released to the media the brothers confessed that they photographed a model [431].

The development of computer analysis has had a devastating effect on UFO photography. However, a small number of photographs have emerged with their credentials enhanced. A very detailed analysis of two photographs by the US Government resulted in the conclusion that

"This is one of the few UFO reports in which all factors, geometric, psychological and physical appear to be consistent with the assertion that an extraordinary flying object, silvery, metallic, disc-shaped, tens of metres in diameter and evidently artificial flew within sight of the two witnesses." [432]

Some photographs do not survive scrutiny for long.

A Brazilian photograph shows a disc flying over a forest area. It has long been published in the UFO literature as classic evidence even though the sun is shining from opposite directions on the disc and the trees below. This has been explained as a UFO space warp effect by diehard UFOlogists.

Claims have been made many times in magazines that astronauts have photographed UFOs.

This is not true.

The negatives of all 'astronaut UFO' photographs have been examined by Professor Hynek and declared to be either hoaxes or misinterpretations.

Thus, apart from a handful of provocative cases photographs and ground markings provide no scientifically acceptable evidence in favour of the theory that UFOs are machines.

What the world awaits is clear physical evidence - a lump of metal of indisputable extra-terrestrial origin or some other evidence.

Close Encounters of the Third Kind

Hynek (433) applies this classification to UFO sightings involving alien creatures. Hence it was used in the 1977 film entitled *'Close Encounters Of The Third Kind'* in which Hynek was given a cameo role.

Many hundreds of such sightings have been reported.

Among the largest number of witnesses to an alien sighting was thirty-eight when discs and humanoids were sighted over an Anglican missionary school in Papua.

An attempt to explain this remarkable event in rational terms has completely failed [434].

Another multiple sighting occurred in Kentucky when eleven witnesses saw - and shot at - a large number of grotesque creatures about one metre tall.

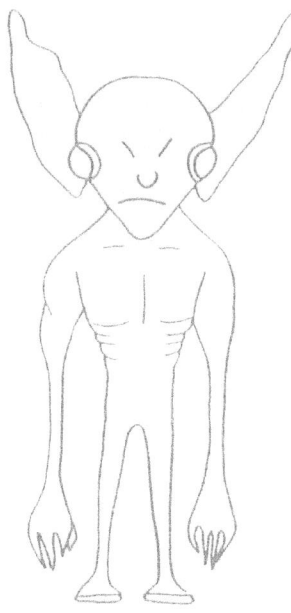

The sketch above shows the alien invaders as drawn by the witnesses.

Very similar creatures were allegedly seen and drawn by a class of infants at Broad Haven School, Dyfed as shown on the next page.

The sketches above have been redrawn by me from some of the drawings by the children made on 4 February 1977. The common features with the Kentucky - and other - creatures are the large eyes, pointed ears and diminutive height [435].

If all sightings were so comparable we would be tempted to believe that we were indeed being invaded from space.

This is not the case.

Above we see heads of a typical cross-section of reported alien creatures. There are many more types in the literature.

Obviously there is no overall pattern.

Either we are being investigated by a large consortium of species or the reports are unreliable.

Barney And Betty Hill Case

In one reported encounter with aliens evidence was claimed to have been presented to the two witnesses which rises above the usual low standard of physical evidence for UFOs being alien spacecraft.

Barney and Betty Hill claimed under hypnosis that they had been abducted into a UFO and subjected to detailed medical analysis. The details of this case are widely known and will not be repeated here [436], [437].

The factor of interest in this case is the evidence of a purported star map shown to the Betty Hill in the UFO. She was told by her alien abductors that this map showed stars local to the sun and that the 'home star' of the aliens was marked.

A three-dimensional model was later made by investigators which included stars near to our Sun. A view into this model was claimed to show a good agreement with the alien star map with the twin star 'home star' in the aliens' map being Zeta Reticuli.

The agreement was claimed to be "remarkable".

The later addition of some faint local stars discovered during a survey by Gliese completed the agreement of the star map with the three-dimensional model of local stars.

This started a long-running cult aimed at contacting the aliens living in the Zeta Reticuli binary star system.

However.....

Carl Sagan went back to the original deposition by Betty Hill and found that her sketched 'star map' was very different from the version widely published in the media.

The original map was inconsistent with the model of local stars to our Sun.

Carl Sagan suggested [438] that better agreement might be obtained by shaking ink from a quill pen over the paper!

Thus, the Betty Hill star map offers no evidence for alien visitors to our planet.

Do we have any more solid evidence?

Many objects purporting to be of alien manufacture were investigated by Dr Condon [439] and Alan Hendry [440] in their detailed investigations in the USA.

For example, an 'alien' metal tube reported to have appeared overnight on a man's lawn and to have burned the grass was identified as the man's lawnmower silencer which had fallen off un-noticed.

A bar of metal given by aliens to a Mr Hermann in the USA carried strange writing of no known language. Its composition was found to be identical to plumbers' lead.

A sliver of metal recovered from the famous Socorro landing incident [441] was reputedly analysed by a Dr Frenkl of NASA.

An alleged telephone message from the NASA laboratory asserted that the metal was certainly of a non-terrestrial alloy but all traces of Dr Frenkl and the metal were later lost.

NASA denies all knowledge of the incident.

Perhaps the most provocative sample was recovered from the beach at the tiny Brazilian fishing town of Ubatuba in 1957. Local fishermen reported seeing a flight of metallic discs pass over the beach and one swept violently upwards and exploded showering pieces to the ground.

Three pieces were recovered and analysed. Two fragments were destroyed during testing. These pieces were reported to be exceptionally pure magnesium with no detectable impurities.

The third piece was analysed by the Condon Committee. Their quantitative report is tabulated below.

Element	Parts per million
Manganese	35
Zinc	500
Strontium	500
Barium	160
Chromium	32

It can be seen that the Ubatuba metal is very pure but magnesium of comparable purity had, or so it has been alleged, been produced, albeit in small quantities, since 1940 by the Dow Chemical Company.

A recent analysis of the third sample [442] concluded that the crust of the magnesium could only have been formed by heating the sample to near melting point in a vacuum and quenching it in air.

The only rational explanation so far proposed is that the metal came from a Soviet spacecraft or aircraft, using very pure magnesium alloy and that a sample was dropped at high speed into the atmosphere from space

Intriguing though this sample may be it cannot be demonstrated to be extra-terrestrial in origin.

The sample could have been made on earth.

Conclusions

After more than three decades of searching there is still no definite proof that alien spacecraft are visiting the earth. Some evidence is very provocative and suggests a covert surveillance operation but the total of evidence is inconsistent with human notions of how such an operation would be conducted. UFOs and reported occupants do not come in consistent forms and no material evidence exists that it is clearly extra-terrestrial.

Perhaps the next thirty years will provide the perfect case so long awaited by UFO enthusiasts. It is a fact that large numbers of sane and sensible people report sightings of huge machines in our skies and creatures are reported to be contacting the human race.

I leave you to decide.

(End Of Article)

Bearing in mind that I wrote the above article over three decades ago, has much changed? Not really. There is still no undisputed physical evidence for UFOs being alien spacecraft.

The origin of Ubatuba 'UFO fragments' are still hotly debated in UFO circles [443] and the sighting of aliens by staff and children of Broad Haven School in Wales are still extant as some of the best evidence available for alien visitations - little has changed since the original report [444], [445].

In my opinion, there has been no shift in the direction of proving that aliens have been paying us visits since I wrote the above article in 1981.

A 2015 book comes to the same conclusion [446].

Maybe one day we will have irrefutable evidence of alien life outside our own Earth.

The odds are stacking up in favour of a universe in which life is scattered widely.

The growing number of discoveries of potential environments in our solar system and on exo-planets makes the existence of some form of life increasingly probable.

However, there is always the worrying 'Fermi Paradox' where the question is posed

"If alien intelligent life exists widely in the universe why have we not yet seen any evidence for it? Why has Earth not been visited?" [447]

The reference listed twenty possible reasons why we are not indisputably seeing intelligent aliens on Earth.

Time will tell which of the reasons is correct.

Lecturing About UFOs

Starting in 1969 I travelled extensively around the United Kingdom giving talks - mostly speaking about astronomy.

The talk I have given more than the other topics put together is entitled

"UFOs - Fact Or Fiction?"

and my title slide is shown on page 14.

I have given this talk approaching one hundred times at venues ranging from the Isle Of Wight on the south coast to Cumbia.

Some of the UFO talks have been given in odd locations.

For a local Naval Wife's Support Group I was directed to a terraced house in the middle of a naval housing estate. I was welcomed by a lady who appeared to be the welfare officer for the group. I was escorted into an upstairs bedroom where about six ladies were sitting on chairs and a bed.

I started to panic.

Could I entertain all these ladies with a single bed with only my slide projector to assist me?

Sometimes the organiser does not stay.

I was booked to talk to the National Housewives Register group in Weymouth. I turned up and the organiser was absent. She had gone to Bournemouth to see 'Hurricane Higgens' playing snooker - much more interesting than listening to my talk.

I was booked to talk on UFOs to Charlton Hawthorne Women's Institute which is somewhere in darkest North Dorset near Sherborne. I drove around the village which had not a single street light.

No sign of a village hall.

It was now 7.30 - the time I was booked to arrive. I abandoned the car and stumbled around in the darkness, squelching over muddy verges and through full gutters - the torrential rain did not help!

And then, like an angelic choir, I heard the faint strains of 'Jerusalem' coming from down the road. I followed it and, on the final triumphant note, I burst into the hall giving dramatic effect to Blake's masterpiece! The entrance of a muddy steaming fallen angel clutching a box of damp slides must have been impressive.

The organiser welcomed me with much relief and invited me to set up my projector.

Oh No!

The hall was fitted out with round pin sockets - which had gone out of fashion two decades ago.

"Oh, Yes" agreed the organiser, *"That is a bit of a problem for speakers"*

With the aid of some borrowed mascara brushes and matchsticks, I managed to get the naked wires from my projector cable into the live socket and did the talk with only the occasional flickering of the bulb.

I was persuaded to talk to the Verne Prison Life Prisoner's Discussion Group several times in the 1980s. I thought this would be a 'bit of an experience' and so it was.

The life prisoners at the Verne on Portland in Dorset in the 1980s were there for murder, rape and other nefarious crimes but, despite their crimes, were not considered high risk.

Indeed, at that time, the life prisoners that I talked to were coming to the end of their sentences and these talks were part of a rehabilitation programme.

Usually I was let in through the double security system with all my paraphernalia of screen, projector and cables.

"Don't worry about the prisoners - they are great chaps. However, don't turn your back on them, don't take your eyes off your equipment - even an otherwise useless lens is currency in here - and I'll leave the door ajar so that you can call for help."

Gee, thanks warder!

In fact, the men were quite pleasant. They were attentive to my talk, mumbled only a little when I made the mistake of urging them to go out and look at the night sky and were polite in offering me tea and biscuits at the end of the talk.

One young lad had assiduously read up on herpetology (study of lizards) and was reputed to have become a bit of an expert consulted by the outside world. I wondered why he was inside but it was very bad form to ask.

I gave my UFO talk on the Isle Of Wight but I was about to start my talk only to find the front row taken up by a nerdy bunch of UFO believers with bundles of the more extreme and nutty UFO magazines on their knees.

Because my talk was a sceptical account of the UFO phenomenon I was expecting trouble and I was not wrong.

I once - and only once! - gave my UFO talk to a local Round Table meeting.

They gave me a meal after which I set up my projector.

They were rowdy and drunk. It was clear that my talk about UFOs was of little interest. Throwing alcohol down their throats was clearly of urgent importance.

Then they started to throw bread rolls around the room.

I decided to notch up my delivery speed to PM90 - that is 90% of the speed that Patrick Moore spoke when excited. Nobody has ever reaching a speed of PM100; not, that is, without dentures and spittle flying around the room.

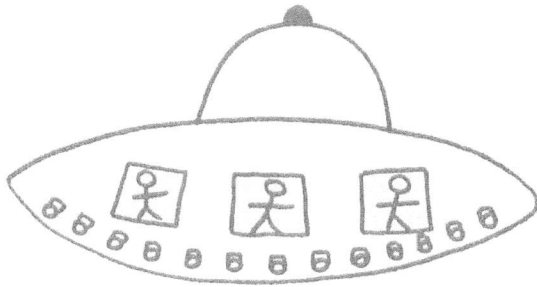

When I projected the above picture they burst into drunken laughter.

"It's a tit!! Yes! Turn it round! IT'S A TIT!!!"

someone yelled and bread rolls rained down on me.

I finished at a talking speed of about PM99 and left hurriedly.

I was booked to give the UFO talk at a local astronomy club meeting.

I started in great form but went to pieces when I accidentally said 'libido' instead of 'albedo' and said 'orgasm' instead of 'organism'.

The first bloop I passed over but, after the word orgasm came tumbling out of my mouth out I stopped dead.

I turned to the audience.

They were staring at me.

My wife had come along for moral support. She had her head in her hands and was looking very hard at the table top.

After that I went to pieces.

Searching For Extra-Terrestrial Intelligence (SETI)

The SETI programme was started in 1999 and continues to the present time with several interruptions due to funding shortfalls [448].

The search for radio signals was just one branch of the general search for evidence of life in space.

The feature film 'Contact' is a wonderful dramatization of SETI with its technical and political problems and (potential) consequences of success [449].

In 'Contact' the heroine, played by my favourite actor, Jodie Foster, is working at the Very Large Array at Socorro. On page 406 you can see me on a visit to this huge facility.

The Seti@Home program has generated two million years of CPU time in analysis of radio 'noise' in the search for a pattern indicating the possibility of an artificial (intelligent?) source for the radiation.

The analysis runs in the background whilst a computer is not otherwise engaged in user tasks.

In my case I left my computer on day and night (except when there was nobody in my house) for several years.

The Seti@Home software automatically downloaded samples of radio noise collected mainly at the Arecibo Radio Telescope, analysed the sample for anomalies which were worth further investigation, uploaded the results to the Seti@Home project and then repeated the cycle.

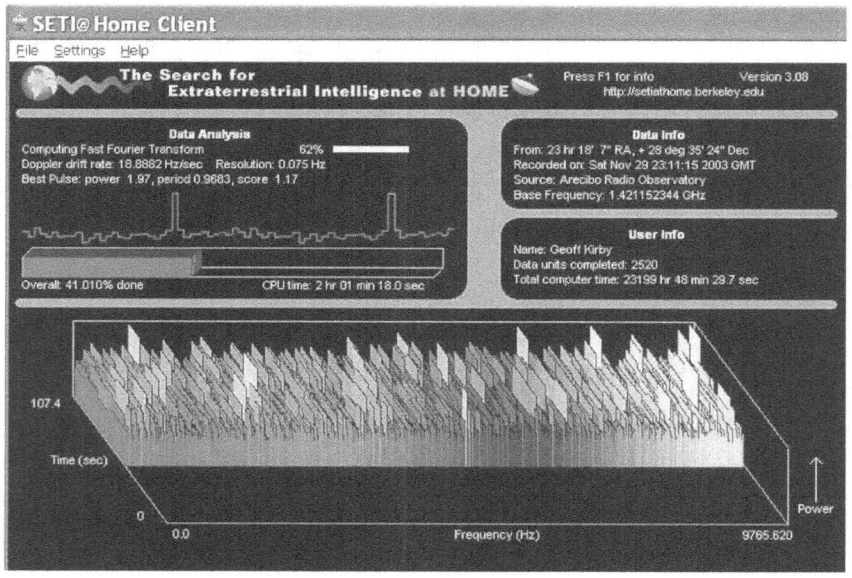

The screen grab above shows a typical set of data being analysed in 2003.

This shows that I had accumulated over 23,000 hours of CPU time at the time this picture was captured. This corresponds to about 2 years and 7 months of CPU processing.

I went on to accumulate another year of CPU processing before I gave up the Seti@Home project.

There were two reasons for abandoning this project.

Although I believed that intelligent life very probably does exist somewhere in the universe, the probability of me being the one to find a signal in the radio noise coming into the Earth's region is vanishingly small. I had accumulated about 3.6 years of analysis out of a total of about two million years of analysis by participants in the Seti@Home project.

That statistic showed that there was a very small chance that the arrival of an intelligent signal would be discovered by me rather than someone else.

In fact, as I write this, Seti@Home has been running for 16 years and nothing has so far been found except for one possible feature in the radio noise back in 2004 that has not repeated.

I decided that I would help to mitigate global warming by switching my computer off when not in use and I left the Seti@Home project.

Another reason for leaving the project was that it became amalgamated into a much wider BOINC program [450]. This is an umbrella project for distributed analysis which covered many dissimilar projects.

In addition to Seti@Home there were programs for investigating cancer drugs, solving mathematical problems and host of other problems to be investigated using distributed computing.

I did not like the BOINC protocols and I found the new set-up relatively unfriendly.

So, I gave up.

Index

References

For your convenience the references can be easily reached by going to my website which is located at

www.geoffkirby.co.uk/References

There you can click on the number of the reference in this book and be taken to that reference website page.

_____

1.
http://www.spacetelescope.org/about/history/servicing_mission_1/

2. *"Make Your Own Telescope from Everyday Materials"* Reg Spry (1978)

3. Spry, F.R. *"A 150 mm From Scrap"* J. Brit. astr. Ass. (86), 4 pp 285 - 287

4. Lloyd, C., Kirby, G.J., James, N.D., Collins, M.J., and Berthold, T. *"GSC 1657.1754: A New Deeply Eclipsing Binary System in Delphinus"* Commissions 27 and 42 of the IAU Information Bulletin on Variable Stars Number 4442

5. http://www.history.com/news/the-day-skylab-crashed-to-earth-facts-about-the-first-u-s-space-stations-re-entry

6. http://www.usno.navy.mil/USNO/about-us/the-26-inch-refractor

7. *"Starlight Nights: The Adventures of a Star-Gazer"*. Reprint of the 1st Ed Pub by Harper and Row, New York Paperback. Leslie C. Peltier (1980)

8. https://www.aavso.org/leslie-peltier-worlds-greatest-amateur-astronomer

9. *"My Lost Childhood - How Hitler and the Freemasons Stole My Childhood"* by Geoff Kirby. Available from Amazon.

10. https://en.wikipedia.org/wiki/Will_Hay

11. http://cdsads.u-strasbg.fr//full/seri/MNRAS/0094//0000085.000.html

12. http://www.hampsteadscience.ac.uk/astro/reports/special/henry_wildey_2003.html

13. http://adsabs.harvard.edu/full/2004JBAA..114...48D

14. http://www.geoffkirby.co.uk/Books/Lost_Childhood/lost_childhood.html

15. https://en.wikipedia.org/wiki/Slide_rule

16. http://www.bonestell.org/Images/Gallery/10_SurfaceOfVenus.jpg

17. http://s451.photobucket.com/user/hush77hush/media/mekono8_zps359f2375.jpg.html

18 . https://en.wikipedia.org/wiki/The_Conquest_of_Space

19. http://www.bonestell.org/#prettyPhoto

20. *"The Exploration of Mars"* by Willy Ley (Author), Wernher Von Braun (Author), Chesley Bonestell (Illustrator) (1956)

21. http://www.imdb.com/title/tt0042393/

22. https://en.wikipedia.org/wiki/Destination_Moon_(film)

23. http://www.hampsteadscience.ac.uk/astro/index.html

24. http://adsabs.harvard.edu/full/1982JBAA...92..137M

25. *"Can You Speak Venusian? A Trip Through the Mysteries of the Cosmos"* Patrick Moore (1976)

26. https://en.wikipedia.org/wiki/Luna_3

27. http://www.scienceforums.net/topic/7741-extracting-phosphorus-from-urine/

28. https://en.wikipedia.org/wiki/Psychokinesis

29. *"Making and Using a Telescope"* by H Percy Wilkins and Patrick Moore. Eyre and Spottiswoode Publishers Ltd (1956)

30. https://www.youtube.com/watch?v=3f_21N3wcX8

31. *"Return to the Far Side of Planet Moore!"* Martin Mobberley, Springer International Publishing, ISBN 978-3-319-15779-5, 2015 page 174 - 179

32. https://archive.org/details/ThePlanetMercury

33. https://en.wikipedia.org/wiki/2001:_A_Space_Odyssey_(film)

34. http://www.morbidofest.com/wp-content/uploads/2013/04/2001.jpg

35. https://www.youtube.com/watch?v=lj-OlW83b6U&list=RDyazu8vaGPwQ

36. https://www.youtube.com/watch?v=yazu8vaGPwQ

37. https://en.wikipedia.org/wiki/Apollo_11

38. http://www.durlston.co.uk/visit-astronomy.aspx

39. https://en.wikipedia.org/wiki/Mu_Cephei

40. https://en.wikipedia.org/wiki/Danjon_scale

41. https://commons.wikimedia.org/wiki/File:C-west-1976-ps.jpg

42. https://en.wikipedia.org/wiki/Brass_monkey_(colloquialism)

43. https://en.wikipedia.org/wiki/PDP-11

44. http://www.pcworld.com/article/155984/worst_tech_predictions.html

45.

http://www.pcworld.com/article/155984/worst_tech_predictions.html

46. https://en.wikipedia.org/wiki/ZX81

47. http://normanlockyer.com/history/

48. https://en.wikipedia.org/wiki/Norman_Lockyer

49. http://www.vogue.it/en/people-are-talking-about/obsession-of-the-day/2011/09/ethel-granger#ad-image113443

50. https://www.staylace.com/textarea/history/ethelgrangerbio2/eg_bio2_01.htm WARNING! This reference leads to very sexually explicit content!

51. http://www.courts.qld.gov.au/__data/assets/pdf_file/0005/337622/cif-uscinski-20141229.pdf

52. http://www.ianridpath.com/stamps/1990.htm

53. https://en.wikipedia.org/wiki/Hoover_free_flights_promotion

54. https://en.wikipedia.org/wiki/Spot_the_ball

55. http://www.solarstorms.org/SS1989.html

56. http://www.solarstorms.org/SRefStorms.html

57. https://en.wikipedia.org/wiki/March_1989_geomagnetic_storm

58. https://en.wikipedia.org/wiki/Comet_Shoemaker%E2%80%93Levy_9

59. http://www.bbc.co.uk/news/uk-england-34659751

60. https://www.youtube.com/watch?v=YjpcZPT1-NA

61. https://en.wikipedia.org/wiki/Baily%27s_beads

62. https://vimeo.com/23364453

63. http://www.geoffkirby.co.uk/Accident2009/

64. *"It Came From Outer Space Wearing An RAF Blazer",*

Martin Mobberley, Springer International Publishing (2013), page 51.

65. *"Return to the Far Side of Planet Moore!"* Martin Mobberley, Springer International Publishing, 2015 page 73.

66. https://en.wikipedia.org/wiki/Axel_Firsoff

67. Obituary of V A Firsoff by Patrick Moore. Journal of the British Astronomical Association, (92), 3, page 139

68. https://en.wikipedia.org/wiki/Fred_Hoyle

69. *"Return to the Far Side of Planet Moore!"* Martin Mobberley, Springer International Publishing, 2015 page 202

70. *"Return to the Far Side of Planet Moore!"* Martin Mobberley, Springer International Publishing, 2015 page 202 - 204

71. *"Return to the Far Side of Planet Moore!"* Martin Mobberley, Springer International Publishing, 2015 page 73.

72. *"Variable Star Observer's Handbook"* John S. Glasby (1975)

73. *"It Came From Outer Space Wearing An RAF Blazer"*, Martin Mobberley, Springer International Publishing (2013), page 332.

74. *"It Came From Outer Space Wearing An RAF Blazer"*, Martin Mobberley, Springer International Publishing (2013), page 293.

75. http://adsabs.harvard.edu/full/1978JBAA...88..280D

76. https://en.wikipedia.org/wiki/Cnut_the_Great

77. *"It Came From Outer Space Wearing An RAF Blazer"*, Martin Mobberley, Springer International Publishing (2013) page 417.

78. Copyright Kate Charlesworth. Published in New Scientist magazine, 30 July 1994, page 45. Reproduced with kind permission of Kate Charlesworth.

79. https://en.wikipedia.org/wiki/Discovery_of_Neptune

80. http://www.mikeoates.org/lassell/adams-airy.htm

81. *"Autobiography of Sir George Biddell Airy"* George Biddell Airy (Author), Wilfred Airy (Editor) (Cambridge Library

Collection - Astronomy), 2010

82. *"Changes In Sexual Attitudes And Lifestyles In Britain Through Life Course And Over Time"* Mercer, C. H. et al. The Lancet, 382 (9907), 30th November 2013, pp 1781-1794

83. *"Sex By Numbers"* Spiegelhalter, D. Profile Books, 2015

84. There is no significant difference between the frequency of sexual activity by men and women nor between straight and gay partners so no distinctions have been made in my calculations.

85. https://www.youtube.com/watch?v=i0ya5kh4_ZM

86. *"It Came From Outer Space Wearing An RAF Blazer"*, Martin Mobberley, Springer International Publishing (2013)

87. *"Return to the Far Side of Planet Moore!"* Martin Mobberley, Springer International Publishing (2015)

88. *"Eighty Not Out"* Dr Patrick Moore, Contender Books, (2003)

89. http://www.stormdunlop.co.uk/

90. *"It Came From Outer Space Wearing An RAF Blazer"*, Martin Mobberley, Springer International Publishing (2013), page 199.

91. *"It Came From Outer Space Wearing An RAF Blazer"*, Martin Mobberley, Springer International Publishing (2013) page 564.

92. *"It Came From Outer Space Wearing An RAF Blazer"*, Martin Mobberley, Springer International Publishing (2013) page 23.

93. *"It Came From Outer Space Wearing An RAF Blazer"*, Martin Mobberley, Springer International Publishing (2013), page 35.

94. *"It Came From Outer Space Wearing An RAF Blazer"*, Martin Mobberley, Springer International Publishing (2013), pages 25 - 32

95. *"It Came From Outer Space Wearing An RAF Blazer"*, Martin Mobberley, Springer International Publishing (2013) pages 74, 77, 82

96. https://en.wikipedia.org/wiki/Gilbert_Harding

97. https://www.youtube.com/watch?v=O5vrNnntcE0

98. https://en.wikipedia.org/wiki/Barbara_Kelly

99. https://en.wikipedia.org/wiki/Bernard_Braden

100. *"My Lost Childhood - How Hitler and the Freemasons Stole My Childhood"* by Geoff Kirby.

101. It has been suggested that the name 'Lorna' for the imaginary fiancée was deliberately similar to 'Lunar' reflecting that his love of the Moon was his driving life's passion.

102. http://www.independent.co.uk/news/arthur-c-clarke-case-investigated-1145468.html

103. *"Return to the Far Side of Planet Moore!"* Martin Mobberley, Springer International Publishing (2015), page 288

104. https://en.wikipedia.org/wiki/Arthur_C._Clarke#Sexuality

105. https://en.wikipedia.org/wiki/Ruth_Westheimer

106 Martin Mobberley has published a comprehensive catalogue of Patrick Moore's TV appearances but this exchange does not appear in his list. I am sure I have remembered the exchange correctly although I have no idea when the programme was broadcast.

107. Patrick Moore has stated that Werner von Braun was *"More sinned against than sinner"*. Over 25,000 slave workers died on the V2 development programme. They came from France, Belgium, the USSR, Poland, etc. They died by starvation, beatings, hanging and torture. Emaciated corpses of slave workers were suspended from gibbets outside the Director's office. Wernher von Braun was a personal friend of the Director and he frequently visited him in the Dora establishment. He obviously knew what was going on. He could hardly have failed to see the starved corpses hanging outside his friend's office - see *"Timewatch - The V1 and V2 Development Programmes"* BBC TV 13th November 1994

108. *"It Came From Outer Space Wearing An RAF Blazer"*, Martin Mobberley, Springer International Publishing (2013) page 516

109. *"It Came From Outer Space Wearing An RAF Blazer"*, Martin Mobberley, Springer International Publishing (2013) page 518

110. *"It Came From Outer Space Wearing An RAF Blazer"*, Martin Mobberley, Springer International Publishing (2013) page 563.

111. *"Eighty Not Out"* Patrick Moore, (Hardcover), Contender Books (2003)

112. *"Eighty Not Out"* Patrick Moore, (Hardcover), Contender Books (2003) Chapter 20.

113. http://www.independent.co.uk/news/uk/crime/black-people-still-far-more-likely-to-be-stopped-and-searched-by-police-than-other-ethnic-groups-10444436.html

114. https://en.wikipedia.org/wiki/Murder_of_Stephen_Lawrence

115. *"It Came From Outer Space Wearing An RAF Blazer"*, Martin Mobberley, Springer International Publishing (2013) page 438.

116. *"Return to the Far Side of Planet Moore!"* Martin Mobberley, Springer International Publishing (2015), page 138

117. https://en.wikipedia.org/wiki/MV_Empire_Windrush#West_Indian_immigrants

118. https://en.wikipedia.org/wiki/Rivers_of_Blood_speech

119. https://en.wikipedia.org/wiki/The_Black_and_White_Minstrel_Show

120. https://en.wikipedia.org/wiki/Blackface

121. https://en.wikipedia.org/wiki/Al_Jolson

122. https://en.wikipedia.org/wiki/The_Dam_Busters_(film)

123. https://en.wikipedia.org/wiki/Bernard_Manning

124. http://www.jokes4us.com/peoplejokes/comedianjokes/bernardmanningjokes.html

125. https://en.wikipedia.org/wiki/Dr._Strangelove

126. https://www.youtube.com/watch?v=kLDOTr_eQY8

127. https://www.youtube.com/watch?v=LuIJqF8av6I

128. https://www.youtube.com/watch?v=yfl6Lu3xQWo

129. 'Viz' Comic No: 167. Dated August 2007

130.
http://www.staylace.com/gallery/gallery31/belt_of_ethel_gran
ger_corsetworld.jpg

131.
https://www.google.co.uk/search?q=ethel+granger&newwindo
w=1&espv=2&tbm=isch&tbo=u&source=univ&sa=X&ved=0ahU
KEwjn25vx2r

132. *"It Came From Outer Space Wearing An RAF Blazer"*,
Martin Mobberley, Springer International Publishing (2013)
page 312.

133.
http://www.staylace.com/textarea/history/ethelgrangerbio2/eg
_bio2_01.htm

134. *"Can You Speak Venusian"* Patrick Moore, WW Norton &
Co.

135. http://www.stellarium.org

136. *"It Came From Outer Space Wearing An RAF Blazer"*,
Martin Mobberley, Springer International Publishing (2013)
page xvii.

137. http://www.binocularsky.com/

138. *"Binocular Astronomy"* (The Patrick Moore Practical
Astronomy Series). Stephen F. Tonkin BSc PGCE (2013)

139. http://www.binocularsky.com/binoc_choosing.php

140. http://www.ncbi.nlm.nih.gov/books/NBK381/

141. http://www.geoffkirby.co.uk/Accident2009/

142. http://www.binocularsky.com/

143. *"Making and Using a Telescope"* by H Percy Wilkins and
Patrick Moore. Eyre and Spottiswoode Publishers Ltd (1956)

144.
http://geogdata.csun.edu/~voltaire/classics/brunnings/BRUN
NINGS.pdf

145. *"Make Your Own Telescope from Everyday Materials"* Reg

Spry (1978)

146 http://www.equatorialplatforms.com/

147. BAA Journal 1981 91 (2) p 167

148. *"Result of the Project for the Comparison of the Performance of Astronomical Telescopes"* Hilton, J. and Bond, J., BAA Journal 92 (6) pp 264 - 266 (1982)

149. *"Comparison of the Performance of Telescopes"* Henbest, N., BAA Journal 93 (2) p 95 (1983)

150. *"Optical quality in telescopes"* Sky and Telescope, March 1992, p 257

151. http://www.cruxis.com/scope/limitingmagnitude.htm

152. https://en.wikipedia.org/wiki/Astronomical_seeing

153. http://weather.gc.ca/astro/seeing_e.html

154. http://weather.gc.ca/astro/seeing_e.html

155. Adaptive Optical systems have now been installed at the Palomar Observatory.

156. https://en.wikipedia.org/wiki/Speckle_imaging

157. http://www.rfroyce.com/pyrex.htm

158. https://en.wikipedia.org/wiki/Harlan_J._Smith_Telescope

159. http://www.amazon.co.uk/Handbook-Telescope-Making-N-Howard/dp/0571046800/ref=sr_1_1?s=books&ie=UTF8&qid=1448979333&sr=1-1&keywords=making+telescopes

160. https://en.wikipedia.org/wiki/Foucault_knife-edge_test

161. https://stellafane.org/tm/atm/test/understanding.html

162. http://www.spacetelescope.org/about/history/servicing_mission_1/

163. *"Unusual Telescopes"* Peter L. Manly. Cambridge University Press (1991) page 8

164. *"Sky & Telescope"* December 1978 page 569.

165. https://en.wikipedia.org/wiki/Ronchi_test

166. http://www.atm-workshop.com/images/ronchi-test9.gif

167. *"A Ronchi Test For Paraboloids"* Mobsby, Eric. Sky & Telescope vol. 48 pages 325 - 330 (1974)

168. http://www.willbell.com/atmsupplies/atm_supplies.htm (Scroll to bottom of page)

169. http://www.amazon.co.uk/Small-Astronomical-Observatories-Professional-Constructions/dp/3540199136

170. http://www.bbc.co.uk/cult/classic/clangers/intro.shtml

171. http://news.nationalgeographic.com/news/2014/04/140412-moon-faces-brain-culture-space-neurology/

172. https://www.youtube.com/watch?v=wLCUKnvhB4E

173. http://www.cloudynights.com/page/articles/cat/binocular-universe/binocular-universe-jack-and-jill-r2561

174. http://astro.ukho.gov.uk/moonwatch/index.html

175. http://astro.ukho.gov.uk/download/NAOTN69.pdf

176. *"The Astronomical Scrapbook"* by Joseph Ashbrook, Cambridge University Press (1984) page 205

177. Purists will dispute this definition of 'New Moon' but there are many definitions and this one is easiest to understand.

178. *"The Astronomical Scrapbook"* by Joseph Ashbrook, Cambridge University Press (1984) page 206

179. http://www.stellarium.org

180. http://www.space.com/30546-supermoon-blood-moon-total-lunar-eclipse.html

181. *"Moon Maps"* H P Wilkins, Faber and Faber (1960)

182. Every effort was made to obtain the permission of the copyright holder to the Wilkins map to reproduce these two pictures. However no response was received. These two pictures are therefore published under the 'Fair Dealing and Review' provisions of the copyright laws - https://en.wikipedia.org/wiki/Fair_dealing_in_United_Kingdom_law#Criticism_or_review

183. *"The Moon; A Complete Description of the Surface of the*

Moon", H P Wilkins and P. Moore, (1955) London, Faber and Faber.

184. *"Amateur Astronomer's Photographic Lunar Atlas"*, Henry Hatfield, Lutterworth Press (1968)

185. Reproduced by permission of the publisher to whom the required fee has been paid.

186. *"The Hatfield Lunar Atlas: Digitally Re-Mastered Edition"* Anthony Charles Cook

187. For example *"The Cambridge Photographic Moon Atlas"* by Alan Chu (Author), Wolfgang Paech (Author), Mario Weigand (Author), Storm Dunlop (Translator) (2012)

188. Phillip's Moon Map by Dr. John Murray which is widely available.

189. *"Return to the Far Side of Planet Moore!"* Martin Mobberley, Springer International Publishing, (2015) pages 65 - 76.

190. *"It Came From Outer Space Wearing An RAF Blazer"*, Martin Mobberley, Springer International Publishing (2013) page 72.

191. *"The Silent War"* BBC TV programme broadcast 5th and 12th December 2013

192. http://adsabs.harvard.edu/full/1937JRASC..31Q.320H

193. http://www.astronet.ru/db/varstars/msg/1200372

194. https://en.wikipedia.org/wiki/Pathological_science

195. *"Forum: Now you see it...Now you don't – A pathological tendency among astronomers"*, Kirby, Geoff., New Scientist 24 February 1990

196. *"Transient Lunar Phenomena"* Peter Grego, Astronomy Now magazine, December 2015, page 67

197. https://en.wikipedia.org/wiki/Mons_Pico

198. https://www.youtube.com/watch?v=1J9YzZ3BlSk

199. http://www.astronomy.com/columnists/stephen%20omeara/2010/05/stephen%20james%20omearas%20secret%20sky%20oneills%20illusion

200. *"It Came From Outer Space Wearing An RAF Blazer"*, Martin Mobberley, Springer International Publishing (2013) pages 71-78.

201. http://the-moon.wikispaces.com/O%27Neill%27s+Bridge

202. *"Return to the Far Side of Planet Moore!"* Martin Mobberley, Springer International Publishing, (2015) page 330

203. http://www.astronomy.com/columnists/stephen%20omeara/2010/05/stephen%20james%20omearas%20secret%20sky%20oneills%20illusion

204. https://the-moon.wikispaces.com/file/view/Wilkins-Bridge2.jpg/138858701/Wilkins-Bridge2.jpg

205. http://lroc.sese.asu.edu/learn/science9

206. https://en.wikipedia.org/wiki/Rupes_Recta

207. https://en.wikipedia.org/wiki/File:Lunar_libration_with_phase2.gif

208. https://en.wikipedia.org/wiki/William_Henry_Smyth

209. https://en.wikipedia.org/wiki/Charles_Piazzi_Smyth#Pyramidological_researches

210. https://en.wikipedia.org/wiki/Great_Moon_Hoax

211. Public Domain engraving

212. http://www.conspiracyclub.co/2015/05/05/moon-alien-base-china-releases-image/

213. http://articles.adsabs.harvard.edu/full/1969JBAA...79..288A

214. "Einstein crater 4188 h2 4188 h3" by James Stuby based on NASA image - Mosaic of two reprocessed Lunar Orbiter 4 images cropped in Gimp. The original image is in the public domain because it is a work of the U.S. Government (NASA). Immediate source: Lunar and Planetary Institute, Lunar Orbiter Photo GalleryLunar Orbiter 4, image 188, h2 [1] and image 188, h3 [2]. Licensed under Public Domain via Commons - https://commons.wikimedia.org/wiki/File:Einstein_crater_4188_h2_4188_h3.jpg#/media/File:Einstein_crater_4188_h2_41

88_h3.jpg

215. http://www.hampsteadscience.ac.uk/astro/index.html

216. http://nssdc.gsfc.nasa.gov/planetary/ice/ice_moon.html

217. http://adsabs.harvard.edu/full/1967JBAA...78...37M

218. https://en.wikipedia.org/wiki/Lunar_dome

219. https://en.wikipedia.org/wiki/Impact_crater

220. *"It Came From Outer Space Wearing An RAF Blazer"*, Martin Mobberley, Springer International Publishing (2013) pages 55/56

221. https://en.wikipedia.org/wiki/Mercury_(planet)#Ground-based_telescopic_research

222. *"The Interior Planets"* V A Firsoff. Oliver and Boyd Publishers. page 29. (1968)

223. https://en.wikipedia.org/wiki/Mercury_(planet)#Ground-based_telescopic_research

224. *"The Interior Planets"* V A Firsoff. Oliver and Boyd Publishers. (1968) plate 3.

225. *"The Interior Planets"* V A Firsoff. Oliver and Boyd Publishers. (1968) page 53

226. https://en.wikipedia.org/wiki/Atmosphere_of_Mercury

227. *"Mercury's Rotation and Visual Observations"* Cruikshank, D P and Chapman, C Sky and Telescope vol. 34 p 24 - 26 (1967)

228. *"The Astronomical Scrapbook"* by Joseph Ashbrook, Cambridge University Press (1984) page 281.

229. https://en.wikipedia.org/wiki/Vulcan_(hypothetical_planet)

230. https://en.wikipedia.org/wiki/Moons_of_Uranus#Spurious_moons

231. *"The Astronomical Scrapbook"* by Joseph Ashbrook, Cambridge University Press (1984) page 107.

232. *"The Astronomical Scrapbook"* by Joseph Ashbrook, Cambridge University Press (1984) page 97

233. https://en.wikipedia.org/wiki/Claimed_moons_of_Earth#Petit.27s_moon

234. http://adsabs.harvard.edu/full/1993AAS...183.2702O

235. This file is in the public domain because it was solely created by NASA.

236. https://en.wikipedia.org/wiki/N_ray

237. Mars image from Hubble Space Telescope, reproduced by permission of NASA.

238. *Villiger, Walther "Die Rotationszeit des Planetan Venus" Neue Annalen der K. Sternewarte in Munich 3:301-342 (1898)*

239. https://en.wikipedia.org/wiki/Linn%C3%A9_(crater)

240. https://en.wikipedia.org/wiki/Vulcan_(hypothetical_planet)

241. *"The Astronomical Scrapbook"* by Joseph Ashbrook, Cambridge University Press (1984) page 281

242. http://c-lab.co.uk/blogs/userdata/BLOGS/sheehan-venus_Page_03_Image_0001.jpg

243. https://en.wikipedia.org/wiki/Percival_Lowell#Venus_spokes

244. http://www.skyandtelescope.com/astronomy-news/venus-spokes-an-explanation-at-last/1/?c=y

245. *"Schröter's Effect And The Twilight Model For Venus"* Mallama, A, Journal of the British Astronomical Association, vol. 106, no. 1, pp. 16-18 (1996)

246. Chambers, R and Taylor, I., Journal of the British Astronomical Association, Vol. 76, p. 310 (1966)

247. *"Investigations Of Schröter Effect In The U.S.S.R"* Bronshtehn, V. A, Journal of the British Astronomical Association, Vol. 81, p. 181 - 185 (1971)

248. https://en.wikipedia.org/wiki/Ashen_light

249. http://www.universetoday.com/94848/the-mystery-of-venus-ashen-light-2/

250. http://www-

ssc.igpp.ucla.edu/personnel/russell/papers/ashen/

251. https://en.wikipedia.org/wiki/Ashen_light#History_of_observations

252. www.stellarium.org

253. http://www.thegreenwichmeridian.org/tgm/articles.php?article=9

254. https://en.wikipedia.org/wiki/John_Harrison

255. https://en.wikipedia.org/wiki/Ole_R%C3%B8mer

256. BAA Handbook published annually by the British Astronomical Association

257. *"Return to the Far Side of Planet Moore!"* Martin Mobberley, Springer International Publishing, (2015) page 361 - 365

258. http://www.ianridpath.com/startales/startales4.htm

259. http://www.lunar-occultations.com/iota/iotandx.htm

260. www.stellarium.org

261. http://www.techradar.com/news/world-of-tech/calling-time-a-history-of-the-speaking-clock-683753

262. https://en.wikipedia.org/wiki/Time_from_NPL

263. https://en.wikipedia.org/wiki/Greenwich_Time_Signal#Accuracy

264. https://en.wikipedia.org/wiki/Chester_Burleigh_Watts

265. Watts, C. B. "The Marginal Zone of the Moon," Astron. Papers Amer. Ephem., 1963, 17, 1-951

266. http://www.geoffkirby.co.uk/Portland/675680/#Eclipse1999

267. http://www.lunar-occultations.com/iota/planets/planets.htm

268. https://www.youtube.com/watch?v=6CuEf2mwONA

269. https://www.youtube.com/watch?v=ZZVPsNXKwyU

270. https://www.youtube.com/watch?v=j0lpfz_S5Js

271. https://en.wikipedia.org/wiki/Hipparcos

272. https://en.wikipedia.org/wiki/Moore%27s_law

273. http://www.skyandtelescope.com/astronomy-news/observing-news/antiope-occultation-yields-double-bonanza/

274. http://occultations.org/

275. http://adsabs.harvard.edu/full/1994JBAA..104...61M

276. *"Mutual Occultations of Planets: 1557 to 2230"* Steven C Albers Sky and Telescope March 1979 p 220-222

277. http://www.bogan.ca/astro/occultations/occltlst.htm

278. https://en.wikipedia.org/wiki/Occultation#Historical_observations

279. https://en.wikipedia.org/wiki/Michael_Maestlin

280. https://www.facebook.com/groups/582495965214715/

281. http://www.theastronomer.org/first_30_years.html

282. http://martinmobberley.co.uk/Alcock.html

283. http://martinmobberley.co.uk/images/AlcockHurst_small.jpg

284. http://www.theastronomer.org/first_30_years.html

285. *"SAO Atlas Omissions"* Kirby, G. J. JBAA, (90) p 79 (1979)

286. https://commons.wikimedia.org/wiki/File:V1500.Cyg.JD2442500-2444500.LightCurve.png

287. http://www.britastro.org/vss/

288. https://www.aavso.org/

289. *"Observing Variable Stars"* (The Patrick Moore Practical Astronomy Series), Gerry A. Good, (2009)

290. *"Binocular Astronomy"* (The Patrick Moore Practical Astronomy Series), Stephen Tonkin (2013)

291. https://www.aavso.org/

292. http://www.britastro.org/vss/aavso_and_the_baavss.htm

293. https://en.wikipedia.org/wiki/Purkinje_effect

294. http://www.britastro.org/vss/EBHandbook11.pdf

295. https://en.wikipedia.org/wiki/Algol

296. http://journals.plos.org/plosone/article?id=10.1371/journal.pone.0144140

297. http://www.as.up.krakow.pl/o-c/data/getdata.php3?BETA%20per

298. http://www.as.up.krakow.pl/o-c/diagram_html/per_beta_small.html

299. http://www.as.up.krakow.pl/minicalc/PERBETA.HTM

300. J.M. Kreiner, 2004, Acta Astronomica, vol. 54, pp 207-210.

301. http://www.geoffkirby.co.uk/EclipsingBinaryPredictions/

302. www.britastro.org/vss/RZCassiopeiae.pdf

303. *"An Unusual Brightening of the Eclipsing Binary RZ Cassiopeiae"* Wayne M. Lowder, JAAVSO Volume 35, 2006 available online as https://www.aavso.org/media/jaavso/2347.pdf

304. http://www.as.up.krakow.pl/minicalc/CASRZ.HTM

305. http://www.geoffkirby.co.uk/EclipsingBinaryPredictions/

306. https://www.flickr.com/photos/astonuniversity/10456705033

307. http://www.royalobservatorygreenwich.org/articles.php?article=1097

308. http://adsabs.harvard.edu/full/1992JBAA..102..343H

309. http://adsabs.harvard.edu/full/1981QJRAS..22...28K

310. *"Observing Earth Satellites"* Desmond King-Hele, Van Norstrand Reinhold Company (1983) pages 86 - 91

311. *"A Tapestry Of Orbits"* Desmond King-Hele, Cambridge University Press (2008)

312. https://en.wikipedia.org/wiki/Desmond_King-Hele

313. http://sattrackcam.blogspot.co.uk/2016/01/in-memoriam-pierre-neirinck-16-aug-1926.html

314. http://www.history.com/news/the-day-skylab-crashed-to-earth-facts-about-the-first-u-s-space-stations-re-entry

315. www.stellarium.org

316. http://www.heavens-above.com/

317. http://www.heavens-above.com/

318. https://en.wikipedia.org/wiki/Maunder_Minimum

319. https://en.wikipedia.org/wiki/Sp%C3%B6rer_Minimum

320. *"The Great Global Warming Blunder: How Mother Nature Fooled the World's Top Climate Scientists." Roy W* Spencer. Encounter Books. (2010)

321. "The Chilling Stars - A Cosmic View Of Climate Change" Henrik Svensmark and Nigel Calder. Icon Books UK (2007)

322. http://articles.adsabs.harvard.edu/full/seri/JRASC/0029//0000361.000.html

323. http://articles.adsabs.harvard.edu/full/seri/JRASC/0029//0000361.000.html

324. http://www.iflscience.com/space/solar-activity-could-cause-lightning-storms-earth

325. http://articles.adsabs.harvard.edu/full/seri/JRASC/0029//0000361.000.html

326. https://en.wikipedia.org/wiki/W._Heath_Robinson

327. Zero longitude is now 102 metres east of Greenwich - see https://en.wikipedia.org/wiki/Prime_meridian_(Greenwich)

328. http://www.solarstorms.org/SRefStorms.html

329. This work has been identified as being free of known restrictions under copyright law, including all related and neighbouring rights.

330 https://en.wikipedia.org/wiki/Moons_of_Mars#Discovery

331. http://www.usno.navy.mil/USNO/about-us/the-6-inch-

warner-swasey-transit-circle

332.
https://en.wikipedia.org/wiki/(There%27ll_Be_Bluebirds_Over
)_The_White_Cliffs_of_Dover

333. https://en.wikipedia.org/wiki/George_Adamski

334. Christopher Mann McKay [GFDL
(http://www.gnu.org/copyleft/fdl.html), CC-BY-SA-3.0
(http://creativecommons.org/licenses/by-sa/3.0/) or CC BY 2.5
(http://creativecommons.org/licenses/by/2.5)], via Wikimedia
Commons

335. https://en.wikipedia.org/wiki/Zzyzx,_California

336. https://en.wikipedia.org/wiki/Shitterton

337. https://en.wikipedia.org/wiki/George_Adamski

338. http://www.noao.edu/outreach/kptour/kpno.html

339. http://www.noao.edu/outreach/kptour/mayall.html

340.
https://en.wikipedia.org/wiki/McMath%E2%80%93Pierce_sola
r_telescope

341. https://en.wikipedia.org/wiki/Tombstone,_Arizona

342.
http://cms.sbcounty.gov/parks/Parks/CalicoGhostTown.aspx

343. https://en.wikipedia.org/wiki/Biosphere_2

344.
https://en.wikipedia.org/wiki/Contact_(1997_American_film)

345.
https://en.wikipedia.org/wiki/Karl_G._Jansky_Very_Large_Ar
ray

346. https://en.wikipedia.org/wiki/Meteor_Crater

347.
https://en.wikipedia.org/wiki/B612_Foundation#/media/File:
Meteor_Crater_(crop-tight).jpg

348. http://meteorcrater.com/

349. https://en.wikipedia.org/wiki/Lowell_Observatory

350. http://www.lowell.edu/

351.
https://commons.wikimedia.org/wiki/File%3APercival_Lowell
_observing_Venus_from_the_Lowell_Observatory_in_1914.jp
g

352. https://en.wikipedia.org/wiki/Edwin_Hubble

353. "100inchHooker". Licensed under CC BY-SA 2.0 via
Commons -
https://commons.wikimedia.org/wiki/File:100inchHooker.jpg#
/media/File:100inchHooker.jpg

354.
https://en.wikipedia.org/wiki/Mount_Wilson_Observatory#10
0_inch_.282.5_m.29_Hooker_Telescope

355. http://www.mtwilson.edu/vir/snow.html

356. https://en.wikipedia.org/wiki/Teide#Future_eruptions

357. https://www.youtube.com/watch?v=yfl6Lu3xQW0

358.
https://en.wikipedia.org/wiki/La_Palma#Tsunami_scenarios

359. http://www.lapalma-tsunami.com/

360. *"100 metre waves are a threat today"*, New Scientist 10
October 2015 p 12

361.
https://en.wikipedia.org/wiki/Jacobus_Kapteyn_Telescope

362. https://en.wikipedia.org/wiki/Nordic_Optical_Telescope

363. https://en.wikipedia.org/wiki/Zodiacal_light

364. http://www.astronomy.com/magazine/2012/07/brian-
may---a-life-in-science-and-music---the-full-story

365. http://www.astronomy.com/magazine/2012/07/brian-
may---a-life-in-science-and-music---the-full-story

366.
https://en.wikipedia.org/wiki/Sir_Howard_Grubb,_Parsons_a
nd_Co

367. By H. Raab (User:Vesta) (Own work) [CC BY-SA 3.0
(http://creativecommons.org/licenses/by-sa/3.0) or GFDL

(http://www.gnu.org/copyleft/fdl.html)], via Wikimedia Commons

368. http://www.astro-travels.com/pictures/La-Palma-Volcan-Teneguia-1971.jpg

369. Global Oscillation Network Group http://gong.nso.edu/

370. https://en.wikipedia.org/wiki/Schroter%27s_Valley

371. By James Stuby based on NASA image [Public domain], via Wikimedia Commons

372. Walker, M., International Dark Sky Association Report (1977)

373. *"Light Pollution Handbook"* Kohei Narisada, Duco Schreuder Section 5.1.3 Springer Publishing (2004) ISBN 978-1-4020-2666-9

374. http://www.cpre.org.uk/what-we-do/countryside/dark-skies/in-depth/item/1676-light-pollution-maps-where-you-live

375. www.stellarium.org

376. http://freestarcharts.com/stars-guides/17-guides/stars/17-how-dark-are-your-night-skies

377. http://news.bbcimg.co.uk/media/images/64659000/jpg/_64659967_0hnqrd5y.jpg

378. http://www.alcyone.de/SIT/bsc/bsc.html

379. https://en.wikipedia.org/wiki/Campaign_for_Dark_Skies

380. http://www.legislation.gov.uk/ukpga/2005/16/section/102

381. http://darksky.org/

382. *"Light Pollution - Responses and Remedies"*, Bob Mizon, Springer (2002) ISBN 1-85233-497-5

383. *"Light Pollution - Responses and Remedies"*, Bob Mizon, Springer (2012) ISBN 1-46143-821-7

384. *"20 years of fighting for the stars"*, Mizon, Bob, Astronomy Now, September 2009. pages 28–31

385. http://www.britastro.org/dark-skies/health.html?7O

386.
http://news.bbc.co.uk/1/hi/in_depth/sci_tech/2003/denver_2003/2766161.stm

387. http://www.britastro.org/dark-skies/dangers.html?7O

388. http://www.britastro.org/dark-skies/economic.html?7O

389. http://www.britastro.org/dark-skies/stars.html?7O

390. http://www.britastro.org/dark-skies/crime.html?7O#noreduction

391. http://www.britastro.org/dark-skies/churches.html

392. https://en.wikipedia.org/wiki/Alan_Yentob

393. *"It Came From Outer Space Wearing An RAF Blazer"*, Martin Mobberley, Springer International Publishing (2013) page 497

394. *"Flim-Flam! Psychics, ESP, Unicorns, and Other Delusions"* James Randi, Prometheus Books (1982)

395. *"A Double-Blind Test Of Astrology"* Carlson, Shawn. Nature (318) pages 419 - 425. (1985)

396. *"Support For Astrology From The Carlson Double-Blind Experiment"* McRitchie, Ken http://www.theoryofastrology.com/carlson/carlson.htm

397. *'L'Influence des Astres'.* Le Dauphin, Paris 1955.

398. https://en.wikipedia.org/wiki/Mars_effect

399. http://metro.co.uk/2014/10/15/second-coming-of-christ-slice-of-toast-bears-face-of-jesus-himself-4906780/

400.
https://en.wikipedia.org/wiki/Perceptions_of_religious_imagery_in_natural_phenomena

401. https://en.wikipedia.org/wiki/Outliers_(book)

402. I would now also add dwarf planets such as Haumea, Makemake and Eris.

403. https://en.wikipedia.org/wiki/Venetia_Burney

404. King George III famously did all those three things.

405. https://en.wikipedia.org/wiki/Bertram_Forer

406. *"It Came From Outer Space Wearing An RAF Blazer"*, Martin Mobberley, Springer International Publishing (2013), page 368.

407. https://www.youtube.com/watch?v=YjpcZPT1-NA

408. https://www.youtube.com/watch?v=H9bVF_f9hmw

409. *"Can You Speak Venusian?"* Patrick Moore. David & Charles (Publishers) Ltd (1972)

410. *"Has The Earth A Ring Around It?"* Frank G Back, Privately published (New York) c1960.

411. *"Can You Speak Venusian?"* Patrick Moore. David & Charles (Publishers) Ltd (1972) page 47.

412. https://en.wikipedia.org/wiki/Hollow_Earth

413. https://en.wikipedia.org/wiki/Leonhard_Euler

414. https://en.wikipedia.org/wiki/Hollow_Earth#Concave_hollow_Earths

415. *"Can You Speak Venusian?"* Patrick Moore. David & Charles (Publishers) Ltd (1972), Chapter 3.

416. https://en.wikipedia.org/wiki/Occam%27s_razor

417. https://en.wikipedia.org/wiki/Cold_fusion

418. https://en.wikipedia.org/wiki/Warminster#UFO_sightings

419. http://news.bbc.co.uk/local/wiltshire/hi/people_and_places/history/newsid_8694000/8694729.stm

420. *"Flying Saucer from Mars"*, Cedric Allingham. Frederick Muller, Publisher (1954)

421. *"Can You Speak Venusian?"* Patrick Moore. David & Charles (Publishers) Ltd (1972)

422. 'Feedback' page of New Scientist magazine August 14, 1986

423. Journal of Naval Science (7) No. 4 pages 228- 235

424. http://forteana-blog.blogspot.co.uk/2011/02/ufology-in-reputable-journal.html

425. *"UFO Handbook"* A. Hendry. Doubleday and Co. Inc (1979)

426. *"UFOs: The Psychic Solution"*, J. Vallee, Panther Books Ltd. 1977

427. UFO Encounter II, Astronautics and Aeronautics, Sept 1971, page 60

428. *"Scientific Study of Unidentified Flying Objects"*, Dr E. U. Condon, Bantam Books Inc. (1969)

429. *"UFO Encounter I"*, Astronautics and Aeronautics, July 1971, page 66.

430. *"The UFO Experience"*, J. A. Hynek, Corgi Books Ltd. 1974.

431. http://www.ufoevidence.org/photographs/section/1960s/photo 337.htm

432. *"Scientific Study of Unidentified Flying Objects"*, Dr E. U. Condon, Bantam Books Inc. 1969.

433. *"The UFO Experience"*, J. A. Hynek, Corgi Books Ltd. 1974.

434. International UFO Reporter, Vol 2, No 11, 1977.

435. http://news.bbc.co.uk/1/hi/wales/6740247.stm

436. *"The UFO Experience"* J. A. Hynek, Corgi Books Ltd. (1974)

437. *"The Interrupted Journey"* J. G. Fuller, Dial Press, New York (1966)

438. *"Brocas Brain"* C. Sagan. Random House Inc. (1978).

439. *"Scientific Study of Unidentified Flying Objects"* E. U. Condon, Bantam Books Inc. 1969.

440. *"UFO Handbook"* A. Hendry. Doubleday and Co. Inc (1979)

441. *"Socorro Saucer"* R. Stanfield, Fontana Books Ltd. (1978)

442. Omni Magazine. November 1979, p 30.

443. http://www.ufocasebook.com/ubatuba.html

444. https://en.wikipedia.org/wiki/Broad_Haven

445. http://news.bbc.co.uk/1/hi/wales/6740247.stm

446. *"How UFOs Conquered the World: The History of a Modern Myth"* David Clarke, Aurum Press Ltd (2015)

447. https://en.wikipedia.org/wiki/Fermi_paradox

448.
https://en.wikipedia.org/wiki/Search_for_extraterrestrial_intel
ligence

449.
https://en.wikipedia.org/wiki/Contact_(1997_American_film)

450. http://boinc.berkeley.edu/wiki/Project_list